Perfect Astrology (Zodiac Sign)

Shri Ram Babu Sao

Author: Shri Ram Babu Sao
Shri Ram Babu Sao is a Mechanical Engineer (1970) from Patna University. He is a meritorious, genius and very intellectual person since childhood. He secured the "National Merit Scholarship" of India during Secondary Board Examination. He has deep interest in Astrology since childhood. He is deeply committed and involved in Astrological research and development work. Sincerity, deep study and his hobby has put him into the use of many new techniques and methods for horoscope prediction. He has studied many books and magazines on astrology. He thought of the necessity of a consolidated book covering all the aspects, topics on subject matters pertaining to a Zodiac Sign Astrology at one place. This book **"Perfect Astrology (Zodiac Sign)"** is the jest of most of the topics on Zodiac Sign astrology. This will result in

improving the general quality levels of the reader to a greater satisfaction. He has taken a lead in upgrading the process of awareness of various matters benefiting the fresh learner in astrology. Any body can avail of practical knowledge on various topics related to astrology and can make predictions in detail of himself or his family member with the help of this Book.

Mobile: 9819506068
Email: rbsao8844@yahoo.co.in

Disclaimer

The book "Perfect Astrology (Zodiac Sign)" is not a writer's whole & sole product. It is a combination of the knowledge and expertise of the author and the data collected from different Books, specially researched to meet the objective and to enhance the knowledge on Astrology. Wherever necessary, the reference of the above books has been given. Because the prediction is not based on the contents of the book but on the skills, purity and perfection of the astrologer, any prediction based on this book shall be the responsibility of the person making predictions.

First Edition: 2015

ISBN-13: 978-1512112962

ISBN-10: 1512112968

Preface

"Money is Prosperity". The book, "Perfect Astrology (Zodiac Sign)", is a unique book, which is very informative and also easy to understand. One book is truly the equivalent of several books on astrology. You can make horoscope of yourself or any member of your family with the help of this single book. This provides some of the elementary and in depth essential elements on complete horoscope predictions. Many of the basics on astrology are explained in detail.

Astrology is not knowing your future, but planning your future by averting
the misshapenness by action in the right Muhurat and at the right time, wearing Gems, wearing Yantras, chanting Mantras and Prayers. It is important to realize that success comes only with the right actions at the right timing. The whole secret of Astrology is "Right Timing". This Book provides the best times for successful ventures such as starting a business, buying a home, or investing in the Stock Market. JP Morgan made a fortune using astrology for "Right Timing". This knowledge is made available to you though this book. By using the book, your life will be more prosperous than ever before. It is important to work "Smartly" but not hard. This Book gives you the followings:

1. The prospective tools to make your life more rewarding.

2. Pictures of your career and love life at its ultimate zenith.

3. Guidelines to ever dream of becoming a Star.

4. Discovering your financial fortune in life.

5. Points out the secrets of looking the "Best you can be every day".

6. An opportunity to start out a professional practice and setting your fees.

7. Making the Horoscope Predictions of yourself or any member of your family.

Many people need to know about their financial status, most important events and their future whether it is going to improve or is it going to get worse. Is now the time to spend or save? Money controls the way we live our lives, and this amazing readings will give you the insight, you need to take control of your financial future. It could be one of the most important readings that you will get through this Book that can truly change your life. This Book will isolate time to capture the situation and reveal its significance. At last, you have discovered a direct channel that will allow you an insight into your own destiny. Many "psychic services" charge you as much as $50, $75, or even $100 for a single reading, but, this book offers you a single instrument for reading as many as you want and that too at the cheapest rate. The technically advanced matters allow you to deliver your reading to you quickly and effectively. Not only will your reading be incredibly accurate but also you will have it available to read and analyse at your own pace. In addition, this book offers you an opportunity to record your own readings and readings of your family members by yourself. Just realise that how much you are going to save on account of Horoscope readings for you and your family members.

This is your journey. You may feel compelled to consult the any Online Psychic frequently, or as often as your circumstances warrant. Through this book, you may record the information you read and so you can review, re-evaluate and reconsider its significance as often as you like. As you continue to consult this book for answers to important issues in your life, an accurate record of your responses will be chronologically archived to allow you to a greater understanding of your self as you relate to the complex world around you. This means that you can make money by using "Right Time" methods by reading your Horoscope to gain knowledge on when and how to do things. Gain Wisdom through knowledge of planetary influences by understanding when and not to act. Gain Power and Status using your own cycles that empower you toward success. Avoid Problems before they happen by knowing the right time to act. Be Comfortable with gifts provided by the universe, as you deserve them.

Enormous books are available in the market, each one covering one or two topics on Astrology. In the developed countries like America, France, Japan, Germany and Britain, plenty of books on Astrology are available, covering one or two topics but it cost much more than our buying capacity. One has to read many books to master on Astrology and also to predict the individual Horoscope/Chart. Our life is speedy. It is ever active and is changing every moment. Each one of us is facing difficulty at every step. This book will facilitate to reach your destination by moving ahead with ease even in the storming situation. This is so much strife and struggle in the present time as it was never before. This is a time of ready-made food and fast food. No body has time to cook the food and then eat. Only this feeling motivated me and necessitated making this book. This is easily approachable and compact. It is full of all information at one place to be referred easily and quickly by any body whether busy in any profession.

Today the eyes of the whole of the world are fixed on India for any kind of development, because our Indian Astrology is the best in the world and covers almost every aspect of the individual life along with the timings of the events to happen in individual life. Now it is a duty of an Indian to come forward to lead and direct the world in this sphere and let the world know how resourceful Indian are in astrology. The need for development has been felt for quite some time back that an authoritative book is written on astrology work which may contain all the aspect of astrology with illustrations so that complete information is conveyed in a simple language. I am confident that this book will help to all in achieving his object and success in every field of astrology. I have tried to make clear about what is the correct astrology work. These are all correct and true facts & figures collected from various books and incorporated here in a single book for the first time for use by the common men. Behind all this, there is our exhaustive study and collections. More than the study is the presentation of the subject matter and even much more than the presentation of the subject matter is long years of experience and association with the astrology work all over India and abroad. This gives the authenticity to the book. This book is a tool for the Jyotish Students, for the Beginners, for the somewhat Advanced Students and for the Professionals too.

The recognition and importance, which this book has received within a short time, is a positive proof of its efforts and impartiality. Hundreds and thousands of persons from our country and from foreign countries are responding and referring this book. There is a great demand created for this book. This is no credit to us but is the genuineness and absolutely authenticity and clarity of this book. We are confident that the readers and experts will consider my effort.

Table of Contents

1.	**Introduction**	**1-24**
1.1	History of Astrology	1-3
1.2	Astrological Terms	3-12
1.3	Astrological Symbols	12-14
1.4	Chart (Kundali)	14-17
1.5	Classification of Kundali Chart	17-18
1.6	Panchang (Hindu calendar)	19-20
1.7	Correction of Birth Time	20-23
1.8	Reference Book	23-24
2.	**Muhurat Predictions**	**25-38**
2.0	General	25-25
2.1	Prediction of Muhurata	25-35
2.2	Yoga for Muhurata and Neutralisation	35-38
3.	**House (Bhava)**	**39-50**
3.1	Concept of House	39-40
3.2	Concept of House & Lord Working	40-41
3.3	Lagna Concept	41-42
3.4	House-to-House Relationship	42-46
3.5	House/Bhava Signification (Karaktva)	46-50

4. Zodiac Sign (Rashi) 51-102

4.1 General 51-52

4.2 Sign Characteristic 52-63

4.3 Predictions by Sign Element 63-66

4.4 Predictions bySigns in Houses 66-102

5. Moon-Sign (Janma Rashi) 103-116

6. Sun-Signs 117-136

6.1 Aries (March 21 - April 20) 117-118

6.2 Taurus (Vrishabh) (April 21 - May 21) 119-120

6.3 Gemini (Mithuna) (May 21 - June 20) 120-121

6.4 Cancer (Karka) (21st June - 20th July) 121-123

6.5 Leo (Simha) (July 21st to August 21st) 123-124

6.6 Virgo (Kanya) (August 22 to September 22) 125-126

6.7 Libra (September 23 to October 22) 126-127

6.8 Scorpio (Vrischika) (Oct. 23 to Nov. 22) 127-129

6.9 Sagittarius (Dhanu) (Nov. 23 to Dec. 20) 130-131

6.10 Capricorn (Makara) (Dec. 21 - January 19) 131-133

6.11 Aquarius (Kumbha) (January 20 - February 19) 133-135

6.12 Pisces (Meena) (February 19 to March 20) 135-136

7. Planet (Graha) 137-192

7.1 Planets' Description 137-141

7.2 Apparent Motion of Planet 141-146

7.3	Planets' Characteristic	146-159
7.4	Judgement of Benefic and Malefic Planet	159-168
7.5	Signification (Karaktva) of Planets	168-192
8.0	**Lords in Twelve Houses**	**193-256**
8.0	Predictions by Combust Lord Effects	193-194
8.1	Predictions by Lagna Lord (Lagnesha)	194-211
8.1.1	Predictions by Lagna Lord (Lagna wise) position	195-199
8.1.2	Predictions by Lagna Lord in Various Bhava	199-208
8.1.3	Predictions by Asc. Lord Combinations (Yoga)	208-211
8.1.4	Predictions by Lord of a Kendra	211-211
8.1.5	Predictions by Lords of Kendra & Kona	211-211
8.2	Predictions by Dhan Lord in Various Bhava	212-216
8.3	Predictions by Sahaj Lord in Various Bhava	216-219
8.4	Predictions by Bandhu Lord in Various Bhava	220-223
8.5	Predictions by Putra Lord in Various Bhava	223-227
8.6	Predictions by Ari Lord in Various Bhava	227-230
8.7	Predictions by Yuvati Lord in Various Bhava	230-233
8.8	Predictions by Randhra Lord in Various Bhava	234-237
8.9	Predictions by Dharma Lord in Various Bhava	237-241
8.10	Predictions by Karma Lord in Various Bhava	241-249
8.11	Predictions by Labh Lord in Various Bhava	249-253
8.12	Predictions by Vyaya Lord in Various Bhava	253-256
9	**Predictions by Planet (Graha) in Sign**	**257-328**

9.1 Effect s of Planet in Sign during Dasha Period 257-261

9.2 Predictions by Planets in Signs 261- 328

10 Sun and Moon 144-Combinations in Signs 329-342

11. Gulika and Upagraha 343-358

11.1 Predictions by Gulika/Mandi 344-352

11.2 Predictions by Upagraha 352-358

12. Bhava Pad, Upa Pad and Pranapad 359-366

12.1 Predictions by Bhava Pad 359-362

12.2 Predictions by Upa Pad 362-365

12.3 Predictions by Dara Pad 365-365

12.4 Predictions by Pranapad 365-366

13. Divisional Charts (Vargas) 367-438

13.1 Shodasa varga (Divisional Charts) 370-384

13.2 Predictions by Hora Chart 384-387

13.3 Predictions by Drekkana Chart 387-389

13.4 Predictions by D-4 (Chaturthamsha) 389-390

13.5 Predictions by (D-6) Saptamsa Chart 390-391

13.6 Predictions by (D-7) Saptamsa Chart 391-391

13.7 Predictions by (D-9) Navamsa Chart 392-403

13.8 Predictions by (D-10); to D-45 Charts 403-411

13.9 Predictions by Ashtakavarga 411-430

13.10	Predictions by Karakans	430-436
13.11	Predictions by Argala	436-438
14	**Planets Transit (Gochara)**	**439-453**
14.1	Planets' Description	439-442
14.2	Apparent Motion of Planet	442-446
14.3	Planet Transit (Gochara)	446-448
14.4	Predictions by Planet Transit (Gochara)	448-449
14.5	Planet Transit over other	449-452
14.6	Planet Transit over Planet in Natal House	452-453

1

Introduction

1.1 History of Astrology

The growth and achievement in life of a Mankind (Individual) depends on so many factors, such as, 1) Astrological Effects, 2) Genetic Effects, 3) Environmental Effects, and 4) Society & Contacts. Astrological Effects can be found out by "Astrology" or "Jyotish" which means the 'science of light' and is related with the Light and Magnetic Field emitted by the planets (Graha). Indian Astrology has been divided into three main branches of study.

1. Siddanta, 2. Samhita, 3. Hora

Siddanta: Siddanta covers astronomical study of celestial bodies.

Samhita: Samhita deals with mundane astrology such as earth quakes, floods, volcanic eruptions, rainfall, weather conditions economic conditions and effects of sunspots.

Hora: Hora Astrology deals with the Phalitha Jyotish, which means predictions about individual's life. Hora Astrology has six sub-divisions, namely, Jathaka, Gola, Prasana, Nimitta, Muhurata and Ganitha.

Vedic Astrology: Vedic Astrology is one of the main systems of Indian Astrology. It is a part of Vedas and is recognized as the Eyes of the Vedas. It predicts future on the basis of the birth-chart. To predict future according to this method the information of the birth-time, place and date are very important.

Jaimini Astrology: Maharishi Parashara and Jaimini both belong to the same period. They introduced separate systems of astrology based on similar principles. Jaimini Astrology is very popular in South India and is quite similar to Vedic Astrology. However, it has distinct rules and principles. The Astrologers who predict future on the basis of Jaimini astrology too are quite successful in predicting the fortunes just like the Parashari astrologers.

Horary Astrology: Horary Astrology is the astrological method based on questions. The person who does not know his birth-place, time and date can use this method to discover what lies in store. In this method the astrologer decides the Ascendant on the basis of the planetary positions at the time the question has been asked by the native. In this method you will get the answers for the questions you ask from the astrologer. It does not use the rigid mathematical methods like Vedic Astrology. It does not include calculations like Vimshottari Dasha, Anterdasha and Pratyantara Dasha.

Lal-Kitab Astrology: Lal Kitab is also one of the methods to predict the future. Lal Kitab is a simple but exotic method of astrology. It gives a lot of importance to houses and remedies. Lal Kitab does not have signs but numbers for each house. It gives results according to the position of planets in a particular house. It provides remedies to increase the auspiciousness of the planets as well as to reduce the inauspiciousness of the planets. In Punjab, this method of Astrology is very popular and is fast gaining popularity all over the country.

'Hindu Astrology' is founded by the Maharashi Aryabhatta, Parasara, Varaha Mihira, Jaimini, Garga, Kalidasa and Kalyan Varma. The origin of this science can go back as old as 4000 years. The astrology fully knows the individual's future as indicated by horoscope but can't certify the same as it is not 100 percent Mathematics or Science".

In classical Jyotish the Moon has equal or greater power to the lagna (Ascendant). Accurate readings cannot occur from the Rashi-Lagna only. Accurate readings require at least two preliminary scans (predictions): first from the Rashi-Lagna, and second from Chandra-Lagna. Then more detailed scans such as from the Mahadasha-Lord and the Karaka are also required. More accurate Jyotish predictions are often produced by reading significations from the Chandra Lagna first, Radical Lagna second, Navamsha Lagna third, and Varga Lagna additionally. Thus it is easier to identify the behaviours and environments that provide the best psycho-emotional support while predictions of a Horoscope. Horoscope is a mirror in which an astrologer can see one's past, present and future.

Horoscope is like a snapshot of a particular place in time and space. For casting the natal horoscope of an individual the time of birth, date of birth and place of birth is needed. There are 12 houses and 12 Signs in a horoscope from which an astrologer can predict about various areas of the life of an individual. This "Perfect Astrology" Book enables the astrologer to know that what the future has in store for the native. For the calculation of the timing of various events indicated in the horoscope the knowledge of impact of major period/sub period and transit is used.

1.2 Astrological Terms

Ascendant (Lagna): The Ascendant or Lagna or Rising Sign is the Sign in which an individual is born.

Affinity: It is a mutual attraction between the planets.

Affliction: Affliction of a planet is formed by, (1) its placement in the 6th, 8th or 12th houses or (2) association with the ruler of the 6th, 8th or 12th lords; or (3) association with natural malefic; or (4) association of Badhaka planet; or (5) its Combustion or placement between two natural malefic. Affliction of a planet is such a bad condition in which his energies or powers are considered to be zero and give an adverse effect.

Angle (Kendra): The Ascendant (first house), Descendant (fourth house), Mid-heaven Seventh house) and I Mum Collie (tenth house) are called Kendra (Angle).

Association: It is a relationship between two or more planets by their position in a house.

Astrologer: The person who practices astrology is called Astrologer.

Astrologist: The professional lecturer, who teaches astrology, is called Astrologist.

Atma Karaka: The Sun is called the natural Atma Karaka. In Jaimini astrology, the planet having the highest longitudinal progression in the Horoscope; irrespective of the sign in which it is placed, is called Atma Karaka planet.

Auspicious Planet: Planet, which gives positive effects and good result to the individual is called auspicious planet.

Benefic Planet: Some of the planets are considered positive for the individual giving good effects and influences and hence they are called Benefic Planet.

Bhava (House): The complete Zodiac is divided into twelve parts for the purpose of complete study of astrology. Each division is called a Bhava (House).

Bhukti: Bhukti is the period, which indicates how many years of Maha Dasa of the planet have passed before the time of birth of the individual.

Birth Time: This is the moment of first breath of a new born individual.

Chart: It is a figure or sketch consisting of 12 houses, in which the position of the planets and the Signs are given.

Conjunction: Two planets situated together in a house or occupying position close to each other within a certain orb or reaching nearer are called in Conjunction.

Cycle: It is a complete revolution or rotation made by a planet around the Sun.

Debilitation (Khala): When a planet occupies its Sign of Fall for Neecha Bhanga, the planet is in Debilitation and is considered weak or Neecha.

Degrees of Maximum Exaltation & Debilitation: Maximum degree of Exaltation or Debilitation is the defined degree of planet position in a Sign.

Derivative House: This is a house related to another house in the Chart, which signify the events of individual. Example: The 3rd house is the house of the brothers and sisters and the fourth house is the mother's house. Accordingly, the third house from the fourth house, i. e. the 7th house in a natal chart will describe the signification of the brothers and sisters of the native mother.

Defeated Planet: When a planet is in association with an enemy planet within a certain degrees or less than specify, then the weakest planet is said to be a defeated planet.

Descendant: The point opposite the Ascendant, i.e. the seventh house is called Descendent.

Detriment: A planet is said to be in detriment or exile, when it is posited in the inimical Sign. When a planet is in detriment, it is not comfortable in that sign and it tends to operate with the least strength.

Dignity: A planet is dignified when it occupies its Own Sign, its Moolatrikona or its Exaltation Sign, aspect by a benefic planet without any aspect or affliction by a malefic, when it is not retrograde or when it is increasing in its light. Such planet gives good results.

Direct Motion: It is a motion of the planet that follows the natural order of revolution cycle. The letter "D" marked against a planet indicates the direct motion of planet.

Dispositor: The lord of the sign is called Dispositor of a planet positioned in that Sign. The Dispositor of a planet is the soul essence of that planet, like prime minister. It dictates the planet to reacts in the way he likes. Example: Venus is the Dispositor of Saturn as because of ruler ship of Libra and Saturn has to act as per choice of Venus.

Domicile (Domes): A domicile (domes) of a planet is a Sign of ruler ship and is regarded as a home. Example: The domiciles of Mercury are Gemini and Virgo as because Mercury is the ruler of them. They are domicile or gaudier of Mercury, where he rejoices.

Exaltation: The planet in a particular Sign is called the planet in Exaltation Sign. Planet is dignified during his Exaltation.

Face: The "faces" arise from a subdivision of zodiac Sign into six equal parts. This is an obsolete term meaning the division of each sign into six equal parts of 5° each. The faces derive from the ancient Egyptian decants.

Friendly Planet: The planets of one group are called friendly planets.

Functional benefic Planets: As per Lagna Sign of the Natal Chart, some planets are defined as Functional benefic Planets even though they are Natural Malefic. Their influence is thought to be positive or constructive for the native. It makes the house strong in which he is sitting. The planet gives good result during his Main Dasa and the Antar Dasa fouling together in the same period.

Functional Malefic Planets: As per Lagna Sign of the Natal Chart, some planets are defined as Functional Malefic Planets whose influence is thought to be negative or destructive for the native. The lord of the 6th, 8th, and 12th are called Functional Malefic (Inauspicious) Planets and will make the house weak in which he is sitting. The planet does not give the good result during his Main Dasa and the Antar Dasa fouling together in the same period.

Ghati: The Ghati is a measure of the Time. One day = 24 hours = 60 Ghati. 2 ½ Ghati =1 hour; one Vighati = 24 seconds; and 60 Vighati = 1 Ghati.

Group of Planets: There are nine planets. They are divided in two Groups, such as,

Jeeva Group: The Sun, Moon, Mars, Jupiter and Ketu and the signs that they rule, are in Jeeva Group with the Jupiter as its leader.

Sareera Group: The Mercury, Venus, Saturn and Rahu and the signs that they rule, are in Sareera Group with the Saturn as its leader. Both groups are equally important. The planets and their Signs (Rasi) of one group are favourable to each other planets and Signs in that kundali. At the same time, the planets and Signs of one group are inimical (having enemy's behaviour) to the other group's Planets and Signs. In special case, when the planets of one group are harmoniously related to the planets of other group due to other governing factors of the astrology in the Kundali, then there will be the Rajya Yoga in the life of that native.

Gochara (Transit): The revolution or usual movement of planets in the Zodiac Sign during the course of revolution around the Sun is known as Gochara (Transit).

Graha Yudha (planet war): Whenever one planet comes within 5^0 orb of another one among the planets Mars, Mercury, Jupiter, Venus and Saturn, as viewed from the Earth, it causes planetary war, or Graha Yudha. One of the two planets involved in this war is said to be vanquished and another is a victor. The victorious planet produces powerful auspicious effects, while the vanquished or defeated one becomes inauspicious. The house in which this phenomenon occurs is destroyed and the individual suffers throughout his life with respect to that house events.
Hora: Each zodiacal sign is divided into two parts of 15 degree each and is called Hora.

Horoscope: It is the Janma Kundali, which depicts the positions of different signs and planets at the time of birth. It also represents the Rising Sign at the place of birth and the location of planets in various signs.

Horizon: It is the visible juncture of Earth and the sky and represented in a horoscope.

House Cusp: This is the zodiacal degree at which a house begins.

Increasing light (waxing): When Moon moves from the position of New Moon to full Moon, the shining portion of Moon grows larger and is called Increasing or Waxing Moon.

Inimical: It means unfriendly and enemy planet or house.

Janma Rashi: Janma Rashi is the Sign occupied by Moon at the time of birth of the native.

Ketu (Dragons Tail/South Node of Moon): Ketu is the dead body of the lusty demon, killed by Vishnu. So, Ketu is the tail of the dragon and called the South Node of moon, which losses and symbolizes the death along with Saturn.

Latitude: This is the celestial angular distance measured north or south of the plane of the ecliptic. This is the distance of a planet from the Equator.

Local Mean Time: This is the actual time in a given location based upon the Sun's position at the Mid Heaven (noon) of the place. It is abbreviated as L M T.

Longitude: It is the distance in degrees or in arc on Earth from 0° Aries eastward to any given point that intersects the ecliptic, such as, 10° Taurus is expressed as longitude 40°.

Masculine Signs: Aries, Gemini, Leo, Libra, Sagittarius and Aquarius are considered Masculine Signs.

Moolatrikona: The position of the planets in the particular Sign is considered the Moolatrikona of planet. Planet is dignified during his Moolatrikona. The Moolatrikona sign is usually a part of its own house with the exception of the Moon, where it behaves almost as favourably and gets more strength.

Moon sign: The Moon sign is the zone of the zodiac in which the Moon is positioned when a person is born. This is also called Rashi too.

Natal Chart: The horoscope cast at the birth time of the individual, showing the position of Signs and Planet with respect to house, is called a Natal Chart or Nativity.

Native: It refers to a person (male or female) for whom a horoscope is cast and studied.

Natural Benefic: Moon (waxing), Mercury, Venus, Ketu and Jupiter are called Natural Benefic Planets in order of increasing Benefic.

Natural Malefic: Sun, Mars, Saturn and Rahu are called Natural Malefic Planets in order of increasing malefic. Moon remains a malefic from 9th day of Krishna Paksha to 7th lunar day of Shukla Paksha.

Navamsa Chart: It is a nine Divisional Chart of a Sign with 3° 20' segments each.

New Moon: It is the beginning of a lunation cycle and of the waxing phase.

Opposite: It is point at a distance of 180° in the Zodiac or the 7th house apart.

Orb: This is the degrees of Longitude in the Zodiac.

Own house: A planet rules a house in Horoscope and that is called his Own House.

Pad: Each Nakshatra is divided into four divisions or Parts. Each Part is called Pad of Nakshatra. Accordingly, there are 27 x 4 = 108 Pad comprising the whole of the zodiac.

Paksha: The 15 days period during which the Moon goes on in its orbit from the Full Moon to Zero or vice versa is called Paksha. The first fortnight of the month is the period during which Moon is waxing and is called Shukla Paksha. The second fortnight during which the Moon is waning is called Krishna Paksha.

Planet: Sun, Moon, Mars, Mercury, Jupiter, Venus, Saturn, Rahu, Ketu, Uranus, Neptune and Pluto are the twelve heavenly bodies which appear to move in the Zodiac and influence the human body and are called Planets.

Part of Fortune: This is the Arabian Part most commonly used by western astrologers. It is the degree, which is calculated by subtracting the Sun's longitude from the sum of the Ascendant and Moon's Longitude. The degree occupied by the Part of Fortune symbolizes good fortune. It is also called Fortuna and Pars Fortune.

Partile: It is the degree at which an aspect is precisely exact (0° orb). An aspect that is within 1° orb is said to be exact but not partile.

Peregrine: It is the point of orbit at which a planet is closest to Earth. It is 0^0 to 5^0 and 25^0 to 0^0 to the eastern Horizon of the Natal Chart

Prediction: Knowing about natives past, present and future with the help of horoscope is prediction. But, in one accident many lives are taken, does that indicate similarity in everyone's horoscope? No, because the horoscope of the place or a vehicle in which passengers are travelling, supersedes the horoscopes of the natives. Hence, in a calamity everyone's horoscope does not necessarily indicate death. However, this point needs further research and views from the readers

Retrograde motion (Vakra/Saktha): When observed from Earth, it appears the apparent backward motion of a planet or moving in reverse direction than its natural direction of travel and is called Retrograde Motion.

Rising Planet: The planets which are positioned in the Rising Sign or Ascendant in the natal chart are Rising Planets.

Rising Sign (Lagna): The earth is rotating once a day around its axis. One of the 12 zodiacal signs is entering the 1st house every two hours. The Sign entering into 1st house at the birth time is known as Rising Sign or Lagna.

Ruling Planet: The planet which rules the Ascendant or Lagna Sign is called the Ruling Planet.

Significator: The planet ruling a house is called the Significator of that house.

Solar Chart: The horoscope Chart with the Sun in the Ascendant is called the Solar Chart.

Termsinal Houses: The houses ruled naturally by water signs, such as four, eight and twelve Sign and known as in Terms. They pertain to symbolize occult interests. Collectively, they are known as the Trinity of Psychism. The assessment of these strengths is useful in predicting the longevity and other results based on the Dasa System.

Trikona (Trine): The Houses 1, 5 and 9 are called Trikona or Trine.

Trikas (Badhakasthana): The Houses 6, 8 and 12 are called Trikas or Badhakasthana. These are the Badhaka houses and considered the evil houses of suffering.

Triplicity: It is a group of three signs belonging to the same element: fire (Aries, Leo, Sagittarius); earth (Taurus, Virgo, Capricorn); air (Gemini, Libra, Aquarius); and water (Cancer, Scorpio, Pisces). The members of a triplicity lie 120° apart in the zodiac, forming a triangular and harmonious relationship with each other and form an equilateral triangle in a horoscopes diagram.

Ugra (Fierce): The nature of some of Nakshatra is fierce as per astrology and is called Ugra Nakshatra. Hitler Nakshatra was Ugra (Fierce).

Unfriendly house (Deena): The house with the lord ship of enemy group of planet with respect to the planet position in that house is defined as the Unfriendly House.

Upachaya: The 3rd, 6th, 10th and 11th houses from the Lagna are called the Upachaya.

Vishnu House: The houses 1st, 5th & 9th stand for Vishnu and hence called Vishnu House.

Waning: The phase of the Moon during which the visible portion of the Moon decreases, is called Waning.

Waxing: The phase of the Moon during which the visible portion of Moon grows larger, is called Waxing.

Winning Planet: When a planet is in association with an enemy planet and its degrees are more than that of enemy planet, then that planet is said to be a winning planet.

Yoga: Nitya Yoga means a 'yoke' or connection or the Panchanga. Panchanga Yoga is formed by the daily connection between of the Sun and Moon. The Moon travels 13º 20' away from the Sun each day. These 13º 20' sections (similar length of a Nakshatra) form Yoga. Yoga is a subtle blending of solar and lunar energies that give special indications every day. There are 27 such yoga formed and each one is linked to a particular Nakshatra. Also, certain planetary combinations are known as Yoga.

Zodiac: It is literally the circle of stars. Zodiac is defined a band of the heaven approximately 14º wide, centred on the Ecliptic, against which the Sun and other planets are seen to move, as seen from the Earth.

1.3 Astrological Symbol

Sign	**Sign Picture**	**Sign Symbol**	**Lord Planet Symbol**
Aries		♈	Mars ♂

Taurus		♉	Venus ♀
Gemini		♊	Mercury ☿
Cancer		♋	Moon ☽
Leo		♌	Sun ☉
Virgo	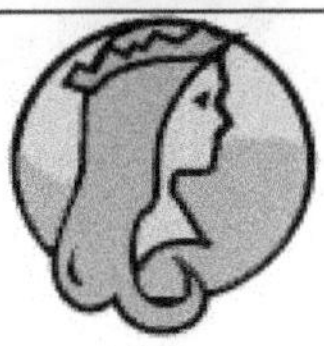	♍	Mercury ☿
Libra		♎	Venus ♀
Scorpio	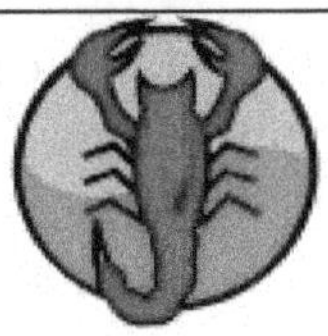	♏	Pluto ♇

Sagittarius		♐	Jupiter ♃
Capricorn		♑	Saturn ♄
Aquarius		♒	Saturn ♄
Pisces		♓	Jupiter ♃

1.4 Kundali Chart

A) South Indian style Kundali (Chart): In South Indian Style Chart, the position of the signs is always fixed and the position of the Ascendant is always changing. The houses are counted in a clockwise direction. The upper top left but one Rectangular box, being denoted by the digit 1 is always Aries. The next Rectangular box right to it, being denoted by the digit 2, is always Taurus and so on as written in the Chart. The digit 1 through 12 indicates the position of the Signs fixed in clockwise direction. It is always fixed in the South Indian style of Chart. The Ascendant (Lagna) falls in one of the Sign depending on rising Sign and is called Lagna or Ascendant of the Chart. The counting of the Houses is always done in clockwise direction from the Ascendant (Lagna) as first house through twelfth house. The planets occupy the House according to their longitudinal position.

12 (Pieces) Venus	**1 (Aries)**	**2 (Taurus)**	**3 (Gemini) Ascendant, Mars**
11 (Aquarius) Mercury, Ketu	**Lagna Chart-1**		**4 (Cancer) Moon**
10 (Capricorn) Sun			**5 (Leo) Rahu**
9 (Sagittarius) Jupiter, Saturn	**8 (Scorpio)**	**7 (Libra)**	**6 (Virgo)**

FIG 1: South Indian Style Lagna Chart

B) North Indian style Kundali (Chart): In the North Indian style Chart, the Houses are always fixed and the rising Sign falls in the 1st House, which is called Lagna or Ascendant, which is at the top in the centre. The Lagna (Ascendant) as well as the other houses are always fixed and are counted in anti-clockwise direction from the 1st House or the Lagna or the Rising Sign. The planets occupy the House according to their longitudinal position. The numbers shown in this format tell us which sign is in the Lagna and other Houses as shown in the Chart.

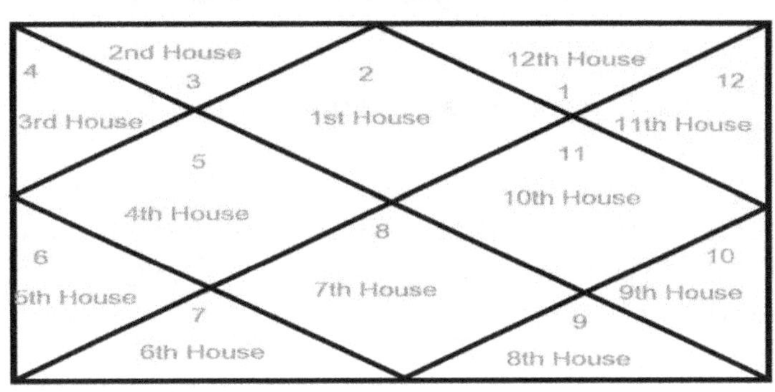

FIG 2: North Indian Style Chart

C) East Indian (Bengali) style Kundali (Chart): In the East Indian (Bengali) style Chart, the Houses are always fixed and the rising Sign falls in the 1st House, which is called Lagna or Ascendant, which is at the top in the centre. The Lagna (Ascendant) as well as the other houses are always fixed and are counted in anti-clockwise direction from the 1st House or the Lagna or the Rising Sign. The planets occupy the House according to their longitudinal position. The numbers shown in this format tell us which sign is in the Lagna and other Houses as shown in the Chart. In Bengali style zodiac is again fixed and ascendant & planets move anti clock wise along the zodiac unlike South Indian System. In western style the ascendant is fixed again and placed on the left hand side whereas

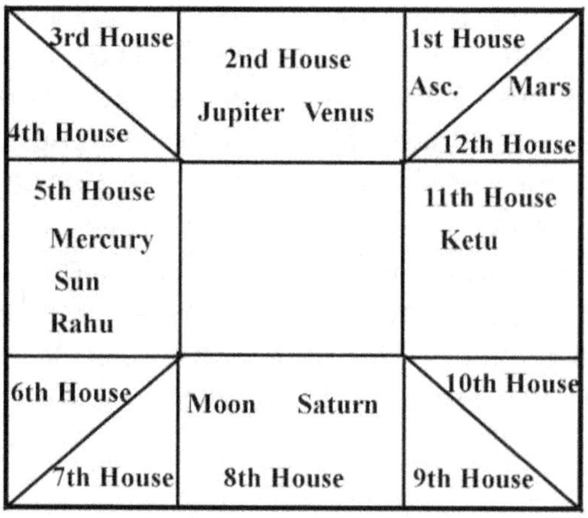

3rd House / 4th House	2nd House Jupiter Venus	1st House Asc. Mars 12th House
5th House Mercury Sun Rahu		11th House Ketu
6th House / 7th House	Moon Saturn 8th House	10th House / 9th House

FIG 4: East Indian (Bengali) Style Chart

D) Western style Kundali (Chart): It is a circular Chart divided in twelve parts in which Sign and Planets position are given as shown below.

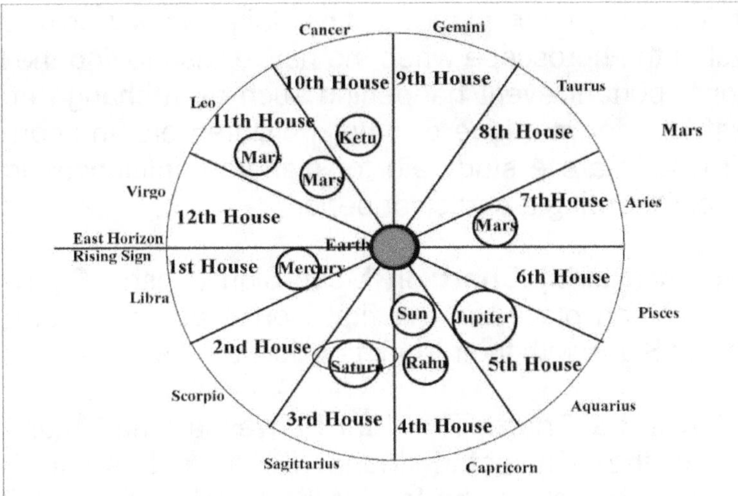

FIG 3: Western (American) Style Chart

1.5 Classification of Kundali Chart

Following different type of Horoscope Charts are prepared to study the different aspects of life:
Natal Chart: This is the Main Lagna-Chart.

Chalit Chart: Chalit Chart shows the actual Zodiac Sign and the actual planet position in a particular house. But Chalit Chart shall not be used for determining the aspects or knowing the sign in which a planet is posited or knowing the strength of a planet. Chalit Chart is the actual calculation of the 12 houses of the Lagna Chart and therefore, actual position of planet by making the house division for corrects predictions, because the planets show their behavior according to the house they occupy.

Transit Chart: The Transit Chart or Progression Chart is prepared by positioning the rising Sign in 1st House and planets in the particular fouling Sign at that particular time. Then, this Chat is compared to find the aspect of transiting planet on Natal chart planet to find daily, weekly or monthly forecast of the horoscope when the native wish to find the best time for important event happening such as a change of job, the best time for marriage or having children etc. In short, the Transit Charts are a study aid for planetary influences in the individual life during a particular period.

Moon-Sign (Rashi) Chart: In Moon-Sign (Rashi) Chart, the Sign with Moon placed in the first House as Ascendant and rests other Signs with their Planets follow the Ascendant.

Tithi Parivesha Chart: The Tithi Parivesha Chart manifests concretely the Vimshotari Maha Dasa and Antar Dasa sequence. In this chart, the lord of the Vimshotari Maha Dasa is placed in the first House as Ascendant. In relation to this, the inner workings of the Antar Dasa are revealed with the help of the Antar Dasa lord planet positioned in different Houses. The Antar Dasa Lord brings the timing of the events that come about as a result in life. Tithi Parivesha Chart distils all this along with the transit of the Moon in the Rashi Chakra (Rashi-Chart).

Divisional Charts: In Divisional Charts, each Sign is divided into various divisions, as per the requirements of the type of the Chart. The Divisional Charts are used for study of (1) strength of the planets, (2) important aspects of life. Each Chart gives a clear history of one of aspects of life of the native, such as (1) The physique of the native is known by Lagna Chart, (2) wealth by Hora Chart, (3) happiness through co-born/sibling by Drekana Chart, (4) fortunes from Chaturthamsa Chart, (5) sons and grandsons from Saptamsa Chart, (6) spouse and planet strength from Navamsa Chart, (7) power and position from Dasamsa Chart, (8) parents from Dvadasamsa Chart, (9) pleasure and adversities through conveyances from Shodasamsa, (10) worship from Vimsamsa, (11) learning from Chaturvimshamsa, (12) strength and weakness from Saptavimshamsa, (13) evil effects from Trimsamsa, (14) auspicious and inauspicious effects from Khavedamsa, (15) all indications from Akshavedamsa and (16) Shashtiamsa charts. The Divisional Charts are discussed in different volume of the book.

1.6 Panchanga (Hindu calendar)

The Panchanga is the "Lunar Calendar" based on Moon's monthly movement. One month is divided into two Paksha, such as, Krishna Paksha and Shukla Paksha. On the average year as 29.53 x 12 = 354.36 days, which is less by 11 days compared to earth's cycle days (365 days). Thus, in every three years there is an Adhika Maah (Extra Month) to cover the gap of these 11 days. Panchanga is used to determine the most ideal or auspicious time for carrying out various activities like getting married, House warming, starting education, laying the foundations of new homes. It gives the Muhurata (time) when a particular task can be undertaken to reap maximum benefits. Panchanga means five organs. These five things are (1) Vara (Day); (2) Tithi (Date), (3) Nakshatra (Constellation), (4) Yoga and (5) Karana.

i) Vara (Day): The day starts from the sunrise and continue till next sunrise. It is the solar day of the week. There are 7 days in a weak.

ii) Tithi (Date): There are 15 Tithis, commencing from next day of Amavasya. The 1st Tithi is the 1st day of bright half of a lunar month to Purnima (Full moon), the 15th Tithi of the bright half. They are called the Tithi of the Shukla-Paksha (brighter phase). Similarly, there are 15 Tithi commencing from next day of Purnima to Amavasya called Krishna Paksha (darker phase). The 15 Tithi are called Pratipada (Prathama), Dwitiya, Tritiya, Chaturthi, Panchami, Shasthi, Saptami, Ashtami, Navami, Dasami, Ekadasi, Dwadashi, Trayodashi, Chaturdasi, and Purnima or Amavasya.

iii) Nakshatra (Naal): There are 27 Nakshatra (group of stars) that Moon occupies. It is also called the constellation.

iv) Yoga: Yoga is an auspicious moment. There are 27 Yoga in a month.

v) Karana: There are one Karana in a one and a half of the Tithi a (Day). There are 11 Karana altogether, which repeats.

Hindu calendar (Maah): The Hindu calendar has twelve months (Maah) in a lunar year corresponded to the following English Calendar Months. The solar month is given against Maah in which they begin and ends. Each Maah starting and the end date are as follows:

S.N.	Maah	Solar Month
1	Chaitra 30 days	March 22 – April 20
2	Vaisakha 31 days	April 21 – May 21
3	Jyaistha 31 days	May 22 - June 21
4	Asadha 31 days	June 22 – July 22
5	Sravana 31 days	July 23 – August 22
6	Bhadra 31 days	August 23 – September 22
7	Asvina 30 days	September 23 – October 22
8	Kartika 30 days	October 23 – November 21
9	Agrahayana 30 days	November 22 – December 21
10	Pausa 30 days	December 22 – January 20
11	Magha 30 days	January 21- February 19
12	Phalgun 30 days	February 20 – March 21

1.7 Correction of Birth Time

The correction of the "Birth Time" is needed before proceeding to the horoscope prediction. Even a slight change in the Birth Time can change its longitude drastically and hence leads to incorrect prediction of the Kundali. The Birth time has to be rectified to have the correct longitude of the Ascendant. The correct degree and the minutes to the extent of 20 minutes of the Ascendant can be found by the following three methods:
1. Navamsa-Dvadasamsa and Kunda Method.
2. Balance of Dasa from the Ascendant Method.
3. Fixing Seven Sub Planets, Gulika, and Pranapad with Sun Position Method

Navamsa-Dvadasamsa and Kunda Method: Each Sign is divided into 9 of 3^0 and 20' each to make a Navamsa Chart. Then, each Navamsa is divided into 12 (Dvadasamsa) equal parts of 16' 40" duration. Thus we get a Navamsa Dvadasamsa, called in short N D. In case of male native birth, the male Signs like Aries, Gemini, Leo, Libra, Sagittarius and Aquarius will fall in the 1st, 3rd, 5th, 7th, 9th, and 11th only in Navamsa-Dvadasamsa (N D). Similarly, Taurus, Cancer, Virgo, Scorpio, Capricorn and Pisces will fall in the 2nd, 4th, 6th, 8th, 10th, and 12th in N D in the female Signs only in case of female births. Each N D takes approximately about 1 minute 6 seconds to rise. Therefore, depending on whether it is a male or female birth, we can easily reject those longitudes of the Ascendant, which are not suitable as per above calculation. But in each Navamsa, six sets of N D show the male births and the other six sets show the female births. We can use the Kunda method to pick up the correct one out of these six sets of N D.

In Kunda Method, we multiply the longitude of the Ascendant by 81 and after expunging multiples of 360^0, it gives the longitude of the birth star or the 10th or 19th star from the birth star. We find that, after multiplying the longitude of the Ascendant by 81 and after expunging the multiples of 360^0, it gives the longitude of either 0^0 to 13^0 20', then the birth star is Ashwini, or between 120^0 to 133^0 20', then the birth star is Makha or between 240^0 to 253^0 20', then the birth star is Moola. Now, if this result does not match with the calculation as told above, then the longitude of the Ascendant should be corrected till the Kunda method conditions are fulfilled or obtained. Thus, by combining the "N D and Kunda" methods we can, in most of the cases, correct the longitude of the Ascendant up to 20 minutes difference in the birth Time.

Balance of Dasa from the Ascendant method: It may happen that within a period of 20 minutes, two or more longitudes of the Ascendant may fulfil both conditions of "N D and Kunda" method. In such case, the longitude of the Ascendant can be corrected by the second method, i.e. Balance of Dasa from the Ascendant method. According to this method, the Dasa balance at the birth is arrived at from the Moon or the Ascendant, whichever is the stronger. But as, even a slight change in the Birth Time, results in the longitude of the Ascendant changing drastically, the Dasa balance arrived at from this longitude becomes unreliable and has, therefore, fall into disuse. But, if we arrive to the correct Ascendant degree and minute, the Dasa progression derived from it will give very accurate result. If after the N D and the Kunda Methods are judged through, we obtain two or more Ascendant longitudes, and then by deriving the balance of the Dasa period from each of the Ascendant longitudes and checking with the past events, it can tell us which of this longitude is correct.

Seven Sub Planets, Gulika, and Pranapad Method: The following thumb rule tests must be applied to accept the Ascendant as the correct one. (a) The Ascendant occupies the 1st, 3rd or any odd position from the Gulika or Pranapad position. (b) The Ascendant is found to be in Trine aspect with the Navamsa of Pranapad. If the result of the above two tests are not found satisfactory, the decision should be taken only after the Seven Sub-Planets, Gulika, and Pranapad Test. (c) In the Seven Sub Planets Test, the five Sub-planets (Invisible Planets) are Dhooma, Vyatipat, Parivesha, Indradhanu and Upketu. Find out these sub-planets as per calculation given in the chapter of Sub-Graha. We must get the Sun longitude back, which shows that the Ascendant is correct. (d) The Gulika Test aims at deciding the correct Rashi. Gulika is very important in judging the correctness of an Ascendant by conforming to any of the following positions, especially, if Moon is not strong in the natal chart. The Ascendant is correct, if, (1) Ascendant will be Trine (the 3rd house) to the lord of the house occupied by Gulika and (2) Ascendant will be Trine (the 3rd house) to the lord of the house occupied by the Gulika in Navamsa. (3) Ascendant will be Trine (the 3rd house) to the Navamsa house of the Gulika itself. In Pranapad Method, the Pranapad test-1 is based on the principle of finding the correct degrees. Multiply the Ghati by 4. Let the product be “P’. Divide the “P” by 15 and let the quotient be “X” and remainder be “Y”. Now, divide the (P + X) by 12. Let this quotient be “Z” and remainder is “R”. Then, the Remainder “R” indicates the Rashi number and “2Y” indicates the number of the degrees in the mean Pranapad.

Pranapad Test-2 is to convert the Ghati into Pala. Divide the result by 15 and let the quotient be "x" and remainder be "y". Now, divide the "x" by 12. Let this quotient be "z" and remainder is "r". Then, the Remainder "r" indicates the Rashi number and "z y" indicates the number of the degrees in the mean Pranapad.

The true Pranapad is obtained as here. Note down the Rashi and the degree of the Sun. If it is not a Movable Rashi, note down the Rashi, which is in the Trine with this and is also movable. Add this Rashi and the Sun's original degrees to the Rashi and the degrees of the mean Pranapad as obtained above. Thus, we get the true Pranapad.

The Accuracy of the Ascendant Test is as here. The Ascendant is considered to be the accurate, if; the Rashi indicated by the Gulika and also by the Pranapad occupy the same or an odd position from the Ascendant. The sum of the degrees of the mean Pranapad and the Sun's longitude must coincide with the degrees of the Ascendant.

1.8 Reference Books

1. Hora, Ganita and Samhita: The great sage Parasara narrated the science of astrology as heard through Lord Brahma in three divisions, viz. Hora, Ganita and Samhita.

2. The Brihat Parasara Hora Shashtra: The great sage Parasara lived at the time of the Mahabharata war, about 3000 BC. The Brihat Parasara Hora Shashtra (a compendium on astrology) is the primary textbook of Vedic astrology.

3. Saravali Translated by R. Santhanam: This focuses on planets in houses and Decanate.

4. Jaimini Sutras: The Jaimini sutras by Rishi Jaimini are a unique classic, and considered as next only to the Brihat Parasara Hora Shashtra.

5. Bhrigu Samhita: Rishi Bhrigu was the first compiler of predictive Astrology. His famous compilation, Bhrigu Samhita, which contains the predictions for thousands of combinations, is popular even today.

6. Bhrigu Sutras: Rishi Bhrigu was the first compiler of predictive Astrology. His famous compilation, Bhrigu Samhita, which contains the predictions for thousands of combinations, is popular even today.

7. Prasana Tantra: By Neelakanta Daivagyna around 1550 AD, is a great classic dealing with the Prasana or Horary Astrology and a must for any astrologer.

8. Brihat Parasara Hora Shashtra Translated by R. Santhanam: The "bible" of Vedic astrology.

2

Muhurata

2.0 General

Panchanga is used to determine the most ideal or auspicious time for carrying out the undertakings based on the Tithi (Lunar day), Vaar (Day), Karana, Yoga and Nakshatra which is known as Muhurata. Tithi (Date) brings prosperity; Vara (Day) prolongs life, Nakshatra removes misdeeds, Yoga gives immunity from diseases, and Karana and Hora leads to success. Sun and Mars rules forceful punishment (danda), Mercury and Saturn rules diplomacy (bheda), Moon rules temptation (Dana) and Jupiter and Venus rules good counsel (Sama).

2.1 Prediction of Muhurata

a) Muhurata by Tithi (Waightage value = 1 Point):

Tithi: The Tithi (Hindu Date) is indicated in Panchang by T. The time next to the T is the ending time of Tithi. If the Tithi ends after midnight but before next sunrise, the end time will have a value greater than 24 hours. To get actual time, subtract 24 from that time. Remember that Indian Tithi does not change at midnight 0:00 hours like English date. Example: If the Panchang for October 23rd is (T: Dwadashi 27:10:11). It means it ends after midnight of that day at 27 hours, 10 minutes and 11 seconds. So it means Dwadashi will end at 3:10:11 AM of the next day (October 24th).

Muhurata by Good Tithi: All Tithi are good for General purpose, Wedding, Buying a New Vehicle, or Starting Business, except the Lunar Days (Tithi), such as, the 4th, 6th, 8th, 12th and 14th (from both Paksha), full and new moon days should be avoided. The Lunar Tithi Ashtami (8th Tithi) in both Paksha is discarded well for all auspicious matters including Travel/Yatra. Please avoid Rikta Tithi 4-9-14 (from both Paksha) and other tithi like K-13th (Krishna Trayodashi), Amavasya, and S-1 (Shukla Paksha Prathama) Tithi. However, other Tithi are auspicious for certain types of undertakings. Tithis 2, 3, 5, 7, 10, 11 and 15th are best for all undertakings.

b) Muhurata by Vaar (Day) (Waightage value = 8 Point):

Good Day (Vara): Monday, Wednesday, Thursday and Friday are good days. Avoid Sunday, Tuesday and day of eclipse. Following Vara (Day) allocated for worship of Devas (God) are, Monday for Shiva Ling, Paravati and Shri Hari Pujan; Tuesday – Shri Ganesh Pujan; Wednesday – Shri Vishnu Pujan; Thursday – Shri guru Pujan; Friday – Ganga & Laxami Pujan; Saturday – Shri Hanuman Pujan; and Sunday – Shri Surya narayan Pujan. All Amavasya – Pitru (all Soul) Pujan and on all Purnima – Shri Shtyanarayan Pujan.

c) Muhurata by Karana (Waightage value = 16 Point):

Karana: The Karana is half a tithi. Karana means to take action, to do. The Karana have the ability to strengthen or mar a tithi. They represent the underlying influences. Karana are underlying energies and have to be studied as subordinate to the tithi not the other way round. Karana are the 6° difference between the Sun and the Moon. They are 60 in all. There are thirty tithi and sixty Karana. There are 11 types of Karana. Four fixed and seven movable. The fixed Karana are Shakuni, Chatuspada, Naga and Kintughna are negative. The seven movable Karana are Bava, Balava, Kaulava, Taitila, Gara, Vanija and Vishti. They start with Bava on the second half of the first tithi and repeat themselves in the same order eight

times in a lunar month. The four fixed Karana start with the 2nd half of the 14th day of the Krishna Paksha, Amavasya and the first half of the 1st Tithi of the Shukla Paksha. The first Karana in shukla paksha is always the fixed Kintughna. The fixed Karana do not repeat themselves. Karana name is indicated by K: There are two Karana per Tithi. Hence there are two lines for Karana, where the time next to the Karana indicates when it is going to end. Example: (Karan: 08:32:33). There are 11 Karana altogether. They are: 1) Bala, 2) Balava, 3) Kaulava, 4) Taitil, 5) Gara, 6) Vanijya, 7) Vishti, 8) Shakuni, 9) Chatushpada, 10) Naaga, and 11) Kinstughna.
Good Karana: All Karan are good for Muhurat except Vishti (Bhadra). Try to avoid Vishti (Bhadra) Karana in all auspicious ceremonies.

d) Muhurata by Yoga (Waightage value = 32 Point):

Yoga (Nitya Yoga): Nitya means daily and Yoga means a 'yoke' or connection or the Panchanga. Panchanga yoga is formed by the daily connection between of the Sun and Moon. The Moon travels 13° 20' away from the Sun each day. These 13° 20' sections (similar length of a Nakshatra) form yoga. Yoga is a subtle blending of solar and lunar energies that give special indications every day. There are 27 such yoga formed and each one is linked to a particular Nakshatra. Some are malefic in nature and others benefic. The yoga at the time of the birth influences the personality and also gives the yogi and avayogi planets. The yoga are connected to the Nakshatra but are different from the planetary Nakshatra. Yoga name is followed by Y: Ganda 08:31:32 indicates it is going to end at 08:31:32.

Good Yoga: All Yoga is good for Muhurat. Avoid Vyatipata, Vaidriti, Parigha, Vishkumbha, Vajra, Shoola, Atiganda, and Vyagata.

Table: List of 'Yoga'

S. N.	Yoga	Nakshatra	Yogi Planet	Avayogi

1	Vishkumbha	Pushya	Saturn	Moon
2	Preeti	Ashlesha	Mercury	Mars
3	Ayusmana	Magha	Ketu	Rahur
4	Saubhagya	P Phalguni	Venus	Jupitar
5	Shobhana	U Phalguni	Sun	Saturn
6	Atiganda	Hasta	Moon	Mercury
7	Sukarma	Chitra	Mars	Ketu
8	Dhriti	Swati	Rahu	Venus
9	Shoola	Vishakha	Jupiter	Moon
10	Ganda	Anuradha	Saturn	Moon
11	Vriddhi	Jyeshta	Mercury	Moon
12	Druva	Mula	Ketu	Rahu
13	Vyaghata	P Ashadha	Venus	Jupiter
14	Harshana	U Ashadha	Sun	Saturn
15	Vajra	Sharvana	Moon	Mercury
16	Siddhi	Dhanishta	Mars	Ketu
17	Vyatipata	Shatabhishak	Rahu	Venus
18	Variyana	P Bhadra	Jupiter	Sun
19	Parigha	U Bhadra	Saturn	Moon
20	Shiva	Revati	Mercury	Mars
21	Siddha	Ashwini	Ketu	Rahu
22	Sadhya	Bharani	Venus	Jupiter
23	Shubha	Krittka	Sun	Saturn
24	Shukla	Rohini	Moon	Mercury
25	Brahma	Mrigasira	Mars	Ketu
26	Indra	Ardra	Rahu	Venus
27	Vaidhrati	Punarvasu	Jupiter	Sun

e) Muhurata by Nakshatra (Waightage value = 60 Point):

Nakshatra 2-4-6-8-9-11-13-15-17-18-20-22-24-26 and 27 is good for muharata. Please avoid 1-3-5-7-10-12-14-16-19-21-23-25 Nakshatra counting from your birth Nakshatra. Avoid Janma-Nakshatra for all good works. Bharani and Krittika Nakshatra should be avoided for all auspicious works. Avoid Ashwini Nakshatra on Tuesday, Rohini Nakshatra on Saturday, and Pushya Nakshatra on Thursday for Griha Pravesha, Marriage, and Travel Muhurata. The 16th Nakshatra (Star) from his/her birth star is prohibited for any kind of muhurata and if performed, it will cause grief. Example: If your birth Nakshatra is Swati, and muhurata's day (today's) Nakshatra is Krittika, then counting from Swati to Krittika it gives us the count of 16th, which is not a good Nakshatra for any muhurata. Hence till Krittika Nakshatra prevails, that time is bad to do any job.

Table: Nakshatra List

Nakshatra	Nakshatra	Nakshatra	Nakshatra
Ashwini	Pushya	Swati	Shravana
Bharani	Ashlesha	Visakha	Dhanishta
Krittika	Makha	Anuradha	Shathabisha
Rohini	Purrva Phalguni	Jyeshtha	Purva Bhadrapada
Mrigashira	Uttara Phalguni	Moola	Uttara Bhadrapada
Aridra	Hasta	P. Ashada	Revati
Punarvasu	Chitra	U. Ashada	

Nakshatra Bala (Tara Bala): Tara Bala is to find a day suitable for Muhurata. To calculate Tara Bala, count the position of muhurata's day Nakshatra from the Janma (birth) Nakshatra (both Nakshatra inclusive). Divide that by 9. The reminder is Tara Bala. If the remainder (Tara Bala) is 2, 4, 6, 8, 9 or 0 then it's very good otherwise, if the remainder (Tara Bala) is 1, 3, 5, 7, then they are not auspicious, but it is bad for muhurata. Example: In one case, the Nakshatra (Birth Star) is Uttar Ashada and today's star / Nakshatra is Ashwini so the Tara Bala will be 8 (very good). But had the birth star was Ashwini and today's star was U. Ashada, so the Tara Bala for today would had been 3 (Bad). Meaning of Tara Bala is explained in following table.

Tara Bala Number, Name & Result			
No.	1	2	3
Name	Janma	Sampatha	Vipatha
Muhurata	Not Good	Very good	Bad
No.	4	5	6
Name	Kshema	Pratyaka	Sadhana
Muhurata	Good	Not Good	Very Good

No.	7	8	9
Name	Naidhana	Mitra	Paramitra
Muhurata	Very Bad	Very Good	Good

f) Muhurata by Chandrabala:

The consideration of the Moon and his position are of much importance in Muhurtha. The best position of Moon is that it should not occupy the 6th, 8th or 12th from the person's Janma Rasi in the election (Muhurata) chart. Example: a person born in Mrigasira (Janma Rasi being Taurus) wants to have his marriage on a day ruled by Bharani which means the Moon will be in Aries. There is neither Tarabala (as Bharani will be Naidhana to Mrigasira) nor Chandrabala (the Moon falls in Aries in the election (Muhurata) day which would be the 12th from the subject's Janma Rasi). Hence, the day is most inauspicious. Hence certain constellations, apart from their being harmonious or otherwise disposed with reference to one's own Janma Nakshatra, should be avoided for this specific purposes on account of their inherent evil natures.

g) Muhurata by Sun in Zodiac Sign:

The next line is SN. This shows which Rashi or Sign the Sun is positioned. After that it is the time when Sun will enter to the next Rashi or Sign on that day. Example: (SN: Makara 4:36). This is also known as Samkaranti. Makara Samkaranti is when Sun enters Makara Rashi. Avoid six hours before and after from the time of Samkaranti for any good work Muhurata. However, this Samkaranti time is good for donation, Satyanaraya Puja but not good for Weddings, Griha Pravesha.

h) Muhurata by Gulika/ Rahukaal/ Yamagandakaal:

1) Muhurata by Gulika: The GK: means the Gulika. Gulika is also called Mandi (i. e. the son of Manda alias Shani). His rising period is also not auspicious. The time frame indicates

Gulika duration. Example: (GK: 09:03-10:06). Please avoid this time for your Muhurata.
2) Muhurata by Rahu Kalam: The RK means the Rahu Kalam. The rising period of Rahu is considered inauspicious in the South as he is considered a malefic for auspicious functions. The time frame indicates Rahu Kalam duration. Example: (RK: 09:03-10:06). Avoid this time for your Muhurata.
3) Muhurata by Yamagandakaal Kalam: The YM: means the Yamagandakaal. Yamagandakaal is the son of Guru and he is also considered inauspicious. The time frame indicates Yama Kalam duration. Example: (YM: 11:09-12:12). Avoid this time for your Muhurtha. Any auspicious work, Travel or Celebration Puja shall be prohibited in Gulika/Rahu Kaal/Yamagandakaal Timing as per the table given below:

Days	Rahu Kaal	Gulika	Yamagandakaal
Sunday	4.30 PM - 06.00 PM	03.00 PM- 04.30 PM	12.00 AM- 01.30 PM
Monday	07.30 AM- 09.00 AM	01.30 PM- 03.00 PM	12.00 AM-1.30 PM
Tuesday	03.00 PM- 04.30 PM	12.00 AM- 01.30 PM	09.00 AM- 10.30 AM
Wednesday	12.00 AM- 01.30 PM	10.30 AM- 12.00 AM	07.30 AM- 09.00 AM
Thursday	01.30 PM- 03.00 PM	09.00 AM- 10.30 AM	06.00 AM- 07.30 AM
Friday	10.30AM- 12.00 PM	07.30 AM- 09.00 AM	03.00 PM- 04.30 PM
Saturday	09.00 AM- 10.30 AM	06.00 AM- 07.30 AM	01.30 PM- 03.00 PM

i) Muhurata by Dur muhurata:

The DM means Durmuhurata (DM: 10:34- 11:22, 15:22- 16:10) means first Durmuhurata starts at 10:34 AM till 11:22 AM and the next one starts at 15:22 PM to 16:10 PM. It should be avoided for any good ceremonies.

j) Muhurata by Varjyam or Vishagatika:

The V: means Varjyam or Vishagatika (V: 30:00-31:37) means Varjyam starts at 6:00 AM of following day and lasts till 7:37 AM of following day. There could be more than one Varjyam line. This time is not a good time. Please avoid in all Muhurata.

k) Muhurata by Amritkaal:

The AK: means Amritkaal (AK: 15:13- 16:53) means Amritkaal starts at 3:13 PM till 16:53 period is a good time. This time is good for Annaprasana Samskara as well as other rituals.

l) Muhurata by Abhijit Muhurtha:

The AJ: means Abhijit Muhurtha (AJ: 12:30- 13:18). Abhijit Muhurata is the time when sun transits in the mid heavens are called. The results of Abhijit Muhurta are beyond the scope of this article. It is Sri Hari Vishnu's time. When you don't have any Muhurata you can use this time if it's not overlapped by Rahu Kalam or Durmuhurata, or Varjyam. On Wednesday Abhijit Muhurtha is bad as it's the same time as Durmuhurata.

m) Muhurata by Panchaka Rahita Vidhi:

The most common method to determine Panchaka is to add, (1) The number of the lunar Tithi, (2) The number of the weekday (like Sunday = 1, Monday = 2, Tuesday = 3), (3) the number of the Nakshatra from Ashwini 1 (like Bharani = 2) and (4) The number of the Lagna Sign from Aries 1 (like Taurus = 2). Now after adding these together, divide the total by 9. The remainder with 3, 5, 7 and 0 are good Muhurata. If the remainder is 1, 2, 4, 6, or 8 then it indicate bad Muhurata, like, 1 indicates (Mrityu Panchaka), danger. 2 indicates (Agni Panchaka), risk from fire. 4 indicates (Raja Panchaka), bad results. 6 indicates (Chora Panchaka), evil happenings. 8 indicates (Roga Panchaka), disease. If the remainder is 3, 5, 7 or zero then it is very good. Example: Let's consider Tithi is Krishna Paksha Dwitiya = (15+2=17), Saturday = (7), Moola Nakshatra = (19), and Kumbha Lagna = (11). Let's add these together. This gives us total of 54. Now divide this total of 54 by 9, which gives us the reminder of zero. Hence it is a good Muhurata.

n) Muhurata by Decanate for Travel:

The Auspicious Functions / Travel / Yatra will be productive and fruitful if the Decanate of the Rising Sign is occupied, owned or aspect by the Benefic denoting flowers or pearls or precious stones, but it will be non-productive and unfruitful or harmful if the Decanate of the Rising Sign is occupied, owned or aspect by the malefic. The Function / Travel / Yatra will be of great strain and will cause the enforced seclusion to him/her, if the Decanate of the Rising Sign happen to be serpent. The Functions/Travel/Yatra will be productive and fruitful, if the Decanate rising at birth is a Quadruped, Nigada (Fetters) or Ayudha (Weapon/Armed) Drekana.

o) Muhurata by Special Yoga:

When a certain weekday coincides with a certain asterism and a certain lunar day, it becomes specialty Auspicious Yoga (Auspicious Muhurata) for good work. Followings are few such special combinations which go under the special distinction of Siddha Yoga for Muhurata:

Sunday: Sunday coinciding with the 1st, 4th, 6th, 7th or 12th lunar day and ruled by the constellations Pushya, Hasta, Uttara, Uttar Asadha, Moola, Sravana or Uttarabhadra gives rise to Siddha Yoga of Muhurata.

Monday: Monday identical with the 2nd, 7th or 12th lunar day and with the constellations Rohini, Mrigasira, Punarvasu, Chita, Sravana, Satabhisha, Dhanishta or Poorvabhadra produces the Siddha Yoga of Muhurata.

Tuesday: Tuesday falling on a day ruled by Aswini, Mrigasira, Chita, Anuradha, Moola, Uttara, Dhanishta or Poorvabhadra and coinciding with Jaya (3rd, 8th and 13th lunar days), gives rise to Siddha Yoga.

Wednesday: Wednesday coinciding with Bhadra and Java and with the constellations Rohini, Mrigasira, Aridra, Uttara,

Uttara Asadha or Anuradha and coinciding with Bhadra (2nd, 7th and 12th lunar days), generates Siddha Yoga.

Thursday: Thursday identical with the 4th, 5th, 7th, 9th, 13th or 14th and on 5th, 10th or 15th (Poorna) lunar day and with the asterisms Makha, Pushya, Punarvasu, Swati, Poorva Asadha, Poorvabhadra, Revati or Aswini gives rise to Siddha Yoga. Thursday falling lunar days constitute Siddha Yoga

Friday: Friday ruled by Aswini, Bharani, Aridra, Uttara, Chitta, Swati, Poorvashadha or Revati coinciding with Nanda (1st, 6th and 11th lunar days) and Bhadra constitutes Siddha Yoga of Muhurata.

Saturday: Saturday falling on a day ruled by Swati, Rohini, Visakha, Anuradha, Dhanishta or Satabhisha and with lunar days Bhadra and Rikta tithi (4th, 9th and 14th lunar days), generates the Siddha Yoga of Muhurata.
Sunday to Saturday respectively coinciding with the constellations Hasta, Sravana, Aswini, Anuradha, Pushya, Revati and Rohini will give rise to Amita Siddha Yoga.

2.2 Special Adverse Yoga for Muhurata and their Neutralisation:

There are 21 great evils (Mahadosha) to Muhurata to be avoided for any auspicious work. These are mentioned below:
1. Panchanga Suddhi: The Lunar Days (Tithi), such as, the 4th, 6th, 8th, 12th and 14th, full and new moon days should be avoided. Tuesday is to be avoided except when it happens to be the 10th, 12th or 16th day of the child's birth when the child's Namakarana (baptising or giving name) may be performed. Sunday coinciding with the 14th lunar day (Aryama); Monday the 8th (Vidhi) and the 12th (Naktanchara); Tuesday the 4th (Prtru) and the 11th (Vahni); Wednesday (Abhijit); Thursday the 12th (Naktanchara) and 13th (Varuna); Friday the 4th (Pitru) and the 8th (Vidhi) and Saturday the 1st

(Rudra) and the 2nd (Ahi). Bharani and Krittika Nakshatra should be avoided for all auspicious works as these two are said to be presided over by the god of death (Yama) and the god of fire (Agni) respectively except in cases of the Lagna is fortified, as the dosha due to Nakshatra may get neutralised. The last parts of Aslesha, Jyeshtha and Revati should also be avoided. The Yoga like the 6th (Atiganda), 9th (Soola), 10th (Ganda), 17th (Vyatipata) and 27th (Vyadhruti) should be avoided. The Karana like Bhadra (Bisti) is unfit for any good work but is eminently suitable for violent and cruel deeds. Therefore, Panchanga Suddhi means a good lunar day, a beneficial weekday, an auspicious constellation, a good yoga and a fertilising Karana.

2. Surya Sankramana: The Surya Sankramana is the solar ingress into different zodiacal signs. When the Sun is about to leave one sign and enter another there seem to occur certain disturbances in the organisation of the solar forces and such times are not recommended for any good work. Sixteen Ghatis (6 hours 24 minutes), both before and after the entry of the Sun into a new sign, should be rejected for all new works.

3. Karthari Dosha: Karthari means scissors. In an election, when two evil planets are placed on either side of the Lagna, the combination goes under the special name of Karthari Dosha and it should be rejected for good work particularly in regard to marriage.

4. Shashtashta Riphagatha Chandra Dosha: The Moon should invariably be avoided in the 6th, 8th and 12th houses from the Lagna rising in an election (Muhurata) chart.

5. Sagraha Chandra Dosha: The Moon's association with any other planet, benefic or malefic, should be avoided. This injunction is especially applicable in case of marriage.

6. Udayasta Suddhi: The Lagna and the seventh should be strong. The Lagna should be occupied by its own lord and the Navamsa Lagna by its own lord or vice versa or lord of Lagna should aspect Navamsa Lagna and vice versa. Similarly the seventh and the lord of the seventh Bhava should be favourably disposed. The strength of Lagna and the seventh is necessary in all elections (Muhurata) but especially so in regard to marriage.

7. Durmuhurtha: Muhurtha technically means 48 minutes or 2 ghatis in terms of time. A sidereal day consists of 30 muhurthas. The 1st fifteen diurnal muhurthas named are: (1)

Rudra. (2) Ahi. (3) Mitra, (4) Pitrii, (5) Vasu, (6) Vara. (7) Vishwedeva, (8) Vidhi, (9) Sathamukhi, (10) Puruhuta, (H) Vahini, (12) Naktanchara, (13) Varuna, (14) Aryama and (15) Bhaga. The nocturnal muhurthas are: (1) Girisa, (2) Ajipada, (3) Ahirbudhnya. (4) Pusha, (5) Aswi, (6) Yama, (7) Agni. (8) Vidhatru, (9) Chanda, (10) Aditi, (11) Jeeva, (12) Vishnu. (13) Yumigadyuti, (14) Thyasthur and (15) Samdram. In regard to the diurnal muhurtha, the 1st, 2nd, 4th, 10th, 11th, 12th and 15th are inauspicious while in nocturnai muhurthas the 1st. 2nd, 6th and 7th are inauspicious. In calculating the muhurtha, the exact length of day and night should be ascertained. Each muhurtha is said to last for 48 minutes (2 ghatis) on the assumption that the duration of day and night is of equal proportion, viz. 30 ghatis, or 12 hours. If the length of day is 28 ghatis, then each muhurtha extends for 1 ghati and 52 vighatis (20h 20s. 8).

8. Gandanthara: The last 2 ghatis (48 minutes) of the 5th, 10th and 15th (Full Moon) and the first 2 ghatis of the 6th, 11th and 1st (dark half) lunar days go under tithi gandanthara and they should be rejected for all new works. Similarly, the last 2 degrees of Cancer, Scorpio and Pisces and the first 2 degrees of Leo, Sagittarius and Aries are inauspicious. The last ghatis of Aslesha, Jyeshta, Moola, Revati and Aswini and the first four ghatis of Makha should be avoided as injurious for good work.

9. Papashadvargs: Malefics is not strong in shadvargas in an election (Muhurata) chart.

10. Bhrigu Shatka: The position of Venus in the 6th is injurious. This is especially so in regard to marriage. Even when Venus is exalted and associated with benefic, such a disposition is not approved.

11. Kujasthama: Mars should be avoided in the 8th house, as it indicates destruction of the object in view. In a marriage election (Muhurata) chart. Mars in the 8th is unthinkable. Even if Mars is otherwise powerful, he should not occupy the 8th house.

12. Ashtama Lagna Dosha: In selecting a time for marriage, the Lagna ascending should not happen to be the 8th from the Janma Lagna of the bride and the bridegroom. Suppose the husband and wife is born in Aquarius and Capricorn respectively. At the time of marriage, the ascending Lagna should be a sign other than Virgo or Leo as these two happen to be the 8th from the bridegroom and bride's Janma Lagnas respectively.

13. Rasi Visha Ghatika: Elsewhere has been given the negative periods of different Lagnas (Lagna Thyajya). They are to be rejected for all auspicious work.

14. Kunavamsa Dosha: The Lagna selected for an auspicious work should not occupy the Navamsa of a malefic.

15. Varadosha: This has already been explained on page 26. Certain weekdays are to be avoided for certain special activities.

16. Grahanothpatha Dosha: The constellations in which the eclipses appear should be avoided, and in regard to marriage, such a constellation should be avoided for six months.

17. Ekargala Dosha: This dosha is powerful only during the daytime. Affects matters started under certain yogas such as Vishkhambam. We need not go into details as it is not of much significance.

18. Krura Samyuta Dosha: The constellation occupied by the Sun at a given moment, and the one immediately preceding and succeeding it have to be deemed unpropitious for all good work and they should be rejected for purposes of marriage.

19. Akalagharjitha Vrishti Dosha: When there is rainfall and thunder, out of season, such days should be deemed unfit for all good work.

20. Mahapatha Dosha: When the Sun and, the Moon are equally removed from the equator upon the same side of it, the aspect is known as Vyatipata, which indicates excess of evil. This is held to be unfavourable for all good work.

21. Vaidhruthi Dosha: This is also an evil aspect (yoga) and should be avoided in all favourable activities. One should always remember that in electing a suitable moment one should try to avoid the major dosha by fortifying the ascendant and taking advantage of the exceptions and ignoring the minor ones. The following combinations are held to neutralise the adverse yoga mentioned above: 1. The lagnathyajya referred to supra prevails only on particular days as per details below. In the first Navamsa, Wednesday and Saturday; in the middle Navamsa, Monday and Friday; in the last Navamsa, Tuesday, Thursday and Sunday; in other days the thyajyam has no significance. 2. Chandrashtama shows no evil when the Moon is waxing and occupies a benefic sign and a benefic Navamsa, or when there is Tarabala. The sting is lost when the Moon and the 8th lord are friends. 3. Tuesday is not evil after midday. 4. The aspects attributed to Vyatipatam, Vaidruti, become defunct after midday. 5. No day of the week is blemished if the lord thereof is strongly placed in the election chart. 6. Venus, Mercury or Jupiter in the ascendant will completely destroy all other adverse influences. 7. Jupiter has the power of dispelling all the evils due to the Lagna, Navamsa and malefic aspects and render the time highly propitious. 8. The mere presence of the Moon or the Sun in the 11th will act as an antidote for other evils obtaining in the horoscope. 9. If the angles are well fortified, evil influences are countered. 10. A planet exalted in Lagna will nullify the other adverse influences. 11. Jupiter or Venus in a Kendra (quadrant) and malefic in 3, 6 or 11 will remove alt the flaws arising on account of unfavourable weekday, constellation, lunar day and yoga. Thus it will be seen that the most important question in Muhurtha is the fortification of Lagna and its lord.

22 Mundana (First Hair Cut) Muhurata: Mundana ceremony should take place on the dates given by the Website: 'Kesha Khandana Muhuratham' or 'auspicious days for first hair cut'

3

House (Bhava)

The Sanskrit name of the House is "Bhava" or "Sthana". The Houses govern all the "events" of our lives. If we look directly east and point our hand at the Eastern horizon where the Sun rises, we are pointing to the first House, which is called Lagna or Ascendant. Each House/Ascendant/Lagna occupies the 30-degree span of space below the Eastern horizon. Directly across from it or 30-degree span of space above the Western horizon is the 7th house of the Natal Chart. From the first to the 7th are the 2nd through 6th houses. Directly over our head is the 10th house, which is called MC. That is the straight up into space where the high noon Sun beams down on us. Straight below our feet is the 4th house. Suppose 9 O' clock position of the clock represent the Eastern horizon. So, the first house governs at 9.00 to 8.00 O' clock position of the clock. The 2nd house is just below it, i.e. at 8.00 to 7.00 O' clock position of the clock and so on. Thus, the 12th house governs at 10.00 to 9.00 O' clock position of the clock, and this brings us back around to the first house of the Natal Chart. Thus, the entire 360 degree circular span of space surrounding us are divided into 12 equal part of 30 degree sections each.

3.1 Concept of House

It is like that we are standing inside a circular room. The circular room is divided into eleven rings up to the wall, starting from the ring nearest to the wall to the ring at the centre of the room. There are nine Planets, one in each ring. The native is in the central ring, which represents the Earth. The ring nearest to the wall has 12 divisions, each representing one Sign. The native is standing at a point in the central ring, i.e. on the Earth which is fixed. The other nine rings containing one Planet each and the tenth ring containing 12 Signs are moving around the native. The Planets are moving at different speeds. The wall of the room is fixed and is divided into 12 parts, each representing one House. When native looks around he sees the 9 Planets and 12 signs rotating within the room. Whenever he looks at any one of the planets, he sees a particular coloured portion of the Sign and a House behind them from his viewpoint. Thus, each planet in a certain Sign is in a certain House always. It does not mean that the planet is actually in those Sign at that time, but simply that the particular Sign is the backdrop far away behind the planet.

3.2 Concept of House & Lord working

In Vedic astrology, one sign always occupy one house. Whatever sign is raising the entire portion of sign is considered in the first house. In this way there is an exact correlation between Sign, Sign Lord and House. The Lord of the House is the Lord of the Sign in that House. The lord of the first House and its position in the other house will together determine the body. The body gets shape of the house in which the First House ruler has come to stay in and that house greatly affects the body. The matter related to the body is affected during periods ruled by the lord of 1st House in the Vimshotari Dasa. Secondly, the planets associated with the 1st lord greatly affect the affairs or events of body of the native. Example: If the lord of the first House is placed in the first house; the body becomes the focus of the life. If the lord of the first house is placed in the second house then the body will be greatly influenced by the second house, i.e. possession and wealth. The native is busy with accumulating wealth whether he is coming from a strong family background or otherwise. If the lord of the first House is placed in the third house, then the body is found to be connected with younger brothers, sisters, and the intelligence, good education, college degrees and other things that the third house rules. If the lord of the first house is placed in fourth house then the person will be greatly concerned with family affairs, home, parents and house & land properties in his life. The native will spend a lot of time at home, rather than on the road or at work. Since the fourth house is tenth from the seventh house, it also means that the native will have some connection with the career of the spouse. This means that the person might work for spouse or work with her spouse or perhaps a home-based business. If the lord of the first house is placed in the fifth house then the children, creativity, romance and helping nature becomes the focus of his bodily activities. If the lord of the first house is placed in the sixth house, then the deaths, diseases and enemies overwhelm the person. If the lord of the first house is placed in the seventh house, then the person is heavily focused on his partner, his spouse during his life. If the lord of the first house goes to the eighth house in a horoscope, then serious physical harm; unforeseen difficulties and serious problems comes to the native. If the lord of the first is placed in the ninth house then there is an overall fortunate protective cover on the native throughout his lifetime. If the lord of the first is placed in the tenth house, then the native is heavily focused upon the attainment of career, status position and success. If the lord of the first is placed in the eleventh house, then the person is heavily focused upon the achievement of desires and certain gainful things in life. If the lord of the first is placed in the twelfth house, since this house rules charity, donations, and losses, the person will be busy in these ways throughout his life. In other words, he may donate himself to these causes. Or, there may be a lot of loss in his life.

Thus, the placements of the lords have their first level meanings. Then the second, third, fourth and deeper and deeper meanings depend on the astrologer's ability to read the complexity of the house relationships with the planets and signs.

3.3 Lagna Concept

Sage Parasara mentioned a few special Lagnas. He wanted special Lagna to be used for clear predictions of the events, which are follows:

Janma Lagna: In Natal chart, the first House is called Janma Lagna. There are Badhaka Graha defined as per the Janma-Lagna Sign. Even though the Badhaka Graha is natural Benefic, he works as the Badhaka Graha and harms badly that house in which he is sitting during his Maha Dasa and Antar Dasa.

Chandra Lagna: It is the Hose, where the Moon is sitting in the Natal Chart. It is also called Moon-Sign or Rashi in Hindu astrology and is very important Lagna for predictions.

Surya Lagna: It is the House, where the Sun is supposed to be sitting in the Natal Chart. It is defined with the date of Birth of the native. It is called Sun-Sign in Hindu astrology and is very important Lagna for predictions.

Indu Lagna: It is calculated as per details given in other Chapter and taken for predictions of wealth.

Hora Lagna: It is taken for predictions of finance & prosperity. Hora Lagna remains in between day and night. From sunrise till the time of birth, Hora Lagna repeats itself every 2½ Ghati (i. e. 60 minutes). Divide the time from sunrise time to up to birth time by 2½ and add the quotient in Rashi, degrees and so on to the longitude of Surya, as at the sunrise. This will yield Hora Lagna in Rashi and degrees.

Ghati Lagna: It is taken for predictions of name and fame. Ghati Lagna changes along with every Ghati (24 minutes) from the sunrise. Find out the birth time in Ghati and Vighati and consider the number of Ghati past, as number of signs, or Ghati Lagna. The Vighati be divided by 2 to arrive at degrees and minutes of arc, past in the said Ghati Lagna. The product so arrived in Rashi, degrees and minutes is added to Surya longitude, as at sunrise, to get the exact location of Ghati Lagna.

Karaka Lagna: It is the Lagna/Ascendant taken from the Lordship of "Karaka Graha" and taken for predictions of the events related to that House with respect to the position of that Karaka Graha.

Varnada Lagna: It is Lagna taken for predictions of social company.

Shri Lagna: It is Lagna taken for predictions of prosperity and marriage.

3.4 House-to-House Relationship

Houses are related to each others. The second house to any of the other house will indicate the wealth or money acquired from that house. The eighth house is second to the seventh house and will indicate the wealth and money of the spouse. The sixth house is the 2nd to the fifth house and will indicate the financial prosperity of the children. The eleventh house is the second house to the tenth house and will indicate financial

gain through the career. Similarly, the twelfth house to any house is the end or loss to the house in question.
Example: The sixth house is the twelfth to the seventh (marriage), so it represents the end of the marriage or wife. The third house is the end of the mother because it is the twelfth to the fourth (mother), so it represents the end of the mother. The third house is our energy, will, and life force and the second house is the twelfth house from our life force and so is the loss of our life force. Thus, Maraka house (2nd) derives its meaning from this principle. It is the twelfth from the third. Similarly the eighth house is our length of life so the twelfth from the eighth, i.e. the seventh house would be the loss of life, i.e. the death house. So, 2nd and 7th houses are called Maraka houses.
The 4th house indicates about "mother". 5th house will indicate the mother's money as it is the 2nd house from 4th house. 6th house will indicate the mother's younger siblings as is the 3rd house from 4th and so on. 1st house will indicate the mother's Career, because this is the 10th house from mother's house. These are called the "compound" or "secondary" houses. The 12th house rules loss, and therefore, the 12th house from any house is the loss of that house. Take the 9th house which rules fortune, religion or Dharma, the father, the spiritual master and guru, and God's grace. The 9th house is 12th from the 10th house. This means that the house of Dharma or religion is the house of loss to the house of career, profession and position. The 10th house rules not only career, but mainly it rules rise and status in material life. It rules standing up tall and straight and getting some position, some fame, and some power in this material life.
Similarly, the 5th house is 12th to the 6th house. Amongst other things, the 5th house rules winning at the lottery and the 6th house rules debts, therefore, it is easy to understand that if we win at the lottery, we can cure all our debt problems. So, the 5th house, which is a money house, puts an end to the 6th house, the house of debts. The 6th house is 12th to the 7th house. The 7th house rules marriage, the spouse and partners in our life. Therefore the 6th house rules the loss of the partner or loss of the spouse.
The 7th house is 12th to the 8th house. The 8th house rules the vital source of energy, which is longevity and so, the 7th house rules the end of vital energy. The 8th house is 12th to

the 9th house, which rules father. The 8th house, therefore, represents loss of father. Therefore, if we find the 9th lord in the 8th house in a chart, it is often found that the person lost his father.

Classification of House/Bhava:

Houses Categories	Houses	Effects of houses
Trikona (Trine) or Kona or Auspicious houses	1, 5, 9	They are the most auspicious or Benefic houses. They give fortune, luck, bring spirituality and well being if, unaffiliated.
Upachaya (Pratipas/Trika)	3,6, 11	Upachaya means "improvement" and are considered auspicious. Life improves and gets better over time with these houses if, unaffiliated.
Dustasthana (Trikas) or Malefic houses	6, 8, 12	Trikas are the most in-auspicious or malefic and deal with suffering, ill health, disease, death, loss and sorrow. The rulers of these houses will inflict this type of suffering.
Bhoga	2, 4, 10	Bhoga are considered auspicious and deal with the pleasure and luxury in all respect such as Cars, furnished sweet home and other part of life.
Kendra	1, 4, 7, 10	Kendra is considered auspicious. The planets in Angles give effects in one's boyhood.
Panapharas	2, 5, 8, 11	The effects of planets in Panapharas are felt in the middle age.
Apoklima (Cadent)	3, 6, 9, 12	The planets in Apoklima give result at the conclusion of the life, i.e. old age.
Maraka	2, 7	Maraka means "killer". The rulers are considered the killer sometimes. They are prominent when death or injury occurs.
Lakshmi	4,	The planets in Lakshmi gives home,

	10	happy life, conveyances, happiness, treasure, lands and buildings, heritage, real estate, good profession or livelihood or honour, living in foreign lands, reputation, business and social activities.
Vishnu (Abodes of Lakshmi, the Goddess of wealth)	5, 9	The planets in Vishnu provide pleasures through children, love affairs and romance, knowledge, royalty or authority, and fun, long distance travel and make fortunate, and gain of spiritual knowledge,
Dharma	1, 5, 9	They relate to our sense of purpose and the spirit that moves us such as self through 1st; creative expression through 5th; and our spiritual beliefs and truths through 9th house.
Artha	2, 6, 10	They relate to our achievements, such as money and material through 2nd; hard work through 6th; and the public recognition and career through 10th house respectively.
Kama	3, 7, 11	They relate to our sense of conveying our ideas, needs, and desires, such as, the need of a life through 3rd; partnership through 7th; and feel connected to every one through 11th.
Moksha	4, 8, 12	They relate to our liberation or freedom of soul through 4th; past essence of the soul through 8th; and about releasing all attachments to the world through 12th.

Weak House: The House/Bhava gets weak under the following circumstances:

When it gets sandwiched between malefic, particularly the natural malefic.

If Bhava is occupied or aspect by the 6th, 8th or 12th lords (lord of Dustasthana).

If the lord of the Bhava is in debility or combust.

If the lord of the Bhava is influenced by the lords of the 6th, 8th or 12th house.

If the lords of the Bhava occupy any of the 6th, 8th or 12th houses (Dustasthana) and is also aspect by the natural or functional malefic.

3.5 House/Bhava Signification (Karaktva)

The Karaka Sign and the most Karaka Planet of the twelve Houses and their Signification are as given below. If the Karaka planet is in the house by position or aspect, that house gives very good effect to the native.

House	Karaka Sign	Interpretation
1st	Aries	Body, head, longevity, health, character and nature, complexion, residence abroad, livelihood, servants, happiness and robes.
2nd	Taurus	Wealth, money matters, speech quality, face, foodhabits, death, primery education, one's deposits, income, sanyas, and security.
3rd	Gemini	Brothers and sisters, sports, throat and singing quality, business, servants, communications, education of higher secondary level, short distance travels.
4th	Cancer	Mother, Property, pleasure, Happiness, Conveyance, Enjoyments.
5th	Leo	Children, Knowledge, Creation, Enjoyment, Games, Gambling, Love and sex.
6th	Virgo	Jobs and Employments, Health and Overall well-being, Service performed for others.
7th	Libra	Close relationships, Marriage and Business partners, Agreements.
8th	Scorpio	Deaths, Rebirth, Sexual relationships,

		Inhrited properties, Finances, Self-transformation.
9th	Sagittarius	Foreign travel, Foreign countries, Long distance travels and journeys, Religion, Law, Higher education, Knowledge.
10th	Capricorn	Business, Ambitions, Motivations, Career, Status in society, Government, Authority, Father, Breadwinner of the household.
11th	Aquarius	Friends, Groups, Clubs and Societies. Higher associations, Benefits and fortunes from career.
12th	Pisces	Places of seclusion such as hospitals, prisons and institutions, including self-imposed imprisonments.

1st House (Thanu/Tanu Bhava): The First house or Lagna called Tanu Bhava represents body looks and soul, head/Brain, personality traits, longevity, health, character and nature, life style, complexion, inherent disposition, vitality, temperament, ego, paternal grandmother's wealth, maternal grandfathers' wealth, residence abroad, livelihood and pride.
2nd House (Dhan Bhava): The Second house called Dhana Bhava represents wealth, money matters, speech quality, family, face, right eye, mouth, foodhabits, charity, death, primery education, one's deposits, income, friends, sanyas, and security. It predicts material and financial resources and the ability to earn money, inflow of finance, wealth possessions, precious stones possessions, domestic life, bank position, and understanding with family members, law suits and domestic comforts in general.
3rd House Sahaja /Bhatru): The Third house called Bhatru Bhava represents brothers and sisters, sibling, co-born, sports, throat and singing quality, voice, music, business, servants and subordinates, communications, education of higher secondary level, talents and skills, short distance travels, neighbour, surroundings, relatives and relations with them; boldness, parent's death, arts such as theatre or filmy arts, filmy direction, painting, drawing, and success through own efforts, competition, hearing and father in law. The third house rules menial.

4th House (Bandhu Bhava): The Fourth house called Matru Bhava represents mother, motherland, house, possession of vehicles/conveyances, education of highest level such as master degrees, professional degree or doctorates, home relatives, office or factory, emotions, domestic and house related happiness and luxury, landed and house property, mental peace, chest and lungs, affairs & home pleasure, treasures, conditions at the old age, matters of the heart, inheritance and false allegations, pleasure trips, savings, cattle and pets.

5th House (Santana/Putra Bhava): The Fifth house called Putra Bhava represents children, intelligence, speculation, love affairs, recreation, romance, knowledge and intellect, creativity, mantra, tantra and pooja, creativity and stomach. It gives professions of teacher, principal and gynaecologists. It predicts abortions, politics, good karma, destiny, lotteries, and self-projection in order to please prominent people or the public or a boss at a job.

6th House (Ari Bhava): The Sixth house called Shatru Bhava represents , enemies, health (digestive system), illness, injuries, loans, sports and opposition, quarrels, court cases, litigation, maternal uncle and aunt, servants, work environment, jobs & service, step-mother, imprisonment, medical profession, food, restaurants, subordinates, obstacles in life, mental worries, calamity, employee and hard work. The sixth house denotes fear from thieves or enemies, fighting, misery and success over enemies, loss of moneys, cheating, danger and calamities (troubles) through women

7th House (Yuvati/Kalatra/ Jaya Bhava): The Seventh house called Kalatra or Juvati Bhava represents spouse, partners, sex, marriage, business, trade, employment in a private firm, reproduction and genital organs, sexual enjoyments, Kundalini Shakti, sexual organs and diseases thereof, death, relationships and signifies kidneys. 7th house represents married life, travel, conjugal happiness, loved one, divorce, honour, residence and reputation in foreign country, interaction with others, attitudes, sexual passions, open enemies, impotency, desire, disputes, relationships with wife, family life, age of getting married, wife age, health and her nature, journeys to distant places, loss through females, relationship and freedom.

8th House (Randhra/Ayusthan Bhava): The Eight house called Ayusthan Bhava represents longevity, destruction, accidents, physical pains, inheritance, legacies, death and reasons of death, underground wealth, historical things, monuments, parental property, longevity, failure, family of spouse, needs, life's secrets, joint resources, anus and sex power. 8th house represents financial windfalls, lottery winnings and physical pains. It represents parental property, worries, finances through unfair means, obstacles, gain from in-law, mode of death, imprisonment, enemies, support from others, the intimacy with other; struggles, Mafia & underworld, bankruptcy, obstacles, surgery, research, intuition, and long-term sickness, monetary gains from partner, misfortunes, trouble from enemies and loss of property. Malefic in the 8th will cause loss of spouse in the Dasa periods of the Lord of the Navamsa occupied by the 8th Lord.

9th House (Dharma Bhava): The Ninth house called Bhagya Bhava represents luck, prosperity, guru, father, religious and spiritual progress and knowledge of the scriptures, sadhana, pilgrimages, long journeys like foreign travel and foreign trade and dealing with foreigners, grandchildren, higher studies like doctorate, knowledge of foreign languages, grandparents, and signifies hips. 9th house represents fortunes, prosperity, writing books, powers of foresight, religious institutions, teaching, lawyers, fame, happiness, and unexpected gain of wealth, gain from lottery and affluence, association with good people, inclination toward God, religious and social work and fame in it, and satisfaction and fulfilment of desires.

10th House (Karma Bhava): The Tenth house called 'Karma' represents profession, business, authority, power, honours and achievements, acts (karma) one does, father, government service, politics, management, career, status, mother-in-law, prestige, reputation and signifies knees. It increases public image and makes the native's parent of greater influence, and provides power or authority like bosses, judges, and big stars. 10th house signifies popularity, status, activities outside house, pleasures, government favour, command, adopted son, worldly activities and moral responsibilities, livelihood, living in foreign lands, debts, reputation, social activities, social position, fame, wealth of the father, earned money, meritorious deeds, hardship in work, service, foreign place of settled life, promotion and number of promotion.

11th House (Labh Bhava): The Eleventh house called Labha Bhava represents gains, sources of income, good news, money, elder brothers and sisters, friends, long distance travels, air plane travel, entertainment, friend's circle, daughter/son-in-law, associations, accumulated wealth, ambition, social life, association and club, emotional attachments, son's wife, quadrupeds and attitude towards them, desire in life and colleagues or co-workers, love affairs and girl friends, honour and social success, clothes, the staple food, gold, and gain of wealth, pleasure, followers, dependent, insight, power of overcoming obstacles, redemption, worth of garments and signifies legs. The 11th lord, planet placed in the 11th house or the planet/planets associated with the 11th lord do give the native financial gains or professional elevation but at the same time they give serious health related problems during their major or sub period. 11th house being 6th from the 6th is an epicenter of serious illness.

12th House (Vyaya Bhava): The Twelfth house called Vyaya Bhava represents losses, waste, expenses, long journeys like foreign countries and residence in foreign countries, imprisonment, death, sadhana and Moksha or final liberation, bad habits, hospital, quests, export and import, feet, sleep, donation, foreign stay, subconscious, psychological issues, secrets disputes, and signifies mental agony, bodily injury. 12th house covers one's own death, good food & comforts, bed and couch pleasure, donations, miseries, sufferings, troubles, betrayals, law suits, imprisonments, hospitalisation, conjugal relations with opposite sex other than wife, contacts, misfortunes, secret enemies, spiritual liberation, sea or ocean travel, interest in arts & films, renunciation and enjoyment. The twelfth house also indicates the foreign trips, number of the foreign trips, benefits or loss due to the foreign trips and settled in the foreign lands, divine favour and travels.

4

Zodiac Sign

4.1 General

The Zodiac is a band of group of stars or the positions of celestial bodies. The Zodiac is divided into twelve divisions of 30 degrees each called "Sign". Each segment is called a Sign (Rashi).

Gender (Male/Female) Sign: The alternate Sign starting from Aries onwards are known as male and female on the other hand. The Gender of the Sign will help the Astrologer in assessing one's children, brothers and sisters in terms of Males and Females, from the horoscope.

Directions of Signs: The four Signs from Aries onwards indicate East, South, West and North, while the remaining Signs repeat in the same way. A journey undertaken by a person towards the direction indicated by the Lagna or the Moon at the commencement of journey yields fruitful results.

Night and Day Signs: Gemini, Cancer, Capricorn Aries, Taurus and Sagittarius are night Signs. Leo, Libra, Scorpio, Aquarius, Pisces and Virgo are day Signs.

Strength of Signs: If a Sign is aspect by its Lord, or by a planet friendly to its Lord, or by Mercury, or by Jupiter, it is said to be Strong Sign. Planets other than the above do not lend strength by aspect.

Lagna Sign: The Rising sign, at a particular time, is the sign of the zodiac positioned on the eastern horizon on the cusp of the first house at birth and is called Lagna. The Lagna lord is Atmakarka and is considered more powerful in Lagna.

Badhaka Signs: Similarly, there are Badhaka Rashi defined as per the Janma-Lagna. It works as the Badhaka Rashi and harms badly that house in which it is occupying.

Gandanta and its Effects: The ending portions, 30th degree of Cancer, Scorpio and Pisces are called Gandanta. It is said, that one born in Gandanta will not survive. He will either lose his mother, or he will end the dynasty, i.e. he is the last of his descent and will not have any children. If, however, he survives, he becomes a king with many elephants and horses.

4.2 Sign Characteristics

Zodiac has twelve signs having different characteristics. They act in different ways and are also known for its different nature. These are given in the following tables:

Table 1: Sign Characteristics

Ascendant	Benefic Planets	Malefic Planets	Most Malefic	Neutral Planets
Aries	Mars, Sun, Jupiter,	Venus, Saturn	Mercury	Moon
Taurus	Venus, Sun, Mars, Mercury, Saturn	Moon	Jupiter	--
Gemini	Mercury, Venus, Saturn	Sun, Jupiter	Mars	Moon
Cancer	Moon, Mars, Jupiter	Mercury, Venus	Saturn	Sun
Leo	Sun, Mars, Jupiter	Moon, Mercury, Venus	Saturn	--
Virgo	Mercury, Venus, Saturn	Sun, Moon, Jupiter	Mars	--
Libra	Venus, Mercury, Saturn	Sun, Moon	Jupiter	Mars
Scorpio	Mars, Sun, Moon,	Venus	Mercury	Saturn

	Jupiter			
Sagittarius	Jupiter, Sun, Mars	Moon, Mercury, Saturn	Venus,	--
Capricorn	Saturn, Mercury, Venus	Moon, Mars, Jupiter	Sun	--
Aquarius	Saturn, Sun, Mars, Venus	Moon, Mercury	Jupiter	--
Pisces	Jupiter, Moon, Mars	Sun, Mercury, Saturn	Venus	--

Table 2: Sign Characteristics

Lagna (Ascendant) Sign	Death inflictor (Maraka Graha)	Raja Yoga Karaka Graha	Neutral Planets	Badhaka Rashi (Signs)
Aries (Mesha)	Mercury, Venus	Jupiter	Venus	Aquarius
Taurus (Vrishabha)	Jupiter, Venus, Moon	Saturn	Sun	Scorpio
Gemini (Mithuna)	Mar, Sun, Moon	Saturn	Moon	Leo
Cancer (Karka)	Venus, Saturn, Mercury	Mars	Venus	Taurus
Leo (Simha)	Saturn, Venus, Moon	Mars	Mercury	Aquarius
Virgo (Kanya)	Jupiter, Moon, Sun	Venus	Moon	Scorpio
Libra (Tula)	Jupiter, Sun	Mercury, Rahu	Mars	Leo
Scorpio (Vrischika)	Mercury, Venus, Saturn	Moon	Saturn,	Taurus

Sagittarius (Dhanu)	Venus, Moon, Mercury	Sun, Ketu	Mercury	Aquarius
Capricorn (Makara)	Mars, Jupiter	Mercury	Mars	Scorpio
Aquarius (Kumbha)	Sun, Jupiter, Moon	Venus	Sun	Leo
Pisces (Meena)	Saturn, Venus, Sun, Mercury	Mars	Mercury	Taurus

Note: In case of following Lagna, some planets act as special death inflictor:
Mesh Lagna: Sani will also inflict death, if associated with an adverse Graha.
Vrishabha Lagna: Mangal will inflict death, if associated with an adverse Graha.
Mithuna & Simha Lagna: Chandra is the prime killer.
Karka Lagna: Surya will also inflict death, if associated with an adverse Graha.
Kanya Lagna: Sukra Yuti with Buddha will produce Raja Yoga. Sukra will also inflict death, if associated with an adverse Graha.
Tula Lagna: Guru will also inflict death, if associated with an adverse Graha.
Vrischika Lagna: Sukra will also inflict death, if associated with an adverse Graha.
Dhanu Lagna: Sani and Sukra will also inflict death, if associated with an adverse Graha.
Makara Lagna: Sani and Mangal will also inflict death, if associated with an adverse Graha. Sukra is capable of causing a superior Yoga.

Table 3: Sign Characteristics

Lagna Sign	**Person's Nature (Guna)**	**Cause of Death**	**Person's Nature**	**Varna**
Aries (Mesha)	Rajasic	High Fever	Violent	Kshatriya (Warrior)

Taurus (Vrishabh)	Rajasic	Fire, Weapon	Auspicious	Sudra (Service Person)
Gemini (Mithuna)	Rajasic	Cataract, Asthma, Mental Deviation, Loss of Appetite	Violent	Vysya (Trader)
Cancer (Karka)	Sathwic	Cholera	Auspicious	Brahmana (Intellectual)
Leo (Simha)	Sathwic	Wild Beast, Fever, Boils, Enemies	Violent	Kshatriya (Warrior)
Virgo (Kanya)	Thtamasic	Women, Venereal Disease, Fall from height	Auspicious	Sudra (Service Person)
Libra (Tula)	Sathwic	Brain Fever, Typhoid	Violent	Vysya (Trader)
Scorpio (Vrischika)	Sathwic	Jaundice	Auspicious	Brahmana (Intellectual)
Sagittarius (Dhanu)	Sathwic	Tree, Water, Wood, Weapon	Violent	Kshatriya (Warrior)
Capricorn (Makara)	Thamasic	Stomach Ache, Loss of Appetite	Auspicious	Sudra (Service Person)
Aquarius (Kumbha)	Thamasic	Cough, Fever, Consumption	Violent	Vysya (Trader)
Pisces	Sathwic	Drowning	Auspicious	Brahmana

(Meena)				(Intellectual)

Table 4: Sign Characteristics

Lagna Sign	**Sign Lord**	**Affecting Disease**	**Affected Part of Body**
Aries (Mesha)	Mars	Bile (Pitta)	Head, Face, & Brain
Taurus (Vrishabh)	Venus	Cold (Sleshma)	Neck, Throat, & Gland
Gemini (Mithuna)	Mercury Rahu	Gas (Vata)	Shoulder, Lungs, Hand, Blood, & Hand's Bone
Cancer (Karka)	Moon	Cold	Chest, Breast, Stomach, Shoulder's Bones,
Leo (Simha)	Sun	Bile	Back, Waist, Heart, Spinal Bones
Virgo (Kanya)	Mercury	Gas	Lever, Back Bones, Tili, Gurda
Libra (Tula)	Venus	Cold	Skins
Scorpio (Vrischika)	Mars Pluto	Bile	Penis, Thighs, Nectar
Sagittarius (Dhanu)	Jupiter Ketu	Gas	Waist, Veins
Capricorn (Makara)	Saturn	Gas	Knees, Bone's joints
Aquarius (Kumbha)	Saturn Uranus	Gas	Feet, Digestive Organs
Pisces (Meena)	Jupiter	Gas	Ankles, Palms

Table 5: Sign Characteristics

Lagna Sign	Aspect on other Signs	Sign Age	Relation
Aries (Mesha)	Leo, Scorpio Aquarius	28 ½ YEARS	Chandra Sadga
Taurus (Vrishabh)	Libra, Cancer Capricorn	18 YEARS	Chandra Sadga
Gemini (Mithuna)	Sagittarius, Virgo Pisces	33 ½ YEARS	Chandra Sadga
Cancer (Karka)	Taurus, Libra Capricorn	40 YEARS	Chandra Sadga
Leo (Simha)	Aries, Scorpio Aquarius	28 ½ YEARS	Surya Sadga
Virgo (Kanya)	Gemini, Sagittarius Pisces	18 YEARS	Surya Sadga
Libra (Tula)	Taurus, Cancer Capricorn	33 ½ YEARS	Surya Sadga
Scorpio (Vrischika)	Aries, Aquarius Leo	40 YEARS	Surya Sadga
Sagittarius (Dhanu)	Gemini, Virgo Pisces	28 ½ YEARS	Surya Sadga
Capricorn (Makara)	Aspect on other Signs	18 YEARS	Surya Sadga
Aquarius (Kumbha)	Leo, Scorpio Aquarius	33 ½ YEARS	Chandra Sadga
Pisces (Meena)	Libra, Cancer Capricorn	40 YEARS	Chandra Sadga

Table 6: Sign Characteristics

Lagna Sign	Sign Gender	Mode	Sign Element (Tatva)	Affected Part of Body
Aries	Male	Odd	Fire	Head, Face,

(Mesha)				Brain
Taurus (Vrishabh)	Female	Even	Earth	Throat, Gland Right Eye
Gemini (Mithuna)	Male	Odd Dual	Air	Neck, Nose, Lungs, Blood, Hand's Bone Ear,
Cancer (Karka)	Female	Even	Water	Chest, Breast, Stomach, Shoulder's Bones,
Leo (Simha)	Male	Odd	Fire	Upper Stomach, Back, Waist, Heart, Spinal Bones
Virgo (Kanya)	Female	Even Dual	Earth	Digestive Organs, Kidney, Lever, Back Bones, Tili, Gurda
Libra (Tula)	Male	Odd	Air	Skins
Scorpio (Vrischika)	Female	Even	Water	Uterus, Ovary, Penis, Nectar
Sagittarius (Dhanu)	Male	Odd Dual	Fire	Thighs, Waist, Veins
Capricorn (Makara)	Female	Even	Earth	Knees, Feet, Digestive Organs
Aquarius (Kumbha)	Male	Odd	Air	Calf Mussels, Feet, Digestive Organs
Pisces (Meena)	Female	Even Dual	Water	Left Eye, Ankles, Palms

Table 7: Sign Characteristics

Planet	Moola-Trikona Sign	Moola-Trikona. Degree	Detriment Sign
Sun	Leo	0 – 20	Aquarius
Moon	Taurus	4 – 30	Capricorn
Mars	Aries	0 – 12	Libra Taurus
Mercury	Virgo	16 – 20	Sagittarius Pisces
Jupiter	Sagittarius	0 - 10	Gemini Virgo
Venus	Libra	0 – 15	Scorpio
Saturn	Aquarius	0 - 20	Cancer Leo
Rahu	Virgo	--	--
Ketu	Pisces	--	--

Table 8: Sign Characteristics

Planet	Exaltation Sign	Debilitation Sign	Max. Exaltn. Debilitn. Degree	Detriment Sign
Sun	Aries	Libra	10	Aquarius
Moon	Taurus	Scorpio	3	Capricorn
Mars	Capricorn	Cancer	28	Libra Taurus
Mercury	Virgo	Pisces	15	Sagittarius Pisces
Jupiter	Cancer	Capricorn	5	Gemini Virgo
Venus	Pisces	Virgo	27	Scorpio
Saturn	Libra	Aries	20	Cancer Leo
Rahu	Taurus Gemini	Scorpio Sagittarius	--	--
Ketu	Scorpio Sagittarius	Taurus Gemini	--	--

Table 9: Sign Characteristics

Lagna Sign	**Element**	**Mode of Expression**	**Positive Quality**	**Negative Quality**
Aries	Fire	Cardinal	Vital	Impulsive
Taurus	Earth	Fixed	Stable	Stubborn
Gemini	Air	Mutable (Dual Signs)	Adaptable	Cursory
Cancer	Water	Cardinal	Protective	Jealous
Leo	Fire	Fixed	Authority	Autocratic
Virgo	Earth	Mutable (Dual Signs)	Detailed	Critical
Libra	Air	Cardinal	Diplomatic	Vacillating
Scorpio	Water	Fixed	Resurgent	Ruthless
Sagittarius	Fire	Mutable (Dual Signs)	Discerning	Moralistic
Capricorn	Earth	Cardinal	Principled	Miserly
Aquarius	Air	Fixed	Liberal	Eccentric
Pisces	Water	Mutable (Dual Signs)	Charity	Anxiety

Table 10: Sign Characteristics

Lagna Sign	Person's Characteristics	Person's Profession
Aries (Mesha)	Hasty, impulsive, restless, short-tempered	Govt. job, surgeon, mechanics, industrialists, athletes, Police, Military Service, Fire Service, Sports, Engineering, arm manufacturing, trade union leader
Taurus (Vrishabh)	Slow in movement,	Musician, singer, actors, banking, tailors, property

	inclined to ease and luxury, faithful & obedient	dealing, Jewellery business, money lending, commission agent, financial institutions, handicrafts, fancy articles, scented materials, five star hotels, drama, cinema, music, poet, story writer.
Gemini (Mithuna)	Good speakers, witty and humorous, inquiring and curious, fond of knowledge, fun seeking,	media and journalism, accountants, translators, writers, Information and broad casting, space department, education department, book publishing , mathematics department, auditors, law and order councillor, ambassador.
Cancer (Karka)	Emotional , forgiving, sensitive	Export and Import, naval and marine, fishing, nursing, interior design, food, petroleum, historians, shipping, transport department, agriculture, hotel business.
Leo (Simha)	Dominative, behaves like ruler	Govt. Job, Politics, Administrator, Social Services, Charitable institutions, Engineering, Industry, religion, investing, diplomacy.
Virgo (Kanya)	Intelligent, good speaker, tactful	Auditing, Accounting, Business, Teacher, writer, retail shops, computing, astrology, media, doctors, healing.
Libra (Tula)	Good talker, judicious in dealings	Shop, commission agents, bank, Life insurance, law department, hotel business, bar and Restaurant, Dancing Hall, Beauty parlour, Music, Dance , Cinema, judges, artists, cosmetics, fashion, receptionists, advertising, interior decorating, prostitutes.
Scorpio	Peevish,	Iron Industries, Engineering,

(Vrischika)	straight forward, likes to hide or run away from people and crowds	and Instrument Manufacturing, raw materials, priest, astrology, mantra and tantra, occult practices, chemicals, drugs, liquids, insurance, doctors, nurses, police, occult.
Sagittarius (Dhanu)	Honest, easy going, even-tempered, kind hearted, gambles	Forest department, law, religion, banking and finance, entrepreneurs, athletes, law department, temple, financial institutions, education department, ordnance depot, military training, social service, charitable institutions.
Capricorn (Makara)	Witty, and changeable, good organizer, cautious, secretive, ambitious, preserving, pragmatic	Hotels, food products, engineer, doctor, business, building work, Granite stone and sand business, Labourer like porters, coolies, drivers, shoe polishing, shoe makers, plumber and mining.
Aquarius (Kumbh)	Studious, philosopher, honest, benevolent	advisors, consultants, philosophers, astrologers, engineers, computer, Psychology, Religion, Teaching, Research and Development, Administration, Service in Space Dept., Defence, Fire, Jail, Bomb manufacturing, tourist guide, central excise CBI Dept.
Pisces (Meena)	Lazy, emotional, timid, honest, talkative, intuitive , psychic, fond of good food and company	doctors, captain, hospital, prisons Education Department, Religious Institutions, Medicine, Financial, Law Department, External Affairs, Bank, Navy, shipping, temple worker, priest

Table 11: Sign Characteristics

Name of Sign	Effects	Direction	Progeny Nature	Feature
Aries	Positive	Northern	Barren	Bestial
Taurus	Negative	Northern	Semi-Fruitful	Bestial
Gemini	Positive	Northern	Barren	Human
Cancer	Negative	Northern	Fruitful	Bestial
Leo	Positive	Northern	Barren	Bestial
Virgo	Negative	Northern	Barren	Human
Libra	Positive	Southern	Semi-Fruitful	Human
Scorpio	Negative	Southern	Fruitful	Bestial
Sagittarius	Positive	Southern	Semi-Fruitful	Half Bestial & Human
Capricorn	Negative	Southern	Semi-Fruitful	Bestial
Aquarius	Positive	Southern	Semi-Fruitful	Human
Pisces	Negative	Southern	Fruitful	Half Bestial & Human

4.3 Predictions by Sign Element

Predictions by Sign Element

Every Zodiac Sign falls into one of four elements. There are four Elements, such as earth represents common sense; fire represents action, air represents thinking and communication skills, and water represents the ability to feel and intuitively know. Each Element is assigned to each sign depending to their orientation in the zodiac. Many astrologers consider the element of each of the planets when determining which of the elements may be more significant in a horoscope.

Fire (Aries, Leo, and Sagittarius): Fire is active and masculine. People of the Fire element are outgoing, quite moral, very creative, courageous, passionate, impulsive, hot, dynamic, progressive, action oriented, and direct. Their essence is spirit. They are enthusiastic, optimistic, confident, naive, self-centred, open, confronting, loyal, tactless, impatient, honest, trusting, and independent and feel free.

Earth (Taurus, Virgo, and Capricorn): Earth is a receptive, feminine sign. People are practical, cautious, and pragmatic approach to life and build solid, 'real' material success, i. e. car, home, career success and have long range planning and strong determination to succeed. They are safe/secure, suspicious, sensual, organised, dependable, introvert, and efficient and strong survival instinct.

Air (Gemini, Libra, and Aquarius): They are active, curious, idealistic, unemotional, conceptual, devoid of feeling, good to communicate, social, objective, impersonal, distant, masculine, intellectual, changeable, and impractical, good speech, and natural communicators, extroverted, social, charming, and logical and air has least obvious bad qualities, theoretical, abstract, needs to socialise, needs to share ideas.. The lack of Air Element in a native birth chart indicates difficulty in the expression of that person. Communication of ideas and the ability to conceptualise may prove difficult.

Water (Cancer, Scorpio, and Pisces): People of Water element dissolve everything in them coolly. They take the shape of who they are with, and are quite emotional, sustaining, emotional, sensitive, imaginative, protecting, compassionate, caring, artistic, moody, soulful, subconscious, irrational, introverted, but strong/powerful, vulnerable to hurt, intimate, defensive, psychic, past, suffering, suspicious initially, self-contained, picks up impressions and associated with healing.

Calculation of Element Strength: The method for evaluating the strength of an element in a birth chart is to assign a value of 4 to the element associated with the Sun; the Moon element is assigned a value of 3; Mercury, Venus, and Mars sign elements are assigned a value of 2 each, and Jupiter and

Saturn each have a value of 1. Uranus, Neptune and Pluto are disregarded because their element is more societal affecting large groups of individuals born during a period. Using this approach, if as many as 8 points are concentrated in one element, it is considered “Preponderance” in that element. If we get less than 8 points with this approach in one element, it is considered “Absence” in that element.

Example: In one chart, the planets are placed as is given below: For Sun in Aquarius, an Air sign, we assign = 4 Points. For Moon in Libra, an Air sign, we assign = 3 Points. For Mercury in Aquarius, an Air sign, we assign = 2 Points. For Venus in Capricorn, an Earth sign, we assign = 2 Points. For Mars in Sagittarius, a Fire sign, we assign = 2 Points. For Jupiter in Pisces, a Water sign, we assign = 1 Points. For Saturn in Fire, a cardinal sign, we assign = 1 Points.

Thus, we have Element strength, such as, 9 in the Air Element, 2 in the Earth Element s, and 3 in the Fire Element and 1 in Water Element in this chart. This shows a preponderance of Air element in this chart. The preponderance reading of Air element would be appropriate in the above chart.

The preponderance readings of all elements are given below:

A Preponderance of the Fire element: A preponderance of Fire Element indicates high spirits, great faith in self, enthusiasm, direct, honesty, intensely assertive, most daring, individualistic, active and self-expressive, good natured, fun loving, natural leader, having a good time than on material possessions, big egos. He believes so strongly in his own powers and abilities that he overlooks and frequently fails to take advantage of the talents and abilities of others. He tries to do it all himself and don't delegate well. He is constantly “out front” or "on stage", such as an Artist and they need to be recognized and admired for his attainment and accomplishments. Appreciation is more important than money in his estimation. Nothing hurts him more than being ignored. The fire sign sense of honesty is straightforward and often child-like. Thus, he believes everyone is, like himself, an open book.

A Preponderance of the Earth Element: A preponderance of Earth Element indicates cautious, conventional, dependable but quite responsible, methodical, organizer, a builder, and a hard-worker. It provides the skills and attitude necessary to succeed readily in the world of business and never gamble or take unnecessary chances. They understand the reality of a situation and value, reliable and steadfast. They are dependent, diligence and a pragmatic, no-nonsense approach to life. Lack of ideas or imagination, dullness, rigid, conservatism, extreme materialism, and blind adherence to rules and regulations are their potential faults.

A Preponderance of the Air element: The preponderance of Air Element suggests a strong emphasis on thought, ideas and intellectual and they communicate and express ideas with mental agility and become the impractical dreamers, constantly thinking, people-oriented, but more inclined toward the group than the individual. Your interests are varied, and you're apt to be a life-long student.

A Preponderance of the Water element: The preponderance of Water Element indicates close emotional relationships, romantic, sentimental, affectionate, secure bond with partner, communicate best in non-verbal ways; emotionally, psychically, or through forms as art, dance music, poetry and photography. They have a natural feel and sense for the arts and are apt to let the heart rule the head, highly impractical and impressionable.

4.4 Predictions by Sign Position in House

General Predictions: If the person is born in the Mesh, he/she is brave and a thief; in Vrishabh wealthy, Mithuna learned; Karka king; Simha respected by king; Kanya learned; Tula minister or adviser; Dhanu sinful; Kumbha businessman; and in Meena he/she is wealthy.

Prediction by Sign in 1st House (Thanu/ Body):

Aries (Mesha): He/She is proud, wealthy, having excessive anger, dependent on others. He/She will be medium height, white complexion, smiling face, clever, and lean. He/She suffers from abdominal problems. He/She relishes helping the poor and has faith in God. He/She thinks very high but implements little. He/She inherits huge paternal property. He/She is prone to drowning and accidents. He/She is selfish and forgets a person who is no longer of use. He/She is prone to be cheated by friends and partners. He/She will improve in life and will earn money and believes in donation. Even born in medium income group, he/she earns good money of his/her own. He/She has a personality that is positive, aggressive, and competitive. He/She has leadership qualities, is bold, and empowered with more physical strength. He/She is best in sports, games, trekking, summer camp, and any other outdoor activities.

Taurus (Vrishabh): He/She is a pleasant talker, scholar and loves all. He/She is tall, luxurious, clean-hearted, strong built, good personality, whitish complexion, smiling face, clever and attractive personality. He/She improves in life and earns money and believes in donation. He/She is prone to Litigation or imprisonment because of personal or property disputes. He/She is financially well off, earning much more. He/She is educated and enjoys a happy married life with educated and glorious progeny. He/She has differences with near relatives and is thus socially unpopular. He/She is prone to accidents. He/She has a tendency toward being heavy by both bone structure and the self-indulgence. Lord Krishna, Mata Amritanand Mayi and also Shri Basaveshwara were born in Taurus, but in Rohini Nakshatra. The second Drekana of Taurus gives the skill of fine arts, music and dance to the native.

Gemini (Mithuna): He/She is proud, loves his friends, charitable, and wealthy. He is annihilator of his enemies and progresses slowly in life. He/She will be medium height, whitish complexion and a faithful friend. He/She is sweet-voiced, jolly and humorous. He/She is well-wisher for everyone and gets cooperation from parents. He/She seldom seeks help and doesn't work under any one. He/She has his own successful business set-up - big or small. He/She is considerate to subordinate and weaker people. He/She has long life, lean body. He/She improves in life; earns money and believes in hard working. He/She, even, born in medium income group, earns money of his/he own. He/She does not get help from his/her spouse. 3rd Drekana of Gemini blesses him/her with fine art, music and dance.

Cancer (Karaka): He/She is religious, handsome, long, has good personality, whitish complexion, and donating. He/She is good-looking, rich and famous. He/She respects his elders and teachers. He/She is prone to head injury during childhood. He/She excels his business away from birth place. He/She is prone to be cheated by partners and close relatives. He/She gains from business abroad in white coloured items. He/She is intelligent and heads an organisation or society. People flock to him/her for advice. He/She may undergo political imprisonment. He/She, even, born in medium income group, earns money of his own. He/She does not enjoy his life due to hardship. His family life is not happy and always difference of opinions between husband and wife. He/She is very protective of those who are close to him/her. He/She is affectionate, emotional, home loving and lovely in their approach.

Leo (Simha): He/She is annihilator of his enemies, has few children, is tall, strong built, good personality, whitish complexion, and smiling face, clever and has attractive personality. He/She is efficient, undertakes tough tasks, and is hard-hearted and always successful. He/She is self-dependent and doesn't trust others. He/She spends as quickly as he earns. He/She crushes his/her enemy, is religious and donating. He/She doesn't forget or forgive his enemy and takes revenge. He/She has differences with father. His wife is long-lived but he/she keeps quarrelling with his/her. He/She is a devoted friend who will remember and repay a kindness. He/She has royal tastes and a sense of luxury. He/She is angry but vents the anger quickly. He/She goes to any authority to prove that he/she is right.

Virgo (Kanya): He/She is endowed with beauty and has a good fortune, medium height, broad chest, whitish complexion, smiling face, clever, very fast in doing the job and is selfish and harm too much out of his selfishness. His/Her young age is very happy. He/She is a successful in politics because he/she has something inside and speaks something else in public. Nobody can measure his political capability. He/She is never crude or coarse. He/She prefers the role of researcher, observer, critic, or teacher. He/She is fault finding type and hypercritical. He/She feels proud in finding the fault and drawback in others.

Libra (Tula): He/She is a scholar, earns his livelihood by virtuous means, wealthy and is respected by everybody. He/She is tall, fair complexion and healthy. He/She is sweet-voiced and benevolent. He/She doesn't stick to one profession and keeps spending too much on research. He/She enjoys little reputation at home but is reputed outside. He/She excels in occupations related to iron. He/She has problem with brothers/sisters. He/She is talkative, is not hard-working and depends on fate, and thus leads insecure life. He/She is unlucky for father's business and gives setback at the age of 12. Initially he/she begins with service but later settles down in own business. He/She leads happy married life with kids.

Scorpio (Vrischika): He/She is wealthy and a scholar. He/She is tall, lean, either very rich or very poor. He/She lives away from home since very early in life and dominates over his family members as well as outsiders. He/She doesn't forget or forgive his enemy. He/She has his own business. He/She has financial problems up to 30 years but later he/she earns money and supports others. He/She cannot sit idle and is world-famous and heads an organisation or society. He/She is prone to injury during fight or accident. He/She gets married more than once. He/She will be rich, famous, popular and smart in love matter. He/She is successful in politics. He/She earns sufficient money in the life but not much savings. This Sign is not good for domestic happiness.

Sagittarius (Dhanu): He/She is an expert in policy matters, religious, important person in his family, medium height, whitish complexion, smiling face, and attractive personality. He/She has strong faith in God, is vegetarian, simple living, believes the people very easily, and is a businessman. He/She does his work with well planning. He/She likes discipline, truth, justice, kindness and independence in his/her life. He/She takes everything and everyone for granted. His/Her reasoning powers are superb. He/She enjoys the good fortune of having thought patterns that remain young and fresh throughout life. He/She makes lots of promises but fail to maintain them.

He/She has a great sense of fairness and adopts only fair means to handle the job.

Capricorn (Makara): He/She is inclined towards evil deeds, is greedy and has many children, but hard working. He/She is tall, strong built, good personality, whitish complexion, clever and but selfish, changing his faces frequently as per situations, and very talkative. He/She works under someone and subjects to heavy ups & downs. He/She spends immediately what he/she earns in bad deeds. He/She dominates spouse and quarrels with other family members on that account. He/She stays away without information. He/She is abusive and short-tempered. He/She fails in business and has to go in for service. He/She is attached to mother and has differences with father. He/She is helpful to brothers and sisters and has more daughters. His/Her expense is more than earning and hence always faces shortage of money. The conjugal life is not happy and there is always difference of opinions between them.

Aquarius (Kumbha): He/She leads a happy and contented life. He/She will be well educated, gentle, peaceful, always ready to help others, having good thinking; tall, whitish complexion and attractive personality and straight forward in nature. He/She successfully tackles early age problems and heads for good time later. He/She would spend any amount of time and money to crush his/her enemy or achieve his/her aim. He/She likes to gossip and interact with women and has interest in astrology. He/She is financially well off and is prone to chest problems. He/She will be very hard working and faces difficulties in life. He/She may suffer with the stomach and heart diseases in old age. He/She is strong willed, detached and unyielding in nature. The child of this Sign is unpredictable regarding his/her behaviour and can change frequently himself/herself to any extent during his childhood.

Pisces (Meena): He/She is wealthy, educated, gentle, peaceful, religious, and always ready to help relatives, medium height, and beautiful curly hair and have self confidence. He/She is famous and heads an organisation or society. He/She studies very hard, but he/she is not a high scorer. He/She helps friends and serves society physically but without spending money. He/She works overtime to finish the work same day. He/She becomes favourable of family members and outsiders. He/She just cannot work under any one and leave his/her service very soon for own business. He/She loses temper beyond control but calms down very quickly. He/She is prone to cheating by partners and should better work alone. He/She is likely to break first marital relation (matured) or otherwise, be unhappy with spouse but happy with progeny. He/She may have great interest in writing, music and acting. He/She is likely to set high goals. Drug, alcohol and false promises attract him/her easily.

Prediction by Sign in 2nd House (Dhan/Wealth)

Aries in 2nd Bhava: The Aries in the 2nd Bhava indicates the uncertainty about his/her earning. Sometimes he/she earns more money and sometimes very less money. He/She spends more than the earning on his/her show business and is sometimes harmed by his/her enemies. His/Her good luck starts after marriage.

Taurus in 2nd Bhava: The Taurus, in the 2nd Bhava, indicates his/her good earning. He faces ups and down in life and is harmed by his/her spouse or partners in the business. His/Her good luck starts after 18, 22, 24, 33 and 35 years of age. He/She has a strong drive to earn money; to build and hold financial worth and material possessions. Venus, the planet of love ruling the second house suggests a real love of money. He/She is good at business affairs. He/She is a natural for making and accumulating money.

Gemini in 2nd Bhava: The Gemini, in the 2nd Bhava, indicates his/her weakness in the earning. He/She wastes money in bad relation with the other woman/man. His/Her enemies sometimes harm him/her financially. His/Her good luck starts in business; the private services or in life insurance, small industries, electrical parts industry etc. The Gemini influence in the second house focuses the mind on material matters and on making money. He/She has an active interest in financial affairs. His/Her resourcefulness in accumulating money may result in holding more than one job simultaneously.

Cancer in 2nd Bhava: The Cancer, in the 2nd Bhava, indicates the less earning as compared to his/her hard work. He/She is miser in nature and spends less than the earning and saves more. He/She faces ups and down in the business and is harmed by his/her partners in the business. His/Her good luck starts after 20, 26, 33, 44 and 54 years of age. The influence of Cancer in the second house shows that he/she is protective of his/her financial assets and possessions.

Leo in 2nd Bhava: The Leo, in the 2nd Bhava, indicates the uncertainty about his/her earning in his middle age. He/She spends very happy childhood and does not face any financial crisis in childhood. He spends more than the earning on his/her show business or in uncertain business and is sometimes harmed by his/her enemies. He/She has good luck always and finishes all his/her work successfully. He/She earns money through political works or government job. The Sun denotes his/her self-esteem, so his/her earning capacity may have a lot to do with his/her sense of self-worth.

Virgo in 2nd Bhava: The Virgo, in the 2nd Bhava, indicates he is financially strong. He/She earns money by hard work in business, especially ready-made shop or fancy store shop. Initially he/she faces financial difficulty due to not taking right time decision or due to hot temper. He/She appears very careful with financial affairs, but often he/she can become somewhat penny-wise and pound-foolish.

Libra in 2nd Bhava: The Libra, in the 2nd Bhava, indicates the uncertainty about his/her earning. Sometimes he/she earns more money and sometimes very less money by business. He/She spends more than the earning on his show business or luxuries life and is sometimes harmed by his/her bad habits. He/She has good luck in business like hotel or restaurants. There is balance and harmony in material affairs. This sign suggests the accumulation of possessions is highly dependent on ventures with a partner, normally the marriage partner.

Scorpio in 2nd Bhava: The Scorpio, in the 2nd Bhava, indicates the uncertainty about his/her earning. He/She spends more than the earning on his/her show business or on his/her big planning and is sometimes harmed by his/her friends or relatives. He/She has good luck in business like small industries rather than service.

Sagittarius in 2nd Bhava: The Sagittarius, in the 2nd Bhava, indicates very lucky and good earning. His/Her good luck starts after 24, 28, 33, 37, 48 and 55 years of age. He/She should be very careful while signing any paper. He/She is lucky in this regard. His/Her personality attracts financial success naturally. He/She is a risk taker who may get burned from time to time.

Capricorn in 2nd Bhava: The Capricorn, in the 2nd Bhava, indicates very lucky and god earning and is financially strong. He/She earns good money by business in mines or stone query as compared to service. He/She may serve in planning commission or plans making organization because he/she is very good in big planning. He/She has prudence and is practical in the handling of money. He/She has a 'poor' complex and does not know his/her earning potential and so he/she continues to live as though he/she had very little. When buying investments, he/she is inclined toward the blue chips and sure bets. He/She is very cautious and practical. He/She may reject luxury or expensive living.

Aquarius in 2nd Bhava: The Aquarius, in the 2nd Bhava, indicates very lucky and good earning and is financially strong. He/She earns good money by many source of income particularly by reporter, writing, publications or businesses as compared to service. He/She does not get benefits from brothers or relatives and gets harmed on good faith by the people. He/She is benefited in partnership and latter life is better than the middle age. He/She is not afraid to take chances financially, and he/she looks for unusual and inventive ways to invest.

Pisces in 2nd Bhava: The Pisces, in the 2nd Bhava, indicates very lucky and good earning and savings and is financially strong. He/She earns good money by many source of income particularly by doctor professions, by sale of medicines, by share business or small industries. He/She controls his expenditures and earns money like anything and does not have peace of mind in life but always busy in earning money. His/Her good luck starts after 24, 28, 33, 37, 48, 55 and 60 years of age. He/She has a sense of timing and intuition that may be an asset financially.

Prediction by Sign in 3rd house (Relations/Sibling):

Aries in 3rd Bhava: The Aries, in the 3rd Bhava, indicates that he/she will be strong mussel and good built, strong built wider shoulders, good personality, very courageous. He/She earns money and believes in making the situations in his/her favours and obedience to his/her seniors. He/She will be talkative and artist and will save money in life. He/She has a strong need to communicate with others, and to communicate forcefully. His mind is active, alert and capable of making quick decisions. He/She is capable in expressing himself/herself. He/She is mentally competitive. He/She is an aggressive learner, always seeking new ideas and new knowledge.

Taurus in 3rd Bhava: The Taurus, in the 3rd Bhava, indicates that he/she will be strong built, good personality, very courageous. He/She earns money by writing, poetry, portrait making and other arts and believes in making the situations in his/her favours and admired by the family members. He/She will be artist. He/She will be able to save money in life. He/She is a slow learner, but once an idea is lodged in his/her brain, it's there to stay. In early education he/she may have been an indifferent student, but as he/she matures he/she accumulates a wealth of knowledge and understanding. He/She has a strong interest in the arts, especially music.

Gemini in 3rd Bhava: The Gemini, in the 3rd Bhava, indicates that he/she will be strong built, good personality, very lucky. He/She earns money and believes in making the situations in his/her favours and honoured by the government. He/She will enjoy the happiness of luxurious vehicles. He/She will be able to save money in life and has very happy family life and good respect and cooperation from spouse. He/She has quickness both in thought and speech. Mercury, ruling Gemini, is the planet of intellect, thought, speech, and wit. He/She is a good conversationalist, a fact collector, and a mentally stimulating person. He/She can be a good diplomat because he/she can agree with divergent views and present rational alternatives.

Cancer in 3rd Bhava: The cancer, in the 3rd Bhava, indicates that he/she will be strong built, good personality, very lucky. He/She earns sufficient money by good business and real estate or construction work and believes in making the situations in his/her favours by his/her noble nature and admired by the society. He/She will enjoy the happiness of friends. He will be able to save money in life and has very happy family life and is religious minded. He/She is very protective of brothers and sisters. He/She has close and emotional ties to the immediate family.

Leo in 3rd Bhava: The Leo, in the 3rd Bhava, indicates that he/she will be strong built, good personality and very courageous and has excellent imaginative power. He/She earns money by writings, poetry, and publication and believes in making the situations in his/her favours and has interest in the great music. He/She will be artist. He/She will face difficulties in education in childhood but finally gets good educations in life and be able to save money in life. His/Her powers of self-expression are outstanding.

Virgo in 3rd Bhava: The Virgo, in the 3rd Bhava, indicates that he/she will be good personality, has very good knowledge of Vedas. His/Her friends are helpful but not the family members. He/She will be short tempered. He/She will be suffering from inferiority complex on communication skills, especially in the early years. He/She can be overly critical of brothers, sisters or neighbours.

Libra in 3rd Bhava: The Libra, in the 3rd Bhava, indicates that he/she will be flicker minded and has relations with low status people. He/She will not be able to take fast decision but too much talkative and does mistake while talking for which he/she repents later on. He/She has differences in family life always. He/She gets along easily with family members because he/she dislikes conflict and argument.

Scorpio in 3rd Bhava: The Scorpio, in the 3rd Bhava, indicates that he/she will be flicker minded and has relations with low status people and is addict of bad habits. He/She will not be able to take fast decision. He/She has differences in family life always and live medium life. He/She is very angry man and does mistake in his/her angriness. He/She does not get help from brothers. There can be friction between him/her and family members, as well as friends and acquaintances.

Sagittarius in 3rd Bhava: The Sagittarius, in the 3rd Bhava, indicates that he/she will lose money in business and will not get success in business. He/She will be able to save money in life in military, police or other government service. He/She is very talkative and has a very cheerful outlook on life, natural exuberance and optimism. He/She has a natural ability to communicate, especially important issues such as politics, education, and religion. He/She has an executive type of mind.

Capricorn in 3rd Bhava: The Capricorn, in the 3rd Bhava, indicates that he/she will be strong mussel and good and strong built handsome body & good attractive face and personality. He/She is very lucky in respect of children and famous among friends and is religious minded. He/She has careful expression of thoughts and ideas. He/She refrains from writing or saying anything unless there is a reason for doing so. He/She has little capacity for small talk, and many people may find it difficult to communicate with him/her. In his early years, he/she may not have good education, but later he/she changes, and he/she is apt to turn to studies to attain his/her ambitions.

Aquarius in 3rd Bhava: The Aquarius, in the 3rd Bhava, indicates that he/she will be peaceful nature and has good relations with brothers but does not get help from them. He/She will get respect in the society and has interest in music. He/She has a sparkling intellect and an inventive mind. He/She is well ahead of time, and sometimes radical. He/She seeks education simply for the sake of learning instead of just for preparing to earn a living.

Pisces in 3rd Bhava: The Pisces, in the 3rd Bhava, indicates that he/she will be good built, handsome body & good attractive face and personality. He/She will be a wealthy man and lucky in respect of children and get full help from children in old age. He/She is famous in society, keeps everybody happy in the society and is religious minded and interested in religious work. In communicating, he/she can become over emotional and he/she may experience problems articulating, especially in his/her youth. He likes to be alone when he/she is performing any type of mental work.

Prediction by Sign in 4th House (Bandhu/Pleasure)

Aries in 4th Bhava: The Aries, in the 4th Bhava, indicates that he/she will be having many cattle. He/She will have relations with many women simultaneously but still he/she enjoys the life happily in different ways and peacefully. He/She is more benefited by agriculture and business. He/She has an aggressive attitude toward the home and family. He/She always takes an active interest in the affairs of the home. He/She has a tendency to force issues and demand too much. The fourth house denotes activities in the latter part of life, the later years will be very active and more daring and outgoing, even youthful personality emerges, as he/she grows older. He/She is more assertive and physically active in the latter part of life.

Taurus in 4th Bhava: He/She is very lucky in respect of children who help him/her in old age and is religious minded. He/She does the social work in big ways. He/She is famous in society, has patient, peaceful, keeps everybody happy in the society and is religious minded and interested in religious work, celebrates worships in big ways. He/She is more benefited by worshipping the lord Shiva. He/She enjoys the happy family life. He/She has strong instincts to provide materially for his/her family. He/She has a very pleasant, easygoing home environment, and harmony, serenity and graciousness in his/her latter years.

Gemini in 4th Bhava: The Gemini, in the 4th Bhava, indicates that he/she will be handsome body & good attractive face and personality. He/She has medium luck in life. He/She is very sexy and happiest person and enjoys the sex in relation with the most beautiful women and spends lots of money on them and fashion and cosmetics. He/She has to work hard to earn money. He/She suffers a lot in his/her old age due to loss of heavy money. His/Her ties to home are not particularly strong.

Cancer in 4th Bhava: He/She will be the luckiest person and has many friends. He/She is very happiest person and enjoys the married life with the most beautiful good nature and fortunate spouse and spends lots of money on fashion and cosmetics. There is a deep attachment to family traditions and home relationships. He/She may be especially protective of his/her parents and assume a role of responsibility for them. He/She may depart his/her early shelter at a young age.

happiest person and enjoys the married life with the most beautiful good nature and fortunate wife and spends lots of money on fashion and cosmetics. He/She is the most angry and irritated person. His/Her relation with the brothers and sisters are not good and does not get help from them. Children are not so happy with him/her. He/She has more female child than male child. He/She has properties and owns a piece of land and lives in a reflection of his/her ego. He/She is apt to be the one that 'rules the roost.' In either sex, this placement often shows the one who is the boss and makes the decisions in the family.

Virgo in 4th Bhava: He/She will be the luckiest person and most fortunate. He/She is very happiest person and enjoys the full and happiest married life with the most beautiful good nature and fortunate spouse and spends lots of money on fashion and cosmetics. He/She gets married in young age and has more male child than female child. He/She has sufficient money to enjoy the life. The childhood is difficult financially but becomes wealthy man after the age of 28 years and is the richest man at 36 years. He/She is learned, educated good nature person.

Libra in 4th Bhava: The Libra, in the 4th Bhava, indicates that he/she will be handsome body & good personality. He/She will be the successful businessman and will spread his business all-around and is most fortunate even though he/she will be poor in childhood. He/She is very happiest person and enjoys the full and happiest married life till old age. The childhood is difficult financially but becomes wealthy man after he/she starts working after marriage. He/She is educated, kind, peace loving and good nature person who helps others. He/She spends some of his/her wealth on religious matters and works and keeps at the distance from the bad people. He/She wants to own his/her home and possessions, free and clear.

Scorpio in 4th Bhava: The Scorpio, in the 4th Bhava, indicates that he/she will be very anxious and worried person; his/her mind will not be peaceful and will face many difficulties in life. He/She is afraid of enemy as they harm him/her frequently. Every time, he/she faces difficulties while starting any job but gets success at the end. He/She starts life from poor childhood but earns money as he/she grows older and saves money to enjoy the family life till old age. He/She has extremely strong feelings about the home and family life. He/She has a strong loyalty and protective tendency shown toward family members. He/She has a sense of royalty, splendour, and space in the home environment.

Sagittarius in 4th Bhava: The Scorpio, in the 4th Bhava, indicates that he/she will be fighting nature and always fighting with people and will be successful in war. He/She will earn money and gets success in business of lending money & interest as compared to service. He/She is involved in litigation and court cases. He/She gets success in military or police job and is his/her own fortune maker and enjoys a medium happy family life. He/She has a strong urge to control and direct family matters. His/Her latter portion of life will be very beneficial. He/She will get wealthier as grow older, both in material ways and in spiritual ways.

Capricorn in 4th Bhava: The Capricorn, in the 4th Bhava, indicates that he/she will be the owner of gardens and vegetations and serve in the same field. He/She has many good and helping friends and is fortunate in respect of friends. He/She has differences with wife but the children are helpful to him/her and beneficial to him. He/She has much responsibility in house of home and family. With the planet Saturn ruling the home, issues of security can be of paramount importance. There is also ambition linked with this placement. He/She dominates his/her home with a practical and no nonsense outlook. He/She is a strong disciplined man, with strict, old-fashioned principles and a lofty code of ethics.

Aquarius in 4th Bhava: The Aquarius, in the 4th Bhava, indicates that he/she will be happiest person and enjoys the married life with the most beautiful, good nature, educated and fortunate spouse who is helpful in making the fortune in the life and spends lots of money on fashion and cosmetics. He/She gets wealth from father in-law. His/Her fortune starts after marriage. He/She has a strong demand for freedom in the affairs of the home. He/She is likely to find it necessary periodically to change his/her address, and he never likes the idea of moving. His/Her home environment is distinctive, perhaps even a bit unusual. He/She has many original ideas and wants the latest innovations as a part of his life style.

Pisces in 4th Bhava: The Pisces, in the 4th Bhava, indicates that he/she will be a captain on the ship or will be in service connected to water transport. He/She will be educated and peaceful person and will get respect in the society due to his/her good nature. He is educated, kind, peace loving and good nature person who helps others. He/She spends some of his/her wealth on religious matters and works and keeps at the distance from the bad people. He/She has sufficient money to enjoy the life. The childhood is difficult financially but becomes wealthy and enjoys the happiest family life. He/She has an emotional tie to the home. He/She is sentimental about his/her family and willing to make sacrifices for his/her loved ones. He/She has a strong need for domestic peace and seclusion.

Prediction by Sign in 5th House (Santana/Children)

Aries in 5th Bhava: The Aries, in the 5th Bhava, indicates that he/she will be very angry and foolish man and does mistake in haste for which he/she repent later. There is always difference of opinions among the husband, wife and children. He/she is always restless and can't take a decision on any matter immediately. He/she never minds spending his/her money for leisure time activities, whatever he/she may be. He/she loves the outdoors and a need to stay on the move. He/she has a sporting attitude that makes him/her fun to be around. Physical activity is necessary for his/her well being and happiness.

Taurus in 5th Bhava: The Taurus, in the 5th Bhava, indicates that he/she will have no issue or will have issue after late or the issue will be weak, unhealthy or mentally retarded. He/she will have better understanding, peaceful, kind and patient. He/she will earn money easily and sometimes from lottery or gambling. He/she will be marrying to a beautiful and fortunate spouse who will be taking all the care of the family. He/she has a very loving nature toward his/her offspring, and has very strong and fixed views regarding their behaviour and the proper upbringing. He/she has personal artistic talents, and he/she may have a natural artistic ability leading to self-expression in some form of the arts.

Gemini in 5th Bhava: The Gemini, in the 5th Bhava, indicates that he/she will have issue and will have all happiness from the issue who will be healthy and educated. He/she will have better courageous, rigid, strong will. He/she will be well educated even though faces difficulties during education. He/she gets angry soon and cooled down soon too. He/she has a cool and intellectual approach to romance. Gemini is the sign of mental energy and the fifth house denotes creativity. He/she is writers and otherwise talented people.

Cancer in 5th Bhava: The Cancer, in the 5th Bhava, indicates that he/she will have more daughters than son in number or the son may take birth at latter age or very late. However, he does not get happiness from the children but pain. He/she will be lazy and has faith on others easily. He/she is very protective of his/her family, and is very maternal toward offspring. His creative work may tend toward the artistic, especially theatre. He/She has good writing abilities and he/she can communicate more easily in writing than in speech.

Leo in 5th Bhava: He/she will have more daughters than son in number or the son may take birth at latter age or very late. He/she will be working away from the native place to earn money for the family. He/she will be courageous, rigid, strong will and hard working. He/She devotes fully to whatever creative activity has his/her interest at the moment. He/She identifies strongly with his/her children and he/she is very proud of his/her accomplishments. He/she is a born gambler and speculator, with more than his/her fair share of luck.

Virgo in 5th Bhava: The Virgo, in the 5th Bhava, indicates that he/she believes in unnecessary show business other than the reality, and does not have issue or does not get help or happiness from the issue. He/she take fast decision and misuse the power immediately after getting it. Relationships with offspring can be strained as he/she lacks the patience for properly disciplining them and understanding their needs.

Libra in 5th Bhava: The Libra, in the 5th Bhava, indicates that he/she will be a gentle man, peace loving, and effective personality, very handsome and educated. Romantically he/she is charming, but very inconsistent and fickle. He/She is lucky at love. Libra is naturally suited to the raising of children, although he/she is somewhat prone to spoiling them. He/She has many child-like qualities in his/her make-up.

Scorpio in 5th Bhava: The Scorpio, in the 5th Bhava, indicates that he/she will be a gentle man, peace loving, effective personality, very handsome, religious minded and educated. He/she will suffer from venereal disease. He/She will get happiness from the children. He/She has a good planning and completes the job taken in hand successfully. He/She has attachments and romantic encounter. He/She is concerned for his/her children, almost to an extreme degree.

Sagittarius in 5th Bhava: The Sagittarius, in the 5th Bhava, indicates that he/she will be interested in horse riding or horse driving like horse race man, or horse cart man. He/She respects the people and enjoys the full happiness from his/her children because all are obedient. He/She has a constant need to show off, and he/she enjoys the thrill of any adventure. He/She is a natural gambler who will speculate on just about anything. He/She has a good understanding of young people, and he/she can get along with them so well because he/she treats them with true respect as individuals.

Capricorn in 5th Bhava: The Capricorn, in the 5th Bhava, indicates that he/she will a bad man, which enjoys the bad work or bad politics, very clever to get job done by the enemies too, principle less. But, if she is a female, she will be a big Guru Matta and worship able by the people, knowledgeable of all the Vedas and religious books. He/She is a hard worker, good with details, and can make a scrupulous teacher and disciplinarian of young people. In romantic affairs, he/she may be prudish, and may even appear cold.

Aquarius in 5th Bhava: The Aquarius, in the 5th Bhava, indicates that he/she will a fixed mind, good in education and get all happiness from the children. He/She talks truth and help and gets respect from others. He/She is courageous and never gets afraid of problems in life and face with courage. He is cool and detached concerning romance. His/Her children are apt to display a rebellious nature, and they need to be taught discipline at an early age.
Pisces in 5th Bhava: The Pisces, in the 5th Bhava, indicates that he/she will be very sexy, popular in a female society and spends a lot for them, always smiling face, sentimental. He/She has differences with spouse and children and hence has medium level happiness from children. He/She has frustration with his/her offspring. He/She may sometimes be disappointed and disillusioned. He/She may dream about creating something so significant that he/she will be remembered for his/her efforts long after he/she is gone. Raising children can be difficult and confusing.

Prediction by Sign in 6th House (Dukha/Sadness)

Aries in 6th Bhava: The Aries, in the 6th Bhava, indicates that he/she will be bad and corrupt man, hard working but always complaining to someone and telling badly about others. He/She is a very hard working member of society. He/She may run into difficulties with co-workers and employees because of his/her critical nature.
Taurus in 6th Bhava: The Taurus, in the 6th Bhava, indicates that he/she will be a frustrated man; hard working but always having differences with family members and his/her children are aggressive type. Teaching is very attractive to him/her because he/she gets the time for his/her varied extracurricular activities.
Gemini in 6th Bhava: The Gemini, in the 6th Bhava, indicates that he/she will be a frustrated man; hard working but always having differences with family members and his/her children are aggressive and fighting nature. He/She does business rather than service but always loses money in life. He/She dislikes repetitious tasks. He/She is likely to succeed in fields related to scientific research, business or finance.

Cancer in 6th Bhava: The Cancer, in the 6th Bhava, indicates that he/she will not be taking rest in life and will not allow his/her relatives or subordinates to take rest. He/She will always be fighting with others and in very much love and affection with his/her children. As a supervisor, he/she is very understanding and concerned about the welfare of his/her employees.

Leo in 6th Bhava: The Leo, in the 6th Bhava, indicates that he/she will be angry-man, infighting and very sexy and will suffer financially due to entangle with other man/women. He/She will suffer and loose relation with spouse and children due to other man's/women's relation. He/She can lose himself/herself completely in work and service dominates over co-workers and subordinates. He/She has a feeling of authority where work and services are concerned.

Virgo in 6th Bhava: He/She will lose huge money due to contact and connection with low status/ level man/women. He/She seldom eats or drinks too excess and he/she is health conscious. History is especially interesting to him/her because of his/her ability to accumulate and relate details.

Libra in 6th Bhava: The Libra, in the 6th Bhava, indicates that he/she will be very rich man but will be selfish and will not believe in god. He/She is cooperative, tactful, and diplomatic in the work place.

Scorpio in 6th Bhava: The Scorpio, in the 6th Bhava, indicates that he/she will be sexy and luxurious. He/She will be hard working and will make money with hard work. He/She will not be so fortunate. He/She has commitment and seriousness about work and gets intensely involved in his/her job. His/Her work often involves matters of investigation such as journalism, research, laboratory science or psychology.

Sagittarius in 6th Bhava: The Sagittarius, in the 6th Bhava, indicates that he/she will be poor and has to work hard and struggle for bread and food whole life. He/She has a tendency to overwork and to over-extend himself/herself in his work. He/She eats or drink too much, or to have an extravagant taste in food and drink.

Capricorn in 6th Bhava: The Capricorn, in the 6th Bhava, indicates that he/she will be poor and has to work hard and struggle for bread and food whole life. His/Her expenditure is more than earning because of children is always sick. As a supervisor, he/she is a disciplinarian. Some Capricornia avoid work altogether because of disinterest in his/her job.

Aquarius in 6th Bhava: The Aquarius, in the 6th Bhava, indicates that he/she will work on ship or in navy. He/She is quarrelling with his/her boss and colleague. He/She has danger of drowning in water in childhood and also at the age of 41 years. He/She can function especially well within group situations.

Pisces in 6th Bhava: The Pisces, in the 6th Bhava, indicates that he/she will be infighting and so does not has good relations with children, spouse and other family members and so feel loneliness in whole life. He/She may be spending in litigation. He/She has excessive worry about the job.

Prediction by Sign in 7th house (Jaya/Wife)

Aries in 7th Bhava: The Aries, in the 7th Bhava, indicates that his//her spouse will be cruel, angry-nature, rigid, very frequently developing bad relation with husband and habituated of making disturbed family life, even though she will be educated. His/Her partner is energetic, aggressive, sexy, innovative, and pioneering.

Taurus in 7th Bhava: The Taurus, in the 7th Bhava, indicates that his//her spouse will be very beautiful, intelligent, sweet-talking, good mannered, better-understanding and talkative but proudly of her beauty. He/She will be good in conjugal life. He/She will be a loyal marital partner. There is much stubbornness in both him/her and his/her partner.

Gemini in 7th Bhava: The Gemini, in the 7th Bhava, indicates that his//her spouse will be beautiful, intelligent, sweet-talking, good mannered, better-understanding and talkative. He/She will be knowledgeable of stitching, dance, music, good cooking. He/She will be fortunate but always unsatisfied in conjugal life.

Cancer in 7th Bhava: The Cancer, in the 7th Bhava, indicates that his//her spouse will be very beautiful, intelligent, attract others due to her good manner, imaginative, thinker, better understanding and excellent in beauty. He/She will be knowledgeable of stitching, dance, music, good cooking and will be fortunate. He/She will be sentimental and good in conjugal life.

Leo in 7th Bhava: The Leo, in the 7th Bhava, indicates that his//her spouse will be cruel, angry-nature, rigid, very frequently developing bad relation with others and stay separately even in the friend's party. He/She will be too selfish and will be attached too much to ornaments and money, even though he/she will be educated, intelligent and good thinker. He/She will have happy married life. His/Her partner is dynamic, dramatic, strong, and vital.

Virgo in 7th Bhava: The Virgo, in the 7th Bhava, indicates that his//her spouse will be very beautiful, intelligent, always helpful to his/her spouse in difficulty and painful time, good mannered, better understanding and excellent in beauty. He/She will have good conjugal life. Marital partner is hard working and effectual, and assumes most of the responsibility of helping him/her and takes care of his/her practical affairs.

Libra in 7th Bhava: The Libra, in the 7th Bhava, indicates that his//her spouse will be very beautiful, intelligent, attract others due to her excellent beauty, helping attitude, interested in religious work, donor and good mannered, imaginative, thinker, better understanding and excellent in beauty. He/She will be helpful to his/her spouse and will be fortunate. He/She will have good conjugal life.

Scorpio in 7th Bhava: The Libra, in the 7th Bhava, indicates that his//her spouse will be less educated or uneducated, does not know stitching or any ladies' art and not even interested in the same, very unfortunate in life since childhood, always facing difficulties in whatsoever work he/she undertake. He/She will be suffering from diseases like headache and stomach. His/Her mate may be inwardly powerful and dynamic.

Sagittarius in 7th Bhava: The Sagittarius, in the 7th Bhava, indicates that his//her spouse will be cruel, angry-nature, rigid, very frequently developing bad relation with others. She/He will be less educated and will have no interest in stitching or any other ladies' arts. He/She will have happy married life, fortunate spouse, who will bring monetary help from his/her parent and will give good children, but reluctance to be hemmed in formal partnership entrapments.

Capricorn in 7th Bhava: The Capricorn, in the 7th Bhava, indicates that his//her spouse will be cruel, angry-nature, rigid, very frequently developing bad relation with others and educated and will have no interest in stitching or any other ladies' arts. He/She will have happy married life. He/She is restricted by the duties of wedlock or of any partnership. He/She needs partners to be his/her nurturing support and a mate as the parent figure to give a solid base, be supportive and responsible.

Aquarius in 7th Bhava: The Aquarius, in the 7th Bhava, indicates that his//her spouse will be very beautiful, intelligent, always helpful to her husband in difficulties, hard working, ready to face strongly in difficulties, fearful of god, respecting elderly people, interested in sacred work. He/She will have good conjugal life and will give good children.

Pisces in 7th Bhava: The Pisces, in the 7th Bhava, indicates that his//her spouse will be beautiful, intelligent, helping attitude, interested in religious work, donor, good mannered, better understanding and very active in social work. He/She will be helpful to his/her spouse and will be fortunate. He/She will be good swimmer and have good conjugal life and deliver intelligent and good children. He/She is willing to make sacrifices to benefit the union. His/Her partner helps him/her expand his/her horizons to new fields.

Prediction by Sign in 8th house (Mrityu/Death)

Aries in 8th Bhava: The Aries, in the 8th Bhava, indicates that he/she will be mostly settled in foreign country. He/She will be suffering from the disease of talking or walking in sleep. He/She will be unhappy by remembering his/her past incidents of life. He/She will be rich and die in foreign country. In joint financial affairs, there can be conflict and disagreement about money. He/She is decisive in this regard. Sometimes such conflicts may result in litigation.

Taurus in 8th Bhava: The Taurus, in the 8th Bhava, indicates that he/she will be seriously suffering from the disease of cough and die due to serious cough congestion of respiratory system. He/She has a good head for business and a sense for sound investments. There is a tendency to marry for money or security, as well as for love. There may be ups and downs in his/her financial affairs, but eventually matters turn out profitably.

Gemini in 8th Bhava: The Gemini, in the 8th Bhava, indicates that he/she will be seriously sick of Prameha and Gurda Roga in old age. He/She will die due to fighting with enemies. He/She is full of ideas about how properly to handle joint finances.

Cancer in 8th Bhava: The Cancer, in the 8th Bhava, indicates that he/she will die due to drowning in water or in the house peacefully at old age. He/She rarely exposes emotional needs and ceases to be the go-getter that usually marks his/her style.

Leo in 8th Bhava: The Leo, in the 8th Bhava, indicates that he/she will die due to snake biting or due to fighting with thief He/She has a large capacity for romance. There is also a much creative energy directed toward business and large-scale investments. Moneymaking becomes a game for him/her to be enjoyed. This position usually promises living to a ripe old age.

Virgo in 8th Bhava: The Virgo, in the 8th Bhava, indicates that he/she will die due venereal diseases or due to consumption of poison. This sign suggests a restrained or constrained sex life.

Libra in 8th Bhava: The Libra, in the 8th Bhava, indicates that he/she will die due to hunger or due to fighting with shoulder or due to anger and blood pressure and has financial gain through marriage and partnerships. He/She gets just everything he/she really wants with a diplomatic approach.

Scorpio in 8th Bhava: The Scorpio, in the 8th Bhava, indicates that he/she will die due to skin diseases or due to consumption of poison.

Sagittarius in 8th Bhava: The Sagittarius, in the 8th Bhava, indicates that he/she will die due to drowning in water or in the house peacefully at old age among the family members. He/She has a natural flare for business and good fortune when it comes to money. He/She has good fortune to benefit from large-scale enterprises that grow and prosper. Jupiter, the ruling planet of Sagittarius, provides good fortune and abundance in the part of the chart it controls. He/She gets financial help throughout his/her life, due to Jupiter influence. He/She can usually get what he/she wants out of life and is good-humoured nature and generous, outgoing demeanour.

Capricorn in 8th Bhava: The Capricorn, in the 8th Bhava, indicates that he/she will die due to old age and worshipping god in the house peacefully among the family members. He/She does not like to borrow and being in debt. He/She would never stoop to cheating or deceptions in business dealing. He/She handles other people's money and does so with serious concern.

Aquarius in 8th Bhava: The Aquarius, in the 8th Bhava, indicates that he/she will die due to burning or due to respiratory problem or due to saviour wound problem. He/She has a strong interest in the spiritual side of life. He/She is very unorthodox on sex, birth control and abortion.
Pisces in 8th Bhava: The Pisces, in the 8th Bhava, indicates that he/she will die due to lever problem or due to blood diseases in old age among the family members.

Prediction by Sign in 9th house (Bhagya/Fortune)

Aries in 9th Bhava: The Aries, in the 9th Bhava, indicates that he/she will be fortunate. He/She is always worried for money. His/Her expenses are more than the earning. He/She makes profit by purchase and sale of animals, agricultural lands and house properties, & product of religious work. He/She loves to travel.
Taurus in 9th Bhava: The Taurus, in the 9th Bhava, indicates that he/she will be fortunate at the age of 28 and 36 years. His/Her child hood is difficult and face problem in education but is successful in completing his/her education. He/She makes profit in business of fancy store but progress slowly in service. He/She earns wealth in second part of life and earns good name & fame and enjoys the life luxuriously. Venus makes travel enjoyable, and lives far from his/her native home.
Gemini in 9th Bhava: The Gemini, in the 9th Bhava, indicates that he/she will be fortunate at the age of 27 years, is simple, vegetarian, and gentleman and talks sensibly. He/She is successful in completing his education and helps poor people. He/She makes profit in business and thinks of always about business. He/She earns wealth in middle part of life and earns good name & fame and enjoys the life as a respected person. He/She especially enjoys travel mostly for observing different people and places.
Cancer in 9th Bhava: The Cancer, in the 9th Bhava, indicates that he/she will be simple, vegetarian, and gentleman and talks sensibly. He/She is successful in teacher, reporter, publisher and writer and helps poor people. He/She suffers from stomach problems and gastric. He/She takes too much time in taking the decisions and due to this, he/she suffers. He/She

earns wealth but his/her middle part of life is full of struggles and loss of wealth.

Leo in 9th Bhava: The Leo, in the 9th Bhava, indicates that he/she will be totally against the religions and will be arguing against the religions. His child hood is difficult and face problem in education but is successful in completing his/her education. Whole life he/she will be struggling and will do hard work for survival. He/She earns wealth in second part of life and earns good name & fame and enjoys the life luxuriously. His/Her nature of job will be touring type.

Virgo in 9th Bhava: The Virgo, in the 9th Bhava, indicates that he/she will be luxurious at the young age and gentleman and talks sensibly. He/She earns wealth in middle part of life but fortune never favours him/her. He/She will be fortunate at the age of 23 years and enjoys his/her life at old age with grand children.

Libra in 9th Bhava: The Libra, in the 9th Bhava, indicates that he/she will be totally religious minded and will be arguing always in favour of the religions. His/Her child hood is difficult and face problem in education but is successful in completing his/her education. Whole life he/she will be struggling and will do hard work for survival. He/She will be more successful in business than service. He/She earns wealth from business and earns good name & fame in the society as a leader and enjoys the life luxuriously in leadership. He/She gets wealth from his/her in-law. He/She will be fortunate after the age of 24 years.

Scorpio in 9th Bhava: The Scorpio, in the 9th Bhava, indicates that he/she will be mean mind. His/Her child hood is difficult and face problem in education, will be struggling and will do hard work for survival. He/She will be more successful after the age of 28 years. He/She earns wealth after 28 year of age and earns good name & fame in the society and enjoys the life luxuriously. His/Her nature of job will be touring type.

Sagittarius in 9th Bhava: The Sagittarius, in the 9th Bhava, indicates that he/she will be clever, peaceful and gentle man. His/Her child hood is good and does not face problem in living. He/She will be struggling and will do hard work for survival at the middle of age. He/She will be more fortunate and successful after the age of 45 years and achieve very high position and wealth. He/She earns wealth after 45 year of age and earns good name & fame in the society and enjoys the life luxuriously. His/Her nature of job will be related to water like captain, sailor or any other service with irrigation department where he/she gets lot of success. He/She will have much more journey in life.

Capricorn in 9th Bhava: The Capricorn, in the 9th Bhava, indicates that he/she will be clever, peaceful and gentle man. His child hood is not good and faces problem in living. He/She will be struggling and will do hard work for his survival. He/She will be more fortunate and successful after the age of 45 years. He/She earns wealth after 45 year of age and achieves very high position and wealth and earns good name & fame in the society and enjoys the life luxuriously.

Aquarius in 9th Bhava: The Aquarius, in the 9th Bhava, indicates that he/she will be clever in politics, a leader. His/Her child hood is not good and faces many problems in living. He/She will be struggling in education and will do hard work for his/her survival. He/She will be more fortunate and successful after the age of 45 years and 36 years and achieve very high position and wealth and earns good name & fame in the society and enjoys the life luxuriously. He/She starts earning wealth after 28 year of age.

Pisces in 9th Bhava: The Pisces, in the 9th Bhava, indicates that he/she will be totally religious minded and child hood will be good but is successful in completing his/her education. He/She will be more successful in business than service, particularly in yellow metals business. He/She will be fortunate after the age of 27 years and earns wealth from business and earns good name & fame in the society as a leader and enjoys the life luxuriously in leadership.

Prediction by Sign in 10th house (Karma/Profession)

Aries in 10th Bhava: The Aries, in the 10th Bhava, indicates that child hood is difficult and face problem in education but is successful in completing his/her education. Whole life he will be struggling and will be busy in more than one work and will do hard work for survival. He/She earns wealth in second part of life but does not earn good name & fame and does not enjoy the life luxuriously. He/She has tendency toward shyness and to disappear on the public stage, or even in public view.

Taurus in 10th Bhava: The Taurus, in the 10th Bhava, indicates that child hood is good but will spend more than his/her earning in his life. Whole lives he/she will be struggling and will be busy in earning and will not be able to save much. He/She will be patriotic to his/her father. He/She chooses career, such as art, theatre, and music.

Gemini in 10th Bhava: The Gemini, in the 10th Bhava, indicates that he/she will be patriotic to his/her father and respect and obey the elders and will be agriculturist and earn money from agriculture or real estate. He/She may earn more money from business rather than service.

Cancer in 10th Bhava: The Cancer, in the 10th Bhava, indicates that he/she will be child hood is good but will spend more than his/her earning in his/her life. Whole lives he/she will be busy in helping poor people and doing social work such as making temples and drinking water facilities for the people and will be busy in earning.

Leo in 10th Bhava: The Leo, in the 10th Bhava, indicates that child hood is difficult and face problem but is successful in completing his/her education. Whole life he/she will be struggling and will do hard work for survival. He/She earns wealth in second part of life but does not earn good name & fame and does not enjoy the life luxuriously. He/She does not get help from the family. He/She is likely to get a position of leadership and authority, and be admired and is happiest running his own business.

Virgo in 10th Bhava: The Virgo, in the 10th Bhava, indicates that he/she will be patriotic to his/her father but does not enjoy the life luxuriously. He/She will be self respected person and works to maintain his/her respect at any cost. He/She will not do buttering of any other person in life, even though he may lose something. He/She will earn good money from business rather than service. He/She finds employment in a large, well-established organization such as civil service, a church, or an educational institution.

Libra in 10th Bhava: The Libra, in the 10th Bhava, indicates that he/she will be patriotic to his/her father and respect and obey the elders and enjoy the life luxuriously. He/She will earn more money from business rather than service. Business will be lucky for him/her and will progress in life by business. He/She will be rich of his/her speech and fulfil whatever he/she will promise. He/She will reach to the highest position in the middle age. He/She will enter a business with a partner, or in cooperation with others. He/She will be an excellent administrator. His/Her public standing may rise well above that of his parent.

Scorpio in 10th Bhava: The Scorpio, in the 10th Bhava, indicates that he/she will be totally religious minded and get popularity and respect in the society and will be clean and honest. His/Her child hood is good but will spend more than his/her earning in his/her life in religious work. Whole lives he/she will be busy in helping poor people and doing social work and will be busy in earning and will not be able to save much. He/She will reach to the highest position in the service. He/She make his/her mark in the world. There has drive to succeed and is naturally able to impress others with the forcefulness of his/her mind.

Sagittarius in 10th Bhava: The Sagittarius, in the 10th Bhava, indicates that he/she will be patriotic to his father and respect and obey the elders and enjoy the life luxuriously. He/She will earn more money from business rather than service. Business will be lucky for him/her and will progress in life by business. He/She may face a lot of problem in service but will be hard working and makes everything favourable in life. He/She will get popularity and respect in the society. He/She has the abilities to be a leader and never hesitates using his/her influence to attain goals. A good deal of travel is likely to be associated with his/her career. He/She may have more than one career in his/her life.

Capricorn in 10th Bhava: The Capricorn, in the 10th Bhava, indicates that he/she will be totally religious minded and get popularity and respect in the society and will be clean and honest. His/Her childhood is good but will spend more than his/her earning in religious work. Whole lives he/she will be busy in helping poor people and doing social work. He/She will be able to make favourable conditions and will reach to the highest position in the middle age and people will be astonished with his/her rise. He/She has a very strong sense of duty, an attitude of dedication. The progress in the career may be slow, but it is consistent. He/She is capable of climbing to the top. He/She can build a solid public image. This is a very achievement oriented placement.

Aquarius in 10th Bhava: The Aquarius, in the 10th Bhava, indicates that he/she will be political minded and get popularity and respect in the society and will be talkative. His/Her child hood is good but will be busy in helping people. He/She has followers in politics. He/She makes favourable conditions and will reach to the highest position. He/She is a team player and function well as a part of the team. He/She may be involved in many humanitarian ventures.

Pisces in 10th Bhava: The Pisces, in the 10th Bhava, indicates that he/she will be patriotic to his father and respect and obey the elders and enjoy the life luxuriously. He/She will earn more money from service related to water. He/She may face a lot of problem in life but will be hard working and makes everything favourable in life. He/She will get popularity and respect in the society. He/She can make money by business too.

Prediction by Sign in 11th house (Gain/Income)

Aries in 11th Bhava: The Aries, in the 11th Bhava, indicates that he/she will be patriotic to his/her father and respects and obeys the elders and does not enjoy the life luxuriously. He/She will be self respected person and will not do buttering of any other person in life, even though he/she may loose something. He/She will be hard working person and does not get help from the father and family. Whole life he/she will be struggling and will be busy in more than one work and will do hard work for survival. He/She is fortunate after 28 years of age. He/She is active in associations, club or other such organizations. He/She attracts a wide and varied circle of friends. He/She makes the most of his contacts. Often, friends are the key to helping him/her attain his goals.

Taurus in 11th Bhava: The Taurus, in the 11th Bhava, indicates that he/she will belong to the medium family and does not enjoy the life luxuriously. He/She will be self respected person and work to maintain his/her respect at any cost. He/She will be hard working person and does not get help from the father and family. Whole life he/she will be struggling and will be busy in more than one work and will do hard work for survival. He/She will be known to the great personalities and will live in his/her surroundings. He/She gets benefited with the opposite sex. His luck starts from his/her middle age and reach to the highest post. He/She will be making money, accumulating a comfortable standard of living, and establishing a secure situation. His/Her interest is not just money and possessions for the sake of wealth. Instead he/she wants to gain a sense of security, and status to overcome a basic insecurity. He/She has well-to-do friends, who may be called on to help him/her attain his/her goals.

Gemini in 11th Bhava: The Gemini, in the 11th Bhava, indicates that he/she will be hand to mouth since child hood. He/She will be self respected person and work to maintain his/her respect at any cost. He/She will be hard working person and does not get help from the family. Whole life he/she will be struggling and will be busy in more than one work and will do hard work for survival. He/She is fortunate in business as compared to service. Up to 42 years, his financial conditions are miserable. He/She will start earning from many sources after 42 years of age. He/She has many personal connections. He/She is naturally attracted to people who are witty, intelligent, and verbal. He/She has a good sense of humour and he can laugh, even at himself/herself. He/She is not a loner, and he/she needs constant, mentally compatible companionship to be happy and fulfilled.

Cancer in 11th Bhava: The cancer, in the 11th Bhava, indicates that he/she will be patriotic to his/her father and respects and obeys the elders and enjoys the life luxuriously. He/She will be self respected person and work to maintain his/her respect at any cost. He/She will be hard working person and does not get help from the father and family. Whole life he/she will be struggling and will be busy in more than one work. He/She will progress in the service as compared to business. He/She will be more attached to his/her family members due to his/her sentiment and he/she may sacrifice for his/her family or friends. He/She can earn money from lottery. He/She will have a successful life.

Leo in 11th Bhava: The Leo, in the 11th Bhava, indicates that he/she will be business minded and calculate everything in term of profit and loss before doing the work. He/She will be self respected person and work to maintain his/her respect at any cost. He/She will be hard working person and does not get help from the family. Whole life he/she will be calculating and will be busy in more than one work and will does hard work for survival. He/She is fortunate in business as compared to service. He/She will start earning wealth in business from many sources and makes lot of money. He/She is kind and respects everybody. He/She will never spend money on luxury and live a simple life. He/She has leadership qualities displayed within groups and organizations. He/She has a very deep need for friends and associations. He/She takes pride in his/her friends and associates; some may be rich and famous. He/She may tend to draw strength from his/her friends as if they fulfilled a special need in his/her life. He/She is always careful to dress in good taste, and notices what others are wearing as well. He/She has a very wide circle of friends, most of who have much respect for him/her.

Virgo in 11th Bhava: The Virgo, in the 11th Bhava, indicates that he/she will be business minded and calculate everything in term of profit and loss before doing the work. He/She will be far sighted and self respected person and work to maintain his/her respect at any cost. He/She will be hard working person and does not get help from the family rather he/she will be helping his brothers and other family members. Whole life he/she will be calculating and will be busy in more than one work and will do hard work for survival. He/She is fortunate in politics as compared to business and service. He/She will start earning from many sources and makes lot of money from politics. He/She is kind and respects everybody. He/She will never spend money on luxury and live a simple life. He/She has a readiness to serve close friends.

Libra in 11th Bhava: The Libra, in the 11th Bhava, indicates that he/she will be calm, quiet, intelligent and gentle person. He/She will be able to identify the right or wrong timing and to do the miracle even in the miserable conditions. Family and friends will be helpful to him/her. He/She has an amiable demeanour with friends and working in group situations. In this context, he is diplomatic and even handed. He/She is likely to be selected to head a group just because of he/she is acceptable to those with divergent interests. He/She has a highly social attitude. He/She really loves to be with friends. Often this sign denotes marriage to a friend of long standing.

Scorpio in 11th Bhava: The Scorpio, in the 11th Bhava, indicates that he/she will be so intelligent that he/she will speak anything suitable to the time, person and place. He/She will be able to do many things at a time and will be successful in all the works undertaken. He/She will be doing the business related to the land and spends a lot for this. He/She will be benefited by agriculture. He/She develops close relationships with intense and aggressive friends who are influential and powerful. He/She never accepts weak individuals as friends.

Sagittarius in 11th Bhava: The Sagittarius, in the 11th Bhava, indicates that he/she will be business minded and calculate everything in term of profit and loss before doing the work. He/She will be self respected person and work to maintain his/her respect at any cost. He/She will be hard working person. Whole life he/she will be calculating and will be busy in more than one work and will do hard work for making lot of money. He/She will have popularity among service class officers and managers and will make use of them to earn more and more money. He/She is fortunate in business as compared to service. He/She will start earning wealth in business from many sources and makes lot of money. He/She is kind and respects everybody. He/She will never spend money on luxury and live a simple life. He/She has a wide circle of friends and associates and enjoys such casual relationships.

Capricorn in 11th Bhava: The Cancer, in the 11th Bhava, indicates that he/she will be a self made person, active and intelligent to take right decision at the right time, right place. He/She will be self respected person and work to maintain his respect at any cost. He/She will be hard working person and get help from the family. He/She is fortunate in service as compared to business. He/She will start earning wealth in service and from many other sources and makes lot of money. He/She is kind and respects everybody. He/She will never talk lye and fulfil his/her promises at any cost.

Aquarius in 11th Bhava: The Aquarius, in the 11th Bhava, indicates that he/she will be a self-made person, active and intelligent to take right decision at the right time, right place. He/She will be self respected person and work to maintain his/her respect at any cost. He/She will be hard working person and does not get help from the family. His/Her childhood is very troublesome but as he/she grows up, he/she starts earning and his/her life is comfortable. He/She is fortunate in service as compared to business. He/She will start earning wealth in service and from many other sources and makes lot of money. He/She is kind and respects everybody. He/She will never talk lye and fulfil his/her promises at any cost. He/She spends money on luxury and lives a simple life. He/She is successful in politics too. His/Her friendships are intellectually motivated rather than sentimentally stimulated.

Pisces in 11th Bhava: The Pisces, in the 11th Bhava, indicates that he/she will be a rich man, self made person, active and intelligent to take right decision at the right time, right place. He/She will be self respected person and work to maintain his respect. He/She will be hard working person and get help from the family. He/She is fortunate and shins in a particular line such as singer or scientist or in arts and in service as compared to business. He/She will start earning wealth in service and from many other sources and makes lot of money. He/She is kind and respects everybody. He/She will never talk lye and fulfil his promises at any cost. He/She spends money on luxury and lives a simple life.

Prediction by Sign in 12th house (Loss/Expenditure)

Aries in 12th Bhava: The Aries, in the 12th Bhava, indicates that he/she will enjoy the life luxuriously since childhood. He/She will earn more money from business rather than service in beginning but will become bankrupt in the middle age and there will be financial crisis. Business will be lucky for him/her in beginning or early age and will progress in life by business up to middle age. He/She may face a lot of problem in business at middle age but will be hard working and makes everything favourable in life. He/She will get popularity and respect in the society. His/Her eyesight will be weak and will get operated in the old age. Though he/she is very slow to anger, when he/she does cross the line, his/her temper can be irrational.

Taurus in 12th Bhava: The Taurus, in the 12th Bhava, indicates that he/she will be a rich man, self made person, active and intelligent to take right decision at the right time, right place. He/She will be self respected person and work to maintain his/he respect in the society. He/She will be hard working person. He/She is fortunate and shins in a particular line such as administrator or tourist or salesman or writer and in service as compared to business. He/She will start earning wealth at the age of 30 years and from many other sources and makes lot of money. He/She is kind and respects everybody. He/She will never talk lye and fulfil his/her promises at any cost. He/She does not spend money on luxury and lives a simple life. He/She has more worry about financial affairs than his/her happy-go-lucky demeanour would imply.

Gemini in 12th Bhava: The Gemini, in the 12th Bhava, indicates that he/she will be imaginative and sentimental type. He/She believes any person and get hurt and cheated in due course of time by that person, particularly by relative, brothers, sisters and others. He/She will never talk lye and fulfil his/her promises at any cost. He/She spends too much money on luxury and lives a luxurious life. Whole life he/she will be struggling and will be busy in more than one work and will do hard work for survival. He/She is fortunate in business as well as service. Throughout, his financial conditions are miserable. He/She will start earning from many sources and will spend too much money. His/Her eyesight will be weak and will get operated in the old age. He/She will find it very difficult to live with or near his brothers, sisters or other close relatives.

Cancer in 12th Bhava: The Cancer, in the 12th Bhava, indicates that he/she will be rigid and angry promise fulfiller type. He/She makes mistake due to his angriness. He/She will never talk lye and fulfil his/her promises at any cost. He/She spends too much money on luxury and lives a luxurious life. Whole life he/she will be struggling and will be busy in more than one work and will do hard work for survival. He/She is fortunate in business as well as service. Throughout, his/her financial conditions are miserable. He/She will start earning from many sources and will spend too much money. He/She believes any person and gets hurt and cheated in due course of time by that person, particularly by relative. He/She is a generous person. He/She gets hurt when those near and dear are beginning to take him for granted.

Leo in 12th Bhava: The Leo, in the 12th Bhava, indicates that he/she will be imaginative and sentimental calm and quite, kind and sweetly spoken type. He/She believes any person and gets hurt and cheated. He/She will never talk lye and fulfil his/her promises at any cost. He/She will not like to spend too much money on luxury and lives a simple life. Whole life he/she will be struggling and will be busy in more than one work and will do hard work for survival. He/She is fortunate in service rather than business. He/She will start earning from many sources and will not spend much money on decorative items. He/She is a power behind the scenes. He/She plays a back room manoeuvring role in matters. He/She is never one to blow his/her horn.

Virgo in 12th Bhava: The Virgo, in the 12th Bhava, indicates that he/she will be sentimental calm and quiet, and sweetly spoken type and successful businessman. He/She will be able to take decision very fast in any matter. He/She will be in business and find out many source of income in it. He/She will not like to spend too much money on luxury and lives a simple life and teach his/her children the same. Whole life he/she will be struggling and will be busy in more than one work and will do hard work for survival. He/She is fortunate in business rather than service. He/She will start earning from many sources and will not spend much money on luxury items. He/She may have to face litigation and charges and due to that, he may face problem. He/She will be self respected person and work to maintain his/her respect at any cost and will do the religious and social work.

Libra in 12th Bhava: The Libra, in the 12th Bhava, indicates that he/she will be a common person, self-made person, active and intelligent to take right decision at the right time, right place. He/She will be self respected person and work to maintain his/her respect. He/She will be hard working person and does not get help from the parent. He/She will be able to make any person in his/her favours and become popular in the society. He/She will have many friends and known personalities and will get benefited from them. He/She is fortunate and shins in service as compared to business. He/She will start earning in service and from many other sources. He/She is kind and respects everybody. He/She will never talk lye and fulfil his promises at any cost. He/She spends money on luxury and lives a simple life. He/She is a loner with some apprehensions regarding reliance on others. His/Her partnership arrangements, including the marriage, may be fated in some way and acting in collaboration with others, rarely works out well for him/her.

Scorpio in 12th Bhava: The Scorpio, in the 12th Bhava, indicates that he/she will be self made person, active, religious minded person and intelligent to take right decision at the right time, right place. He/She will be self respected person and work to maintain his/her respect in the society and will do the religious work. He/She will be hard working person. He/She is fortunate and shins in a particular line such as writer or publisher and in service as compared to business. He/She will start earning wealth at the age of 16 years and from many other sources. He/She is kind and respects everybody. He/She will never talk lye and fulfil his promises at any cost. He/She does not spend money on luxury and lives a simple life.

Sagittarius in 12th Bhava: The Sagittarius, in the 12th Bhava, indicates that he/she will be intelligent. He will be able to take decision very fast in any matter. He/She will be in business and find out many source of income in it. He/She will not like to spend too much money on luxury and lives a simple life. Whole life he/she will be struggling and will be busy in more than one work and will do hard work for survival. He/She will start earning from many sources and will not spend much money on luxury items. He/She will be self respected person and work to maintain his/her respect at any cost and will do the religious and social work and will be popular in the society. He/She will be able to make any person in his/her favour and become popular in the society. He/She will have many friends and known personalities and will get benefited from them.

Capricorn in 12th Bhava: The Capricorn, in the 12th Bhava, indicates that he/she will be a common person, self-made person, active and intelligent to take right decision at the right time, right place. His/Her childhood will be troublesome. He/She will be self respected person and work to maintain his/her respect. He/She will be hard working person and does not get monetary help from the parent. He/She will be able to make money by his/her own hard work and become popular in the society. He/She will start earning wealth at the age of 30 years and from many other sources. He/She will have many friends and known personalities and will get benefited from them. He/She is fortunate and shins in service as compared to business. He/She will start earning in service. He/She is kind and respects every body. He/She will never talk lye and fulfil his promises at any cost. He/She spends money very cautiously and makes balance between income and expenditure and lives a simple life. He/She has subconscious feelings of limitation and inadequacy deeply ingrained in the psyche.

Aquarius in 12th Bhava: The Aquarius, in the 12th Bhava, indicates that he/she will be intelligent, a common person, self-made person, active, religious minded and intelligent to take right decision at the right time, right place. He/She will be self respected person and work to maintain his/her respect. He/She will be hard working person. He/She will be able to take decision very fast in any matter and will be in service and finds out many source of income in it. He/She will spend too much money and does not have control over it and lives a simple life. Whole life he/she will be struggling and will be busy in more than one work and will do hard work for survival. He/She will start earning from many sources and will not spend much money on luxury items. He/She will be self respected person and work to maintain his/her respect at any cost and will do the religious and social work and will be popular in the society. He/She will be able to make any person in his/her favour and become popular in the society. He/She will have many friends and known personalities and will get benefited from them.

Pisces in 12th Bhava: The Pisces, in the 12th Bhava, indicates that he/she will be a common person, self-made person, active and intelligent to take right decision at the right time, right place. His/Her childhood will be simple. He/She will be self respected person and work to maintain his/her respect. He/She will be hard working person and does not get monetary help from the parent. He/She will be able to make money by his/he own hard work and become popular in the society. He/She will start earning wealth at the age of 32 years and from many other sources. He/She will have many friends and known personalities and will get benefited from them. He/She is fortunate and shins in service as compared to business and reaches to the highest position by his/her hard work. He/She will start earning in service. He/She is kind and respects everybody and will never talk lye and fulfil his promises at any cost. He/She spends money very cautiously and makes bank balance and lives a simple life. He/She will be helpful to others.

5

Moon-Sign (Janma-Rashi)

Predictions by Moon-Sign (Janma Rashi)

The waxing (rising) moon is a very auspicious planet, capable of causing Neechabhanga of other planets by his mere aspect. The position of Moon in a Sign is at the time of birth is called the Moon-Sign or Birth Rashi or Rashi. **Example:** If the Moon is in the Sign of Mesh (Aries), the Moon-Sign or the Birth Rashi or Rashi (Janma Rashi) is Mesh. It is also called the Moon-Sign Chart, in which Moon in the first House. Accordingly, the distance of the house of all the planets from the Moon is accessed, which is essential to predict the effects of the Maha Dasa and Antar Dasa of planet.

Aries Rashi (Moon in Aries) (Aswini, Bharani, Kritika 1st quarter):
Name starts with phonetic (Chu, Hey, Cho, La, Li, Lu, Ley, Lo, Ae): The Moon in Aries is not congruent to her nature. He/She is restless, eyes, inflicted with diseases, unfaithful, gives pleasures to his wife, fears of drowning into water, hard working and is full of tranquillity in his old age. He/She is adventurous and is too changeable, moody, whimsical and flirtatious. He/She is likely to meet with all sorts of disappointments and even disillusion.

Taurus Rashi (Moon in Taurus) (Kritika last 3 quarters, Rohini, Mrigasira 1st half):
Name starts with phonetic (Ee, U, Aye, Oh, Va, Ve, Vo, Vay, Vo): The Moon in Taurus is a natural domicile. He/She is charitable, pious, virtuous, wealthy, and full of radiance, good

health and long lived. He/She has determination, loyal friend and is emotionally very strong and seldom changes his/her mind. He/She attracts opposite sex for strong romance. He/She takes up occupations of real estate, property, art, design, jewellery and business. He/She is an ambitious, selfish, has a goal for a luxurious home, plenty of money and intellectual but has a very practical, astute, shrewd ability to judge the average conditions in life.

Gemini Rashi (Moon in Gemini) (Mrigasira last half, Ardra, Punarvasu first 3 quarters):

Name starts with phonetic (Ka, Ke, Koo, Gha, Jna, Cha, Kay, Ko, Haa): The moon in Gemini is considered weak. He/She has a melodious voice, talks sweetly, is kind hearted, very lusty and is prone to throat diseases, famous, wealthy, fair complexioned, tall, clever, genius, of firm resolution, efficient in work, and remains judicious in every situation. He/She thrives on communication. He/She prefers job in media, travel as well as sales. He/She chooses his/her partners. He/She is highly imaginative, educated, and is both, a good teacher and a sharp student. He/She cannot limit to one activity and business at a time and would do well with a strong & practical partner. He/She will be entrusting the partner with most of the decision-making in business. He/She is very successful as news person, advertising agency, writers, authors or any other creative field. He/She is likely to make plenty of money, and enjoy great popularity. He/She is good in the business world. He/She will retain a youthful look and behaviour and succeeds at "staying young forever."

Cancer Rashi (Moon in Cancer) (Punarvasu last quarter, Pushya, Aslesha):

Name starts with phonetic (He, Hoo, Hey, Ho, Daa, Dee, Doo, Dey, Do): The Moon is considered royal in her own Cancer. He/She is wealthy, has patience, serves his teacher, very clever, lives in a foreign land, keeps good company, and has a high degree of intelligence. He/She is sympathetic, kind, compassionate and sensitive to others' feelings. He/She is with excellent memory, fond of home and parents, peaceful, gentle, affectionate, and romantic. He/She may get delayed marriage. He/She may be excellent artists, musicians, and poets and home life is very important to him/her. As a parent, he/she is quite nurturing and lavishes loved ones. He/She does not hesitate to use tears or a self-sacrificial attitude to get his/her point across. Real estate is an especially good area to invest in for him/her. He/She would also do well in a business run from home and has natural tendency to put on weight.

Leo Rashi (Moon in Leo) (Magha, Poorvaphalguni, Uttaraphalguni 1st quarter):

Name starts with phonetic (Ma, Me, Moo, May, Moo, Ta, Tee, Too, Tay): The Moon in Leo is the natural domicile of the Sun. He/She forgives easily, loves to travel, likes to eat non-vegetarian food, is full of fear, keeps good company, is humble, has excessive anger, devoted to his parents and achieves fame. He/She is self-sacrificing, generous, conservative, discriminating, encouraging, romantic, optimistic, brilliant robust, strong, and decisive and a natural leader and exudes energy and drive. He/She loves excitement and action. He/She may sacrifice everything in the cause of righteousness and justice. He/She makes others dependent on him. He/She cannot be convinced against his/her will, nor be swayed against emotions. He/She is strong-minded and determined. He/She is warm, loving and outgoing. He/She doesn't settle for less than what he/she wants and is too much commanding. He/She will be smothered with love and care, and may be henpeck.

Virgo Rashi (Moon in Virgo) (Uttaraphalguni last 3 quarters, Hastha, Chitra 1st half):

Name starts with phonetic (Too, Pa, Pee, Pu, Sha, Na, Tha, Pay, Poe): The Moon in Virgo is unimpeded mind, passion and flesh. The keyword is "criticism". He/She is a sensualist, respects the virtuous people, religious, clever, charitable, poet, follower of the Vedas, lover of humanity, interested in dance and music, likes to travel, and is troubled by his/her Spouse. He/She has no confusion in the mind, strong social conscience, good communication and is sociable, logical, back-seat drivers and clear-headed. He/She has a strong personal code of conduct and set high standards. He/She frequently takes nursing, dietetics, teaching, and secretarial work as careers. He/She is considered to be cold-blooded and overly ambitious. He/She is very conscious of keeping fit, both physically and mentally. He/She also enjoys taking part in national politics or social issues. He/She makes well, stable business partners and a successful professional. He/She is grave, sexless people with no sexual curiosity, and doesn't understand the meaning of sex. He/She is exceedingly active, and will put more energy into house cleaning, attention to business, and personal doctoring than any other type of person.

Libra Rashi (Moon in Libra) (Chitra last half, Swati, Visakha first 3 quarters):

Name starts with phonetic (Ra, Ree, Ru, Ray, Tha, Thee, Thoo, They): The Moon in Libra is the best positions. The keyword is "decision". The moon in this position gives artistic temperament, creative ability, good mental understanding, but no executive ability. He/She gets angry unnecessarily, talks sweetly, has restless eyes, mixed fortunes, authority inside the house but powerless outside, a devotee and likes to travel. He/She is charming, creative, and diplomatic. He/She may be romantically amorous, notoriously fickle and wavering in romance. He/She is financially motivated, can be reckless, careless, and/or squandering. He/She is known for charm and social grace, and presents an image of total balance and harmony, and has problem of the kidneys and allergies. He/She takes professions as law, architecture, politics, the arts, and even homemaking. He/She is attracted to very gracious partner and finds a good one in life, although it may not be the first one. He/She is voluptuous, deceptive in habits, with a voracious physical appetite. Usually he/she is so pretty and has so much charm that marriage is a foregone conclusion.

Scorpio Rashi (Moon in Scorpio) (Vishakha last quarter, Anuradha, Jyestha):

Name starts with phonetic (Tho, Na, Nee, Noo, Ney, No, Yaa, Yee, Yoo): The Moon in Scorpio is favourable position. The keyword is "Ulterior Motivation". He/She is a traveller from his childhood, has yellow eyes, lusty, proud, behaves roughly with his relatives, acquires wealth through hard work, and is wicked towards his mother. He/She is intelligent, cold, sensual, emotional, materialistic and secretive. He/She can become superb occultists and astrologers. The position is favourable for jobs in medicine, surgery, chemistry and investigative work.

He/She can be very possessive, jealous, very cruel and vindictive. He/She never forgets a wrong that someone has done, and will plot and plan for years or decades, if necessary, to seek revenge. He/She is intensely emotional but projects a perfectly cool exterior at all times. His/Her frenzied desires are never satisfied within home. He/She is constantly seeking outside satisfaction. There are few women who have a good deal of scandal running through the life.

Sagittarius Rashi (Moon in Sagittarius) (Moola, Poorvashada, Uttarashada 1st quarter):

Name starts with phonetic (Yey, Yo, Ba, Bee, Bu, Dha, Pha, Dha, Bay): The Moon position in Sagittarius is risk-taking. The keyword is “enthusiasm”. He/She is pious, wealthy and virtuous, has a loving nature, knowledge of fine arts, likes drawing and painting, has a wife full of good qualities, sweet-talker, has a heavy physique and in some rare cases, a destroyer of his family. He/She is eternal students with an urge to higher education, impulsive, blunt and outspoken, magnetic and forceful, actively philosophical and believes in justice and fair play and helps out anyone who is in need. He/She likes astrology and prophecy. He/She hates anything hidden or secret. He/She has a strong sense in gambling and believes that he/she simply cannot lose and therefore take unconsidered risks. If it is necessary to terminate a relationship, he/she does so quickly and cleanly without looking back. He/She knows something better is waiting just around the next corner. He/She is great spenders and enjoys a jolly, pagan sort of social life, uninhabited and full of romantic interest. He/She is incurable opposite sex partners' chasers and has put a great deal of enthusiasm into his/her dashing love affairs.

Capricorn Rashi (Moon in Capricorn) (Uttarashada last 3 quarters, Sravana, Dhanishta 1st half):

Name starts with phonetic (Bo, JA, Je, Ju, Jay, Jo, gha, Ga, Gee): Moon in Capricorn is one of the least desirable positions. The key word is “Management”. He/She has values in his family, is under the influence of his wife, scholar, undertakes charitable work, respects his mother, is wealthy, has obedient servants, is kind hearted, has a large family, and lot of worries also. He/She lacks sympathy, and has innate selfishness, self-preoccupation, dignity, tenacity and a realistic vision of the world. He/She has defeated ambitions and dreams, misfortunes, occupational and financial troubles, credit difficulties and all sorts of other misfortunes. He/She will have positions of executive, administrative, public and organizational positions and commercial pursuits. He/She has a natural desire to rise to a position of power and fame and is willing to work hard for accomplishments. He/She is not fortunate enough to achieve fame and fortune and becomes terribly frustrated and may even develop ill health as a result. He/She is very selfish, cautious, and thrifty and lays own aims and ambitions.

Aquarius Rashi (Moon in Aquarius) (Dhanishta last, Satabhisha, Purvabhadra 1 to 3 quarters):

Name starts with phonetic (Goo, Gay, Sa, See, So, Say, Da): The Moon in Aquarius is fixed. The keyword is "disinterestedness". He/She is lazy, owns the most expensive vehicles, wealthy, is blessed with beautiful eyes, and has a simple nature. He acquires wealth and knowledge, achieves fame on account of his virtuosity and kindness, is fearless and enjoys his wealth. He/She is idealistic, caring for the global village, a little detached to home and paradoxically. He/She possesses integrity and honesty, and is not likely to ask for help when in trouble, but ready to help others. He/She is well liked and have strong religious and philosophical instincts coupled to a humanitarian urge. He/She possesses absence of jealousy and possessiveness, and favours all forms of humanitarian, political, and educational pursuits, exploration in all fields, authorship, and astrology too. He/She has a tendency to gossip and spread rumours. He/She loves the business and professional world, and has plenty of patience, kindness, intelligence and understanding.

Pisces Rashi (Moon in Pisces) Poorvabhadra last quarter, Uttarabhadra, Revathi):

Name starts with phonetic (Dee, Du, Tam, De, Do, Cha, Che): The Moon in Pisces is not favourable position. The key word is "anxiety". He/She is brave, talks cleverly, but often has excessive anger, loved by his family, a devotee, a very fast walker, efficient in charity and knowledge, virtuous, sacrificing and receives affection and love from his friends and family members. He/She has strong creativity, powerful imagination and impressionability. He/She is natural worshippers of beauty, very loyal to friends and is more inclined to romantic attitudes. He/She succeeds best in intuitive judgement, discretion, assiduity, and detailed work. He/She does well as entertainers, dealers with liquids of all sorts, promoters, seafarers, and detectives. He/She has strong intuitive and psychic qualities. He/She should generally follow his intuition, and cares a great deal about others and seek to serve society as a whole in own way. He/She gives love freely to others, and may get deceived by others. He/She may feel like rejecting the world altogether. His/Her life probably will not be one of the Cinderella or Prince Charming of his/her dreams. He/She is not afraid to work hard and is subjected to wild swings in mood. He/She can give pure, unselfish, transforming love and compassion. He/She can often find satisfaction and relief in religion and art.

6.2 Naming Children and Vehicle Number

Give your Child a Correct Name: The name carries much value in life. Once, a person of around 30 years of age visited an astrologer telling that he has a lot of personal as well as financial problems. He has a dull and sombre face without lustre because of Saturn (Sani). Astrologer asks his name. His first name is Mandeswara Rao but changed to Ravi Kumar as per advice of an astrologer. Second astrologer is shocked. Ravi Kumar means “son of the planet Sun”, which is none other than Saturn. Then second astrologer realizes that the person is completely under the clutches of Lord Saturn, and it is not so easy to come out of the influence of Lord Saturn.

It is very difficult for any one to select a name filled with complete positive vibrations and may not be able to change his name for good or bad at middle age. But we can give a better name to our kids. There are 3 steps to choose a better name, filled with positive vibrations, as advocated by Indian Astrology and Numerology combined.

Step I: Set the positive vibration in the name starting with phonetic according to the birth star (Nakshatra) of the child. Astrology prescribes the starting sound to be used in a name depending on the Birth Star (Nakshatra) and or the Rashi (Moon- Sign) of the child, as shown below in Table:

Nakshatra	**name starting with phonetic (sounds) letters combination**
Ashwini 1,2,3,4 quarters	Chu, Che, Cho, La
Bharani 1,2,3,4 quarters	Li, Lu, Le, Lo
Krittika 1st part or quarter	Aa
Krittika 2,3,4 quarters	Ee, Vu, Ae
Rohini 1,2,3,4 quarters	Vo, Va, Vi, Vu
Mrigashira 1 and 2 quarters	Ve, Vo
Mrigashira 3, 4 quarters	Ka, Ki
Aridra 1,2,3,4 quarters	Ku, Kham, Jna, Cha
Punarvasu 1,2,3 quarters	Ke, Ko, Ha
Punarvasu 4 quarter	Hi
Pushya 1,2,3, 4 quarters	Hu, He, Ho, Da
Ashlesha 1,2,3,4 quarters	Di, Du, De, Do
Makha 1,2,3,4 quarters	Ma, Mi, Mu, Me
Purva Phalguni 1,2,3,4 quarters	Mo, Ta, Ti, Tu
Uttara Phalguni 1 quarter	Te
Uttara Phalguni 2,3, 4 quarters	To, Pa, Pi
Hasta 1,2,3,4 quarters	Pu, Sha, Na, Dha
Chitra 1 and 2 quarters	Pe, Po
Chitra 3, 4 quarters	Ra, Ri
Swati 1,2,3,4 quarters	Ru, Re, Ro, Tha
Visakha 1,2,3 quarters	Thi, Thu, The
Visakha 4th quarter	Tho
Anuradha 1,2,3,4 quarters	Na, Ni, Nu, Ne
Jyeshtha 1,2,3,4 quarters	No, Ya, Yi, Yu
Moola 1,2,3,4 quarters	Ye, Yo, Bha, Bhi
Purva Ashada 1,2,3,4 quarters	Bhu, Bha, Dha, Pha
Uttara Ashada 1st quarter	Bhe
Uttara Ashada 2,3,4 quarters	Bho, Ja, Ji
Shravana 1,2,3,4 quarters	Khi, Khu, Khe, Kho
Dhanishta 1, 2 quarters	Ga, Gi

Dhanishta 3,4 quarters	Gu, Ge
Shathabisha 1,2,3,4 quarters	Go, Sa, Si, Su
P. Bhadrapada 1,2,3 quarters	Se, So, Dha
P. Bhadrapada 4th quarter only	Dhi
U. Bhadrapada 1,2,3,4 quarters	Dhu, Tha, Jha, Da/Tra
Revati 1,2,3,4 quarters	Dhe, Dho, Cha, Chi

Note: If we could not find a name starting with the sound mentioned against the birth star, then we should use the sounds of the other star in the same Moon-Sign group. Example: Purva Phalguni born child name should be with the sounds: Mo, Ta, Ti, Tu or the star Makha sound: Ma, Mi, Mu, Me or the star uttar Phalguni sound of the same Moon-Sign sound: Te, To, Pa, Pi. Similarly Revati born child name should be with the sounds: Dhe, Dho, Cha, Chi or for the same Moon-Sign sound: Dhi, Dhu, Tha, Jha, Da/Tra.

Step II: Let us now set a harmonious vibration between the name and the birth date Number according to Numerology. Remember, every letter is assigned a numerical value as shown below:

Table: Letters with assigned Numerical Value

A, I, J, Q, Y	1
B, K, R	2
C, G, L, S	3
D, M, T	4
E, H, N, X	5
U, V, W	6
O, Z	7
F, P	8

Calculation: Now, take the name of the child, and find the total numerical value of the name using the above digits. Example: RAVI BABU is the name chosen.

Name	Value	Compound Number	Single Number
RAVI	2 + 1+ 6 + 1	10	1
BABU	2 + 1 + 2 + 6	11	2

Name compound number: 10 + 11 = 21 and name's Single Number Value: 1 + 2 = 3.
Now, take the birth date of the child, and make it a single digit Number. Example: the child is born on 27th April, 2006. Take only the birth date number, i.e., 27 = 2 + 7 = 9. The following numbers are having harmonious vibrations:

Table: Digits with assigned Harmonious Vibrations of Single Numerical Value of Date of Birth

Name Single Number	**Harmonics**
1	1,2,9,3
2	2,1,5
3	3,1,2,9
4	3,6,2
5	5,1,6
6	6,5,8
7	7,1,2,9
8	5,6
9	9,1,2,3

If we observe the name's Single Numerical Value of RAVI BABU, it is 3 and his birthday number is 9. From the Table above, Number 9 is harmonic for 3. Hence we can give RAVI BABU name to the child who was born on 27th.
Note: Once, the child is named as RAVI BABU; we should take care to call him with this name only. If others call him, RAVI only, then his name single number is 1, which is a harmonic to his birth number 3. Similarly, if he is called BABU only then also his name single number is 2, which is a harmonic to 3. So even if he is called RAVI BABU or RAVI or BABU, there is no problem.
Step III (Check with defined Favourable Compound

Number): According to Numerology, the following compound numbers have good vibrations (others are bad): 10, 15, 17, 19, 21, 23, 24, 27, 32, 33, 36, 37, 41, 42, 45, 46 and 50. The compound number 'RAVI BABU' is 21, which is falling in the above favourable numbers. So we can confirm the name RAVI BABU is the best suitable name for the child.

Note: If the name compound number is more than 52, then add the digits again. For example, take 53; add 5 + 3 = 8. Take it as 8 only. Similarly, 54 are equal to 9, 55 = 10 and 56 = 11. By following the above steps, we can avoid any obstacles and chaos in the child's future and the child will become a happy and successful person.

Give your Vehicle a Correct Number: People often say that they got good luck after purchasing the vehicle and most of the people throw huge amount of money to get a lucky number for their new vehicle. In India, many people are seeking number '9999' for their vehicle. There is a high demand for this number than any one else. People who choose this number would be shocked if they know that this number is prone to accidents more often than any other number. Yet, they choose such numbers out of knowledge, ignorance and misguidance.

Alphabet	Numerology No.
a, i, j, q, y	1
b, k, r	2
c, g, l, s	3
d, m, t	4
e, h, n, x	5
u, v, w	6
o, z	7
f, p	8

Example: Take a vehicle number AP 9 UL 9902 to understand the Numerology behind Vehicle Number. Numerology assigns numerical digits for each alphabet as shown in the table. Calculate the Vehicle Number representing each alphabet in the form of digits. Add all the numeric digits of the number to make a single digit. Thus, it is A P 9 U L 9 9 0 2. 1 + 8 + 9 + 6 + 3 + 9 + 9 + 0 + 2 = 47 = 4 + 7 = 11 = 1 + 1 = 2. So, the Vehicle Single Digit

(VSD) number is 2. Now see the positive and negative traits of this number in the table below:

VSD No.	Planet	Positive traits	Negative traits
1	Sun	Good for works with government and forestry.	You appear to be egoist.
2	Moon	Good for short trips and picnics.	Emotional driving like sudden changes in the speed of vehicle.
3	Jupiter	Good for political, administrative and advisory works.	You are proud.
4	Rahu	Sudden monetary gains.	Wrong decisions and separation.
5	Mercury	Good for any commercial activity.	You are talkative and appear to be childish.
6	Venus	Good for artistic works and cinema-industry.	Too much indulgence in liquors and ladies.
7	Ketu	Spiritual thinking and pilgrimages.	Material benefits become less.
8	Saturn	Mobilizing masses and unions.	More work and less profits.
9	Mars	Quick decisions and dynamic actions.	You drive too fast.
0		A turning point in your life.	May be sudden downfall.

Since the major influencing digit in this vehicle number is 2 from the table, we can understand that you would enjoy short trips and picnics. Hence your car will be favourable for trips with your family. The negative part is you would drive your vehicle emotionally; some times very fast and some times too slow and you would be engrossed in changing thoughts. This may lead to minor accidents.
Similarly, take another vehicle number 'UP 12 BD 2204'. This gives 6 + 8 + 1 + 2 + 2 + 4 + 2 + 2 + 0 + 4 = 31 = 4. So, the Vehicle Single Digit (VSD) is 4. This digit 4 shows separation, which means the person, may be separated from his family (death) or he may lose a limb.
Lucky number for the vehicle: There are 3 simple principles that we follow to set a lucky number for the new vehicle. **Step 1:** The number should be harmonious with the vehicle purchaser birth date number. For this purpose, take the date of birth of the purchaser and make it a single digit. Example: If he is born on 16th of some month, his Birth Single Digit (BSD) will be: 1+ 6 = 7. From the table below, we can see that the Vehicle Single Digit (VSD) can be 7 or 2. Thus, suppose the vehicle number is 'AP 20 Y 1202' = 1 + 8 + 2 + 0 + 1 + 1 + 2 + 0 + 2 = 17 = 8.

Birth single digit (BSD)	Favourable Vehicle Single Digit number (VSD)	Avoid/Unfavourable Vehicle Single Digit number (VSD)
1	1, 3, 9	6, 8
2	2, 7	8, 9
3	3, 1, 9	5, 6
4	4, 3, 6, 8	1, 2
5	5, 6	2, 9
6	6, 5, 8	1, 9
7	7, 2	6, 8
8	8, 5, 6	1, 2
9	9, 1, 3	5, 8

Step 2: Safe-guard the vehicle number from harmful influences. The vehicle number should not sum up to the single digits shown in the last column of the preceding table, i.e. the Unfavourable Vehicle Single Digit number (VSD). Example: If the Birth date Single Digit (BSD) is 7, then avoid 6 and 8 as the vehicle single digit number. Thus, the vehicle number AP 20 Y 1202 = 8 becomes unfavourable.

Step 3: Avoid Always the combination of accident-prone digits. Numerology identifies 9, 8 and 4 as highly dangerous numbers with respect to accident. Try to avoid too many 9 or 8 or 4 in the vehicle number as well as their combinations, such as, the number like 'AP 9 D 9908' because it will be very harmful even though the sum or VSD Number becomes 3. There are three 9's and one 8 which are 100% accident prone digits.

Note: The digit '0' is considered a wheel of fortune and it represents the ups and downs in life. Single zero will not affect in any way. So, if a zero comes in the vehicle number, there is no problem. But avoid repetition of zeros in the vehicle number too.

6

Predictions by Sun-Sign

Predictions by Sun-Sign

The date of birth determines the Sun-Sign and gives special attributes as mentioned below:

6.1 Aries (March 21 - April 20)

Who works from morning to evening, and never likes to be outdone?
Who is outspoken, alert, and ambitious and whose walk is almost like a run?
Personal Quality: The Arian is vital, impulsive, born leader, adventurous, brave, fearless, highly dominating, and full of energy, aggressive, argumentative, good athletes and soldiers and makes enthusiastic lovers. He/She is outspoken, alert and quick to act and speak. He/She is always willing to help the persons in need. He/She is not a follower. He/She is large hearted and speak straight forward for what he/she feel about the person. He/She is childishly egocentric, extremely demanding and liable to throw tantrums if denied. He/She is quick to anger and known for his impatience, and is prone to be arrogant. Under planetary afflictions he is subject to brain fever, dizziness, nosebleed, neuralgia, inflammation of the cerebral hemispheres, and diseases of face.
Positive Quality: He/She is generous, a lover of justice, and wishes to earn by own efforts and never looks at others wealth. He/She gives time, effort, money and sympathies to others. He/She likes to be challenged and enjoys solving any obstacle. He/She has both moral and physical courage.

Negative Quality: He/She is not very tactful in communicating and will never bend, has strong feeling of admirations and is not diplomatic and is sharp tongue and shows anxiety. He/She has a spending nature to maintain the image. He/She gets nervous when things are not moving his/her way. He/She is quick-tempered, violent, impatient, egotistical and intolerant.

Physical Appearance: He/She is angular, slim in early life, although may fill out later. He/She has sharp elbows and knees.

Relationships: He/She is possessive, jealous, faithful and idealistic, passionate and incurably romantic. He/She tends towards joyous sex and close relationships throughout the lives. His/Her life partner will be Leo and Libra. Aquarius and Sagittarius might be very helpful to him/her in business.

Career: His/Her income and social status will rise at the age of 48-52 and will get promotion at the age of 30, 36 and 45. He/She will deal with 'futures' on the money market or hacking through the Amazonian forest. He/She can be Dentist, Director, E M T, Entertainer, Entrepreneur, Landlord, Lawyer, and Make-up artist, Optometry, Producer, Sports person and Stockbroker.

Health: He/She is always in a tearing hurry and often has a fast metabolism, which keeps the weight down. He/She is prone to stress and suffer from tension headaches.

Ideal Partner: He/She vibes best with one of Gemini, Leo, Sagittarius, Aquarius

Compatible Zodiac Signs: Gemini, Leo, Sagittarius, Aquarius

Incompatible Zodiac Signs: Taurus, Cancer, Virgo, Scorpio, Capricorn, Pisces

Lucky stone: Coral & Pearl. The glittering Coral will gives all the courage, makes him/her rich and gives a comfortable future. Topaz and Moonstones will be auspicious too. Coral is to be worn as ring in middle finger.

Lucky Number: The number 1, 8, 6, 9 & 14 can bring luck in life.

Lucky Colour: Blue, Blue-green. Revel in the magic of peacock blue or Shades of Red will be lucky.

Lucky Day: Friday, Tuesday and Saturday. Dig gold on Tuesday.

Lucky Flower: Sweet Pea.

6.2 Taurus (Vrishabh) (April 21 - May 21)

Who loves good things and smiles through life except when crossed?
Who is stolid, tenacious and determined and thinks he knows the most?

Personal Quality: The Taurus is stable, stubborn, generous, highly reliable, practical, ambitious, and good in the position of managers and achieves almost everything in life. He/She makes friendships very rarely but, once made, he/she is faithful. He/She is reliable, responsible, affectionate and loyal. He/She is easy to get along with and good team player. He/She can reach to the desired height with hard work, devotion and patience. He/She is practical, reserved and is possessing tremendous willpower and self-discipline. He/She is incredibly and uncompromisingly loyal. His/Her economic position will be good from the age of 35-46 and becomes a rich man in the society. His/Her early part of life is very struggling. His/Her children are generally intelligent and bright and he/she has a pleasant married life.

Positive Quality: He/She is helpful and does a lot of things for family and considers it as a sacred duty. He/She has ability to concentrate and never leaves anything unfinished and does not believe in shortcut of anything. He/She is warm, loving, gentle and charming most of the time. He/She is honest, reliable and loving.

Negative Quality: He/She very seldom changes the mind once made-up and does not bother at all for the result. He/She is expressed in dullness, stubbornness and resistance to change. He/She is very suspicious and is afraid of getting deceived. He/She believes to the person so easily that any body can cheat him/her. He/She cannot forget or forgive people so easily.

Physical Appearance: His/Her stature varies from short to medium to stocky. His/Her eyes are bright and soulful and he/she carries himself gracefully.

Relationships: He/She is intense and passionate. He/She demands perfection from mate and is exacting. This makes him/her ardent and fascinating lovers. He/She makes charming company and is loyal and devoted.

Career: He/She has a wide spread of potential careers, right from banking to the fine arts. He/She will hold on to one job for

the rest of his/her working lives. Here are some occupations that a he/she might consider such as Advertising director, Antique dealer, Business person, Cashier, Clothing designer, Financial advisor, Florist, Patron of the arts, Perfumer, Real estate agent, Singer, Venture capitalist and Woodworker.
Health: Traditionally, he/she is endowed with a vigorous constitution and splendid health. If there is a weak point, it is usually his/her throat or neck. He/She is hopelessly addicted to food and alcohol.
Ideal Partner: He/She vibes best with one of Taurus.
Compatible Zodiac Signs: Cancer, Virgo, Capricorn, Pisces
Incompatible Zodiac Signs: Gemini, Leo, Libra, Sagittarius, Aquarius, Aries
Lucky stone: Diamond, Emerald & Blue Sapphire. Wearing Diamond or Emerald or Blue Sapphire can bring wealth and makes a better person and can give him/her strength. Sapphire or emerald, and it is to be worn as a ring in index finger
Lucky Number: Number 1, 2, 3 & 8 are best for good fortune.
Lucky Colour: Lotus pink or Shades of verdant green will do a world of good.
Lucky Day: Friday & Monday is the day of new beginnings
Flower: Daisy

6.3 Gemini (Mithuna) (May 21 - June 20)

Who oscillates, communicates and changes often and who is fond of life, fun and pleasure?
Who has a free soul and loves others attentions and exchanging of an intellectual nature?
Personal Quality: The Gemini is adaptable, dual natured, affectionate, courteous, kind, generous, scientists, and talented, bright, witty, entertaining army personnel, thoughtful towards poor and sufferer, adaptable and adjusting nature. He/She has unpredictable temperament. He/She is very well with Aquarius. He/She has wealth in later half of life. He/She will completely change his/her mind like Chameleon.
Positive Quality: He/She is usually quite, creative and has a strong self-confidence. He/She is often the centre of attraction in the gathering and is versatile, adaptable, and inquisitive and always moves along with the times.
Negative Quality: He/She is sharp-tongued and sometimes boring. He/She cannot concentrate well at one point and hence

does not finish the job. He/She becomes cynical, biting, moody and quickly angered. He/She is superficial, restless, nervous, lacks concentration and conniving. He/She does not keep their promises.
Physical Appearance: He/She has small, narrow hands and feet and slim. He/She is generally tall. He/She is highly energetic and exudes oodles of charisma.
Relationships: He/She is emotionally undemonstrative but enjoys being in a lively family, and seeks a partner with strong opinions. He/She is keen to make relationships but tends to be too egocentric.
Career: He/She is particularly suited to media work, in sales pitches and can be writer, Broker, Commentator, Concierge, Correspondent, Debater, Impersonator, Journalist, Librarian, Linguist, News commentator, Novelist, Orator, and Playwright.
Health: He/She has the ill effects of smoking since he/she has delicate lungs.
Ideal Partner: He/She is best with Aquarians. Virgo, Libra and Aquarius people may help him/her.
Compatible Zodiac Signs: Aries, Leo, Libra, Aquarius.
Incompatible Zodiac Signs: Taurus, Cancer, Virgo, Scorpio, Capricorn, Pisces
Lucky Gem: Blue Sapphire, Diamond & Emerald. He/She can count on the Emerald, which will bless him with all the intelligence he needs. Emerald or Yellow sapphire, which is to be worn as a ring in middle finger
Lucky Number: Number 3, 5, 7 & 9 will bring him good news.
Lucky Colour: Yellow and green, & sky Blue; reach for the skies with sky blue.
Lucky Day: Thursday and Wednesday; all his dreams come true on Thursday.
Lucky Flower: Rose

6.4 Cancer (Karka) (21st June - 20th July)

Who cannot stick to any adhesion and changes like a season? Who most perplexing character and let's go without any reason?
Personal Quality: The Cancer is protective, jealous, sensitive, full of suspicion, not easy to understand for his/her moods, often fluctuate from sweet to cranky and lacks faith in others. He/She can be untidy, sulky, devious, and inclined to self-pity

because of an inferiority complex but always ready to cooperate. He/She is very fond of food and is usually hearty eaters. He/She gets ancestral and sudden properties after the age of 35. He/She is the most family-centred persons and fiercely protective of loved ones. He/She possesses strong paternal and maternal instincts.

Positive Quality: He/She is loving, kind, faithful, loyal, honest, hard working and sensitive and leaves his unforgettable impression on others.

Negative Quality: He/She is sulky, devious, moody, clinging, manipulative, overly emotional and insecure and inclined to self-pity and prone to a sense of personal inferiority and believes his/her views, opinions and behaviour to be impeccable, and beyond question or criticism.

Physical Appearance: He/She is small with round faces and possesses a tendency to gain weight. He/She has abundant shiny hair, expressive eyes and is economical with his/her gestures.

Relationships: He/She treasures emotional bonds and doesn't severe tie easily. He/She clings on to a failing relationship and finds it difficult to let go. He/She wears his/her hearts on his/her sleeve, and is prone to emotional excesses.

Career: He/She is best suited for counselling and charity work, good journalists, writers or politicians, archaeologist, caterer, dairy farmer, deep-sea diver, dietician, hotel worker, manufacturer, merchant small businesses. He/She works well with people and often adopts the role of a mediator.

Health: He/She has a weak digestive system and constantly suffers from heartburn and ulcers. He/She tends to become hypochondriacs.

Ideal Partner: He/She seeks steady, stable and practical partners, and usually bonds best with Capricorn.

Compatible Zodiac Signs: Taurus, Virgo, Scorpio, Pisces

Incompatible Zodiac Signs: Aries, Gemini, Libra, Leo, Sagittarius, and Aquarius

Lucky Gem: Yellow sapphire, Pearl & Coral. If he suffers with sleepless nights, a pearl could be the perfect cure. Wearing the pearl ensures peace of mind, and brings all the good luck in the world. Pearl, this is to be as a ring in small finger.

Lucky Number: 2, 7 and 9. Number 4 & 6 are his/her pick for good fortune.

Lucky Colour: White, cream, red and yellow. Soak in the elegance of white.
Lucky Day: Sunday, Monday. Wednesday; it is time to look ahead on Wednesday.
Lucy Flower: Larkspur.

6.5 Leo (Simha) (July 21st to August 21st)

Who praises all his kindred and expects others to praise them too?
Who possess grace, dignity and generosity but cannot see their senseless view?
Personal Quality: He/She is creative, strong willed, self-confident, generous, warm hearted, loving, broad-minded and faithful. He/She has a powerful presence of mind and power of success in the conquests. He/She is called the true kings of the zodiac. He/She is manager because he dislikes subordination. He/She can be good as chairman or director in business because he is excellent organizers. He/She has self-confidence, alertness and hard struggling power. He/She never forgets the goal and tries hard to achieve it with patience, wisdom and hard labour. He/She is a dominant, always busy with some planning and work and cannot seat idle even for an hour. He/She is able to attain the top most position. He/She achieves full success after the middle age. He/She prefers position and honour to money. For position, he/she can forget any money. He/She always cares for others and other's interest or benefits except, where it is necessary to take care for him/he. He/She is born to lead. He/She is sometimes great saints. Many great Saints have born under the Leo sign. There is a reality in his/her love and he/she can do anything to please that he/she likes most. He/She will keep improving day by day after 30 years of age. He/She has a small family and their children are very brilliant and intelligent. He/She is straightforward and uncomplicated individuals. He/She is stubborn, and may suffer from short bouts of depression. He/She walks forward always, head held proudly and face turned towards the sun. He/She accumulates good amount of money, wealth and properties. His/Her fortune is good and he/she will never lack money in life.
Positive Quality: He/She is a good leader, generous and kind hearted, and confident, ambitious, loving, and honest and

creative minded. He/She possesses a strong positive nature and doesn't shrink from any adverse circumstances. He/She can never bear dishonesty and injustice in life. He/She is witty and sets examples for others to follow. He/She is direct and to the point.

Negative Quality: He/She sometimes thinks himself/herself to be the only capable person in the total world. He/She believes in commanding only and does not care for others feelings. Some Leos think for money and profit and forget their other duties. It is very difficult to face his/her furies. He/She can be too sensitive to personal criticism, and when his dominance is threatened he can go into a sudden rage. He/She is conceited and arrogant.

Physical Appearance: He/She has a distinguished stature and attracts attention easily. He/She normally possesses healthy skin and a well-sculpted body.

Relationships: He/She makes wonderful social companions and is passionate and faithful lovers, but very sensitive and gets hurt easily.

Health: He/She suffers from back and spinal problems, has tendency to be overweight by lack of exercise. He/She gets easily stressed and can suffer from heart ailments.

Ideal Partner: Hence, he/she gets best with Aquarius, but Taurus, Gemini and, Sagittarius women are ideal partner for him/her.

Compatible Zodiac Signs: Aries, Sagittarius

Incompatible Zodiac Signs: Taurus, Scorpio, Aquarius

Lucky Gem: Ruby and Topaz. The ruby is a miracle stone, and wearing it will bring him/her health and good fortune. He/She will also acquire the power to make instant decisions. The Ruby and Topaz can give him/her success and power in life. Ruby, this is to be worn as a ring in ring finger

Lucky Number: Number 1, 3, 4, 5, 6 & 9 will pull him/her out of trouble.

Lucky Colour: : Orange, Gold, Red. Saffron. Let saffron lift his/he spirits.

Lucky Day: Friday, Tuesday and Wednesday. Preen, for he/she will rake in accolades on Friday.

Lucky stone: Peridot.

Flower: Gladiolus

6.6 Virgo (Kanya) (August 22 to September 22)

Who criticizes all she sees and would even analyse a sneeze? Who is observant, shrewd but hugs and loves her own disease?

Personal Quality: The Virgo is critical, precise, easygoing, reliable, steady, helpful, intellectual, studious, logical, methodical, communicative, sciences, languages, takes a romance to new heights, good followers and the best employee one can ever have. He/She often dislikes delegating. He/She knows how to please the persons in power and position to get his/her work done easily and so, gets promotion very fast. He/She has a pleasant nature, colourful personality and sharp mind with a great sense of responsibility. He/She is precise, refined, and a lover of cleanliness, hygiene and order. It is not so easy to measure the depth of his nature. He/She knows to mould others in his shape by his clever activity. If some body offends him/her, he/she does not show his real feelings on his/her face but act secretly to take revenge and hit back all his/her might when he/she gets the opportunity. The early part of his/her life is spent in struggles. The luck favours him/her at the age of 24, 36 to 42 years of age. He/She gets properties after the age of 40.

Positive Quality: He/She is good planners, practical and hard working, trustworthy and able to do perfect work. He/She is a hard worker, conscientious and perfectionists. He/She is plain spoken and is able to express well. He/She leads a moderate life and does not like excesses.

Negative Quality: He/She is sometimes very critical and thinks himself/herself all in all. He/She has a penchant for turning molehills into mountains, difficulties into stress and cleanliness into obsessive behaviour and a capacity for endless worry. He/She is miser and selfish and wish others to follow them.

Physical appearance: He/She has good bone structure and is often highly photogenic. He/She is attractive, with beautiful eyes that sparkle with intelligence.

Relationships: He/She is truthful and loyal. He/She is tense in close relationships, which could badly upset his/her sex lives and makes it hard for him/her to become truly intimate with those he/she loves.

Career: He/She is usually happy working in a job that calls for precision, a shrewd mind, and logic. Dogged, analytical and intellectual, he/she is makes skilled and inspired research scientists, analysts or even literary critics. He/She can be at the best as a manager, a secretary, a lawyer and a trader.

Health: He/She is often martyrs to stomachs, and may suffer from Irritable Bowel Syndrome, from food allergies. Virgo rules the abdominal region, intestines, and the lower lobes of the liver, the spleen, the duodenum, and the sympathetic nervous system.

Ideal Partner: He/She gets attracted to opposites and vibe well with dreamy Pisces.

Compatible Zodiac Signs: Taurus, Capricorn

Incompatible Zodiac Signs: Gemini, Sagittarius, Pisces Earth

Lucky Gem: Pearl, Topaz and Emerald. He/She should wear Emerald to crack a tough problem, to help him. The stone will bless him/her with all the intelligence he/she needs. Emerald, this is to be worn in small finger

Lucky Number: 2, 3, 5, 6, 7 & 15are for all the good luck.

Lucky Colour: Green, white and yellow. Get close to Nature. Wear earthy browns.

Lucky Day: Wednesday. His/Her quest for perfection pays off on Friday.

Lucky Flower: Lavender

6.7 Libra (September 23 to October 22)

Who is easygoing, sociable and keeps you waiting for half the day?

Who puts you off with promise gay and compromises all the way?

Personal Quality: The Libra is diplomat, impartial, sociable, cheerful, charming and sensitive to the needs of others. He/She is affectionate, polite in behaviour, cooperative, peace loving and sacrificing. He/She is natural arbitrators and diplomats, justice, honest and hard worker. He/She can win over his/her staunchest enemies with the help of his/her sweet voice. He/She has an idealistic and generally peace loving nature. He/She is the most civilized of the twelve signs. He/She has financial stability in the life. He/She is sure to have properties of his/her own but keep away from others' properties. He/She is more interested in making friends than enemies, and is willing to go along with others and do whatever it takes to maintain a relationship. He/She is a sensual lover and does not like any interference in the matters of love and marriage. Discord makes him/her totally insecure, and uncomfortable. He/She likes harmony in his/her life, and will do whatever it takes to have it.

Positive Quality: He/She always maintains a sweet relation with others. He/She is usually sympathetic, kind, loving nature and artistic. He/She does not like injustice, quarrels and disagreements. He/She is fair minded and loyal and have reach taste or good sense of humour. He/She does not hurt other person's feelings and likes to help the person in need.

Negative Quality: He/She is insincere and jealous and likes self admirations. He/She does not have argument power well even he is right. He/She tries to keep away from truth and painful experience. He/She can be frivolous, flirty and quite shallow. He/She is fickle minded, dependent, indecisive, sulking, and likes peace at any cost.

Physical Appearance: He/She is smart and attractive. He/She has sweet open faces with laughing eyes, and a devastating smile. He/She tends to be plump rather than angular or skinny.

Relationships: He/She highly understands his/her companions and he meets her with his own innate optimism. His/Her married life is delightful with the Gemini and Cancer girls. He/She gets special co-operation from Gemini, Aquarius, Sagittarius and Leo natives. He/She is not very good at handling relationships.

Career: He/She gets success in life as a businessman, engineer, lawyer, chartered accountant or a doctor. He/She is attracted to careers in the luxury trades, including fashion, beauty and design. He/She also makes good diplomats and counsellors. He/She is successful as writers, composers, fashion designers, interior decorators, critics, administrators, lawyers, and in civil services.

Health: This sign rules the kidneys, the lumbar region of the spine, the skin, the urethras, which are the tiny ducts running between the kidneys and the bladder, and the verso-motor system. He/She tends to suffer from nervous stomachs and ulcers. He/She needs to drink plenty of water in order to flush out the toxins from kidneys.

Ideal Partner: He/She gets along with the best ones and the gentle that captivate attention for life.
Compatible Zodiac Signs: Gemini, Aquarius
Incompatible Zodiac Signs: Aries, Cancer, Capricorn
Lucky Gem: Diamond and Blue Sapphire. He/She does love the real thing, and wearing a diamond can bring him/her wealth and can make him/her a better person. Opal, which is to be worn as a ring in index finger
Lucky Number: Number 1, 2, 4, 7 & 21 will fill with joy.
Lucky Colour: Royal Blue, Orange and white. He/She is can rule over the world with royal blue.
Lucky Day: Tuesday, Sunday and Monday. Pack your bags for a holiday on Tuesday.
Lucky Flower: Aster.

6.8 Scorpio (Vrischika) (October 23 to November 22)

Who has an intense and powerful nature and keeps ready an arrow in his bow?
Who is self centred, wilful, proud, and detective and if you prod him, lets it go?
Who is a fervent friend, a subtle foe?
Personal Quality: Scorpion is ruthless, mysterious, magnetically, attractive and emotional intimacy and is the most intense and passionate. He/She also has immense degree of willpower and is highly tenacious. He/She is of a secretive, timid, retiring nature, one that does not talk of his affairs. He/She is honest and independent nature person with hard struggles in life. He/She does not believe in accumulating the illegal wealth. He/She has strong will power and a natural quality of leadership. He/She does not believe in empty promises. He/She can be vindictive, dangerous enemies and possesses a strong streak of venom. He/She starts good earning at the age of 24 and after 40 years of age acquire properties. He/She is self contained and self centred and seethes and doesn't give up the enmity. He/She may burst any moment. He/She is too demanding, too unforgiving of faults in others. He/She is very jealous. He/She is the symbol of sex and passionate lovers.

Positive Quality: He/She is brave, courageous, sincere and loving, subtle, imaginative, powerful, generous, loyal, passionate, exciting, and magnetic. He/She is the person with the fixed mind and achieve goal directly by his deed. He/She is not afraid of obstacles because he has a strong will power.
Negative Personality: He/She is proudly, over sensitive and careless. He/She is jealous, resentful, obstinate, compulsive, obsessive, brooding, secretive, revengeful, possessive, and extremist and can appear cold and impassive.
Health Concerns: He/She is prone to ailments of the liver and kidneys, stones and gravel in the bladder or genitals, and other genital ills such as pianism, abscesses, boils, carbuncles, fistulas, piles, ruptures and ulcers.
Physical Appearance: He/She is always striking and has a magnetic face and dress elegantly. There is a strange mysticism and magnetism in his personality, which is enchanting to the beholder.
Relationships: He/She likes to stay away from his/her family and leads an independent life. His/Her marital life with Pisces, Taurus, and Virgo and Cancer girls will be pleasant.
Career: He/She is traditionally associated with jobs such as mining and detective work. He/She makes demanding bosses. He/She might consider job such as Analyst, Criminologist, Detective, Doctor, Enforcer, Hypnotist, Insurance agent, Investigator, Lab technician, Private investigator, Psychiatrist, Psychologist, Researcher or Scientist.
Ideal Partner: Scorpios fits best with Taurus. This sign gives him the material and emotional security he craves.
Compatible Zodiac Signs: Cancer, Pisces
Incompatible Zodiac Signs: Taurus, Leo, Aquarius
Lucky Gem: Coral, Opal. He gathers courage from the coral. The stone could also make him/her rich and also gets confident.
Lucky Number: Discover the magical powers of Number 2, 3, 7, 8 & 9.
Lucky Colour: Midnight Blue, Yellow, red, orange, and Maroon. Kiss those blues away with midnight blue.
Lucky Day: Sunday, Tuesday and Thursday. There will be sudden windfall on Sunday.
Lucy Flower: Chrysanthemum.

6.9 Sagittarius (Dhanu) (November 23 to December 20)

Who loves the dim, religious light and always keeps a star in sight?
Who is versatile enterprising and an optimist, both gay and bright?

Personal Quality: The Sagittarius is moralistic, impulsive, full of versatility and eagerness, and has a positive outlook towards life. He/She enjoys travelling and exploration. He/She is ambitious and optimistic, honourable, honest, trustworthy, truthful, generous and sincere with a passion for justice and truth. He/She has very charming voice and benevolent personality. He/She is 'Yes' man and never says 'No' to anyone. He/She doesn't get demoralized in his failure. He/She does not like to harm any person and does social work too. He/She is noted for longevity, intuitiveness and original thinkers. He/She cannot gain or be successful in Gambling, Races and Stock-exchange businesses. He/She can be successful in business of white and artistic gift items or textiles and metal. He/She earns wealth after 36 year of age and also gets parental property.

Positive Quality: He/She is honest, tolerant, and friendly and trusts and respects people. He/She is kind and forgives the people easily is never proud.

Negative Quality: He/She is indiscipline and never learns even from his/her mistake in the past. He/She likes gambling and loose money. He/She does not keep his promises and does not have foresightedness and hence, he/she is unsuccessful. He/She has a quick temper and a biting tongue. Physical Appearance: He/She has darting and piercing eyes that are always likely to flash with laughter. While not particularly fashion-conscious, but he/she looks trendy.

Relationships: He/She has a happy family life and Gemini or Arian girls will be suitable as a partner and also for success in his life. Leo and Libra can be helpful for him. He/She is tactless and can hurt with his/her brutal remarks.

Career: He/She is successful in social administration, in public relations, as scientists and as musicians inquisitive. He/She works best with a tactful, organized business partner. Here are some occupations that he might consider, such as, Academic,

Adventure travel guide, Advisor, Astronaut, Consultant, Entrepreneur, Inventor, Humanitarian work, Market researcher, Senator.

Health: He/She often fails to look before where he leaps, and as a result suffer quite a few bruises, pulled muscles and broken bones.

Ideal Partner: He/She needs to spend his life with organized and tolerant people, so his vibe best with Aquarius or Libra.

Compatible Zodiac Signs: Aries, Leo

Incompatible Zodiac Signs: Gemini, Virgo, Pisces

Lucky Gem: Topaz. If he/she is feeling invincible, thank the Hessonite for it. This is a stone of power, and the world is for you to conquer.

Lucky Number: He should be sure of success with Number 2, 3, 5, 6 & 8.

Lucky Colour: Red, Light Blue and cream. Wear Reds for warmth and energy.

Lucky Day: Thursday, Wednesday and Friday. An old friend will brighten up an otherwise dreary Thursday.

Flower: Holly

6.10 Capricorn (Makara) (December 21 - January 19)

Who takes what's due and climbs and schemes for wealth and place?

Who is confident, a resourceful and morns his brothers fall from grace?

Personal Quality: The Capricorn is miserly, very ambitious, self-confident, loyal, snobbish, true workaholic and accepts hard work and reaches to greater heights. He/She makes superb administrators, and often raises to very high positions in his/her careers. He/She wants to get everything with money power. He/She can be temperamental and moody. He/She acquires wealth, dominant position and what else due to his/her smooth talking and hard working. He/She never gives up the thing what he/she had decided to get and takes rest only after completion of the work. He/She knows the value of money only and is always ready to face the consequences to grab the money. He/She saves money and does not expend without need. He/She has very wide circle of friends. He/She is resourceful, practical manager, works well in a disciplined environment. He/She can be frugal, possessing the ability to achieve results with minimum effort and expense. He/She manages several projects simultaneously. His/Her bright carrier begins after the age of 24 and between the ages of 35 to 48 years. He/She is among the wealthy persons. He/She moulds himself/herself as per the situation and hence gets success in life.

Positive Quality: He/She is honest, very practical person and can be relied on. He/She does duty sincerely and finishes the work with great responsibility. He/She is goal oriented. He/She has great control and authority. He/She is loyal to intimates. Never impetuous, he/she considers business and personal relationships carefully before becoming involved. He/She is family person, and family usually comes first, except where business is his/her primary concern.

Negative Quality: He/She is sometimes too bossy and very narrow-minded and thinks highly of him. He/She doesn't function well in subordinate positions. He/She can spread gloom and tension in a minute and is quite capable of depressing everyone else around him. Never really up, but often down, he/she needs a positive environment to enliven his/her spirits. He/She thinks as the wisest man in the world and likes to show the others. Sometimes, he/she is very proudly. He/She is stubborn, overbearing, unforgiving, inhibited, fatalistic, condescending.

Physical Appearance: He/She tends to be tall and have sharp, angular features. He/She generally has serene expressions and an air of tranquillity. He/She takes great care of appearance.

Relationships: He/She can be surprisingly passionate behind closed doors. However, he/she takes a while to warm up and is very cautious about relating to others. He/She is very unhappy with emotional scenes and upheaval, gets hurt easily, but thaws just as quickly if he/she finds that his/her partner is genuinely repentant. Cancer and Libra girls will be ideal for his/her life partnership. Taurus, Virgo and Libra people will be helpful to him.

Career: He/She is usually determined and ambitious, with a strong sense of discipline and a good head for business. He/She sees duty and law enforcement as of paramount importance. Here are some occupations that he/she might

consider such as Administrator, C E O, Coach, Commissioner, Economist, Governor, Industrialist, Leader, and Manager, Mountain climber, Office manager, Official, Operations manager, Organizer, President, Professional, Programmer, Proprietor or overbearing.
Health: Traditionally, Capricorn problem areas are their joints, bones and teeth. He/She needs to ingest extra calcium, cod-liver oil and evening primrose oil supplements to help his/her flexibility.
Ideal Partner: Always seeking a loving and strong partner, he/she gets along very well with Taurus. Capricorn needs a strong, loving partner and bond best with Taurus.
Compatible Zodiac Signs: Taurus, Virgo
Incompatible Zodiac Signs: Aries, Cancer, Libra
Lucky Gem: Blue Sapphire, Garnet and Diamond. This blue sapphire promises him/her all the luck he/she could wish for. He/She could soon be in for a lot of money, so better brush up on his/her financial skills.
Lucky Number: Number 6, 8, 9 & 18 are his/her secret weapon for the success.
Lucky Colour: White, red and Peacock Blue. Wear peacock blue for luck.
Lucky Day: Friday. Watch things falling into place perfectly on Friday.

6.11 Aquarius (Kumbha) (January 20 - February 19)

Who gives to all a helping hand but bows his head to no command?
Who are inventor, genius and superman and higher laws doth understand?
Personal Quality: The Aquarian is eccentric, inventor, technical wizard, and computer and has strong convictions. He/She possesses ill health and likes to make the world a better place. He/She has spiritual bent of mind. He/She is friendly, humanitarian and original thinkers, but can be rather eccentric. He/She is far sighted and innovative. He/She will go out of his/her way to help when needed, but never gets involved emotionally. He/She gets respect and the dignity in the society due to his/her kind nature and service to the

people. He/She is broad-minded and expects others the same. He/She is known by his own name and is not follower but makes his own path. He/She often adopts a life style that goes against the trends, because the odd and unique fascinate him/her. He/She is an active man who is always busy with some kind of mental and physical work. He/She can achieve masterly in artistry, writing, medical, management, police or intelligence work. He/She enjoys his/her own company and is recharged by this quiet time.

Positive Quality: He/She is kind hearted, honest, kind, tolerant, cool, clear, logical people. He/She always thinks for welfare of everyone. He/She is a dedicated person and never bears injustice. He/She does not like to hurt other's feelings and are very helpful to the people in need.

Negative Quality: He/She is, sometimes, not efficient planners and the work undertaken is seldom completed. Sometimes, he/she thinks himself very clever and intelligent than others. He/She is an enigma. He/She is quite aloof people and, sometimes, does not accept his/her fault.

Physical Appearance: He/She has clean-cut good looks and a ready smile that shows off his/her excellent teeth to advantage. He/She can eat any amount of junk food and still remain slim. He/She likes to wear unusual outfits.

Relationships: Physical relationship is not so important for him/her but he/she is very emotional lover and has in depth love. Gemini, Libra, Leo and Aries are the persons who will be helpful to him/her.

Career: He/She might excel in technical fields linked with electrical and radio industries. Most are hard working, even driven in his chosen field. Many choose careers like Astrology. Here are some occupations that he/she might consider such as Academic, Adventure travel guide, Advisor, Astronaut, Consultant, Entrepreneur, Inventor, Humanitarian work, Market researcher and Senator.

Health: His/Her lungs are particularly sensitive to cigarette smoke and air pollution.

Ideal Partner: The most suitable match for Aquarians is Capricorns.

Compatible Zodiac Signs: Gemini and Libra

Incompatible Zodiac Signs: Scorpio, Cancer, Sagittarius

Lucky Gem: Amethyst, Blue Sapphire and Hessonite. If he/she is feeling invincible, think of hessonite for him, which is

to be worn as a ring in middle finger. This is a stone of power, and the world is for him/her to conquer.
Lucky Number: He/She can be sure of success with Number 2, 3, 7 & 9.
Lucky Colour: : Yellow and Red. Wear reds for warmth and energy.
Lucky Day: Thursday. An old friend will brighten up an otherwise dreary Thursday and Friday.
Lucky Flower: Violet

6.12 Pisces (Meena) (February 19 to March 20)

Who possesses a gentle, compassionate, sensitive and spiritual nature?
Who are friendly and respond to suffering, which others encounter?
Personal Quality: The Piscean is charity, anxious, self-sacrificing, gentle, patient, malleable nature and has strong intuitive powers. He/She has superb observation, concentration while listening and good grasping power. He/She has instinct for nature, beauty, travelling, luxury and pleasure. He/She is good in subordinate positions and heads of small business. He/She is the kindest and most charitable of all the signs. He/She will make many sacrifices for other people. He/She lives life in lonely. As a lover, he/she is faithful and love to dabble in the art of sexual fantasy. He/She has many generous qualities and is friendly, good-natured, kind and compassionate, sensitive to the feelings of those around them, and respond with the utmost sympathy and tact to any suffering. He/She gets ancestral properties but he/she wishes to make money by his/her own efforts. Horseback riding, dancing, skating, swimming or sailing are favoured activities.
Positive Quality: He/She is versatile, intuitive and has quick understanding. He/She observes and listens well, and are receptive to new ideas and atmospheres. He/She readily adapts to change.
Negative Quality: His/Her dominant keyword is “I believe”. His/Her nature tends to be too otherworldly for the practical purposes. He/She also dislikes disciple and confinement. The nine-to-five life is not for him/her.

Relationship: He/She tends to bond romantically well with Aquarians. He/She is never egotistical in personal relationships and gives more. He/She can be loving and affectionate partners for life.
Health Concerns: He/She can be threatened by anaemia, boils, ulcers and other skin diseases, especially inflammation of the eyelids, gout, inflammation, heavy periods and foot disorders and lameness. He/She is prone to all kinds of allergies and crippling headaches.
Partner: Aquarians are the best match for Pisceans and two tend to bond romantically.
Compatible Zodiac Signs: Taurus and Cancer
Incompatible Zodiac Signs: Aries, Gemini, and Leo
Lucky Gemstone: Cat's eye, which is to be worn as a ring in ring finger
Lucky Stone: Yellow Sapphire, Topaz & Coral.
Lucky Color: Red and yellow
Special Flowers: Water Lily, White Poppy & Jonquil
Lucky Numbers: 1, 2, 3, 4 & 6
Lucky Day: Thursday,Sunday and Friday

7

Planet (Graha)

7.1 Planets' Description

There are nine Graha, namely, Sun, Moon, Mars, Mercury, Jupiter, Venus, Saturn Rahu and Ketu. But after further studies, the Uranus, Neptune and Pluto have been added to the astrology to make it more fascinating subject. Moon, Mercury, Jupiter and Venus are benefic by nature and others are malefic.

Introduction of Grahas:

Surya: Surya's eyes are honey-coloured. He has a square body. He is of clean habits, bilious, intelligent and has limited hair (on his head).

Chandra: Chandra is very windy and phlegmatic. She is learned and has a round body. She has auspicious looks and sweet speech, is fickle-minded and very lustful.

Mangal: Mangal has blood-red eyes, is fickle-minded, liberal, and bilious, given to anger and has thin waist and thin physique.

Buddha: Buddha is endowed with an attractive physique and the capacity to use words with many meanings. He is fond of jokes. He has a mix of all the three humours.

Guru: Guru has a big body, tawny hair and tawny eyes, is phlegmatic, intelligent and learned in Shashtra.

Sukra: Sukra is charming, has a splendours physique, is excellent or great in disposition, has charming eyes, is a poet, is phlegmatic and windy and has curly hair.

Sani: Sani has an emaciated and long physique, has tawny eyes, is windy in temperament, has big teeth, is indolent and lame and has coarse hair.

Rahu (The Dragon Head/ North Node of Moon): The Moon's orbit and the earth's orbit intersect and these two intersecting

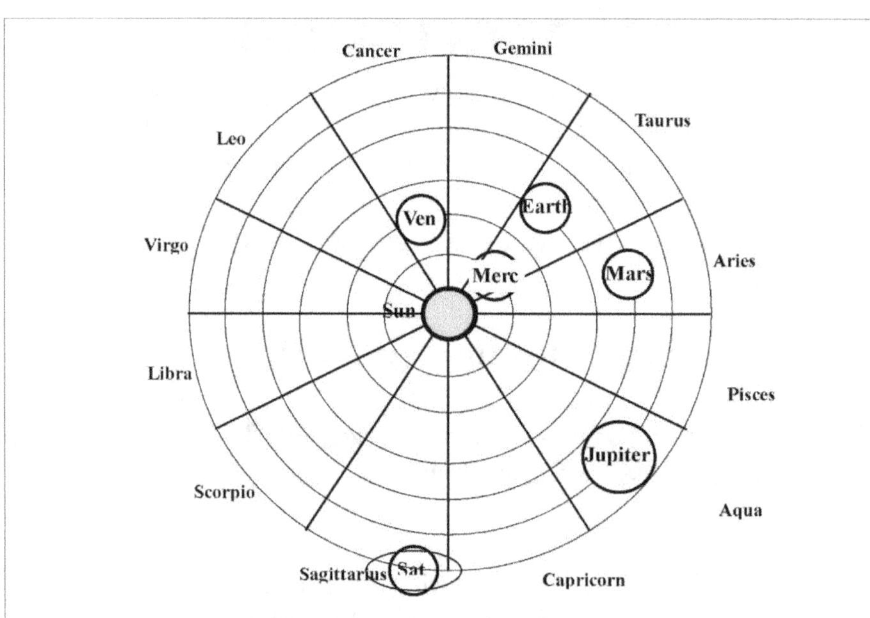

Fig 1: Actual Planet Movement in Zodiac

Points are known as North Node (Rahu) and the South Node (Ketu). These orbits differ by 8 degrees. They are mathematical points which influence human behaviour. Hence great importance has been assigned to these Nodes in Vedic Astrology. Rahu is also known as the dead body of the lusty demon, killed by Vishnu. So, Rahu is the head of the dragon and symbolizes the trouble. Rahu has smoky appearance with a blue mix physique. He resides in forests and is horrible. He is windy in temperament and is intelligent.

Ketu (Dragons Tail/South Node of Moon): Ketu is the dead body of the lusty demon, killed by Vishnu. So, Ketu is the tail of the dragon and losses and symbolizes the death along with Saturn. Ketu has smoky appearance with a blue mix physique. He resides in forests and is horrible. He is windy in temperament and is intelligent. Ketu is akin to Rahu.

Note: Each planet has been allotted some points, such as Sun - 48; Moon - 49; Mars - 39; Mercury - 54; Jupiter - 56; Venus - 52; Saturn - 39. Thus it is totalling 337 points all together.

Body Parts (Sapth Dhatu) represented by the Graha: Bones, blood, marrow, skin, fat, semen and muscles are, respectively, denoted by Surya, Chandra, Mangal, Buddha, Guru, Sukra and Sani.

Abodes represented by the Graha: Temple, watery place, place of fire, sport-ground, treasure-house, bed-room and filthy ground are, respectively, the abodes for the seven Graha from Surya onward.

Tastes represented by the Graha: Pungent, saline, bitter, mixed, sweet, acidulous and astringent are the tastes governed, respectively, by Surya, Chandra, and Mangal, Buddha, Guru, Sukra and Sani.

Deities of Graha: Fire (Agni), Water (Varuna), Subrahmanya (Lord Shiva's son, following Ganesh), Maha Vishnu, Indra, Shachi Devi (the consort of Lord Indra) and Brahma are the presiding deities of the 7 Graha in their order starting from Sun.

Gender of the Graha: Buddha and Sani are neuters. Chandra and Sukra are females, while Surya, Mangal and Guru are males.

Primordial compounds of the Graha: The Panchabhutas, space, air, fire, water and earth, are, respectively, governed by Guru, Sani, Mangal, Sukra and Buddha.

Yogi Point: Yogi Point is an extremely positive degree (beneficial Longitude) in the natal chart. When benefic planet transits over the Yogi Point, native can expect major material success.

Ava-Yogi Point: Ava-Yoga Point is equal to Yogi Point's longitude plus 186 deg 40 sec.

Yogi planet: The exact degree of the 'Yoga' becomes the 'Yogi Point'. The ruler of the Nakshatra where this point is called 'Yogi Planet' and the ruler of the Rashi of the Nakshatra is known as the 'Duplicate Yogi'. The 6th Nakshatra from the yogi Nakshatra is called the 'Avayogi Nakshatra' and its ruler is the 'Avayogi Planet'. The house where the yogi planet is placed in the chart gets enhanced. Yogi planet is a planet for prosperity and Avayogi give the opposite results. Avayogi Planet causes trouble to the house where the Avayogi planet is placed. The duplicate yogi is also a planet that helps the native. There is no duplicate avayogi. The dasha of Yogi Planets give good results and avayogi will give negative results. In relationships compatibility, if your partner's ascendant ruler is your yogi planet, it bring good fortune to your relationship whereas if it is your avayogi planet, it can bring debt, pain and incompatibility. It is important not to get carried away by the principle of yogi and avayogi planets but to view them as supplementary information that enhances the natal analysis. While they have the ability to alter the final judgement, they are not the only factor to consider.

Yogi Duplicate Planet: The planet which rules the Rashi containing the yogi point is known as the duplicate yogi planet. The duplicate yogi planet also enhances (under certain conditions) the prosperity promised by the yogi, and reduces any destitution that the avayogi might forebode.

Ava-yogi Planet: The graha which rules the Nakshatra containing the ava-yogi point is known as Ava-yogi Graha. The avayogi planet obstructs prosperity

Table 1: Astronomical Detail of Planets

Planet	Mean Distance from sun in '000 km	Sidereal (Orbiting) Period
Earth	14,94,56.180	365.25 days
Mars	22,77,21.610	687 days
Mercury	579,36.240	88 days
Jupiter	77,77,94.020	11.86 year
Venus	10,78,25.780	224.7 days
Saturn	142,60,36.100	29.46 year
Uranus	286,9453.000	84.01 year

Neptune	449,48,86.600	164.43 year
Pluto	589,98,40.4000	248.43 year

6able 2: Astronomical Detail of Planets

Planets	Axial Rotation Period	Equatorial Diameter in 000 km	Max. Surface Temperature (0F)
Earth	23 hr 56 min.	12.756	(+) 1400
Mars	24 hr 37 min.	6.759	(+) 850
Mercury	88 day	4.667	(+) 7700
Jupiter	9 hr 51 min.	142.748	(-) 2000
Venus	N. A	12.392	(+) 8800
Saturn	10 hr 14 min.	120.861	(-) 2400
Uranus	10 hr 48 min.	47.153	(-) 3100
Neptune	14 hr	44.579	(-) 3600
Pluto	6 days 9 hr	5.794	Not Available

7.2 Apparent Motion of Planet

Planets in Apparent Motion (Graha's Transit or Bhraman):

Transit refers to the movement or revolution of planets across the Sun. Transit Chart is the position of the Planets at a particular time and place while orbiting the Sun because of their daily motion. The transits Chart show the effect of planets, the ups and the downs brought about by circumstances in our life that we face on daily or monthly basis. The planet's power increases when it is retrograde or stationary. The following table represents the average daily angular motion of the planets and the approximate time they spend in one sign and they need to complete a full circle of the Sun. The symbols in the brackets indicate the direction of their movement. In Primary Directions the apparent motion of the planets and the

House-cusps is clockwise, resulting from the counter-clockwise motion of the Earth's periphery. Direct The true motion of the planets in the order of the Signs is counter-clockwise within the Zodiac.

Table 3: Astronomical Detail of Planets

Planet	Average Daily Motion	Time spent in a Sign/House	Cycle Duration to cover 12 Signs
Sun (D)	0059'	30 days	1 year
Moon (D)	130 10'	2 1/4 days	1 month
Mars (Θ)	00 31'	1 1/2 months	18 months
Mercury(Θ)	40 5'	27 days	1 year
Jupiter (Θ)	00 5'	1 year	12 years
Venus (Θ)	10 36'	28 days	1 year
Saturn (Θ)	00 2'	2 1/2 years	30 years
Rahu (R)	00 3'	1 1/2 years	18 years
Ketu (R)	00 3'	1 1/2 years	18 years

Legend: D = Always Direct Motion; R = Always Retrograde Motion & Θ = Mostly Direct Motion but sometimes Retrograde Motion planets.

Motion of the Planets:

There are eight kinds of motions to planets from Mangal to Sani. These are Vakra (Retrograde), Anuvakra (entering the previous Rashi in retrograde motion), Vikal (Stationary, i.e. fixed or devoid of motion,), Mand (slower motion than usual), Mandatar (slower than the previous motion), Sama (increasing in motion), Char (faster than Sama motion) and Atichar (entering next Rashi in Accelerated motion). The strengths, allotted due to such 8 motions are 60, 30, 15, 30, 15, 7.5, 45 and 30 respectively.

Daily Forecast Analysis (Moon Transit): Moon rotates once every 2.25 days. This makes daily forecast analysis based on the current positions of the Moon in the sky and how he aspects on other planets, i. e. the angles between the Transit

Moon and Planet and the Signs in Natal Chart. For example, when, the transit Moon makes a Trine (a 120 degree angle) to the Moon in the native Natal Chart, he makes intuition strong. He/She identifies and act upon his/her gut instincts about things easily.

Table 4: Astronomical Details of Planets

Planet	Angular Distance from Sun		No. Of Days Stationary	No. of Days Retrograde
Mars	2280	1320	3	80
Mercury	14-200	17-200	1	24
Jupiter	2450	1150	5	120
Venus	290	260	2	42
Saturn	2510	1090	5	140

Weekly, Monthly & Yearly Transit & Forecast Analysis: The weekly and Monthly forecast analysis is based on the current positions of the planet in the sky and how they aspects between them, i. e. the angles between the Transit planet and Planet in Natal Chart and the Signs in that weak of predictions. Like that the different aspects of individual life are highlighted each week to get the most advantageous and beneficial results. This is how a weekly or monthly horoscope is forecast.

Saturn Transit (Sadesati): Transit Saturn gives significant results with respect to Moon. When Saturn crosses over Moon, it gives Sade Sati effects. Sade Sati normally gives lots of tension. Similarly Saturn crossing over 8th house from Moon also gives tension and losses. Saturn transit over Jupiter or over 7th house predicts marriage time and Saturn over Rahu gives break in business partnership. During the first Sade Sati of Saturn, it gives a drop in academic results and in 2nd it gives loss in business. When Saturn crosses over 8th house from Moon it gives break in partnership and brings the income to the rock bottom level. It also gives bad health to mother and loss of life of one of the close relatives. Transit of Saturn from Moon is very important.

Jupiter Transit: If transiting Jupiter is conjunct to Venus in the birth chart, it gives enjoy a time of tremendous financial or romantic opportunities. Jupiter transiting over 4th house, over Rahu or over Saturn, it gives finance to buy a new property. In a nut shell, transits seem to be showing their results very effectively; even more than Dasha. Major transits bring major changes in the life of the native. Jupiter and Rahu transit over Lagna as well as Moon give results. If we superimpose the result of Dasha over that of the transit we can predict better.

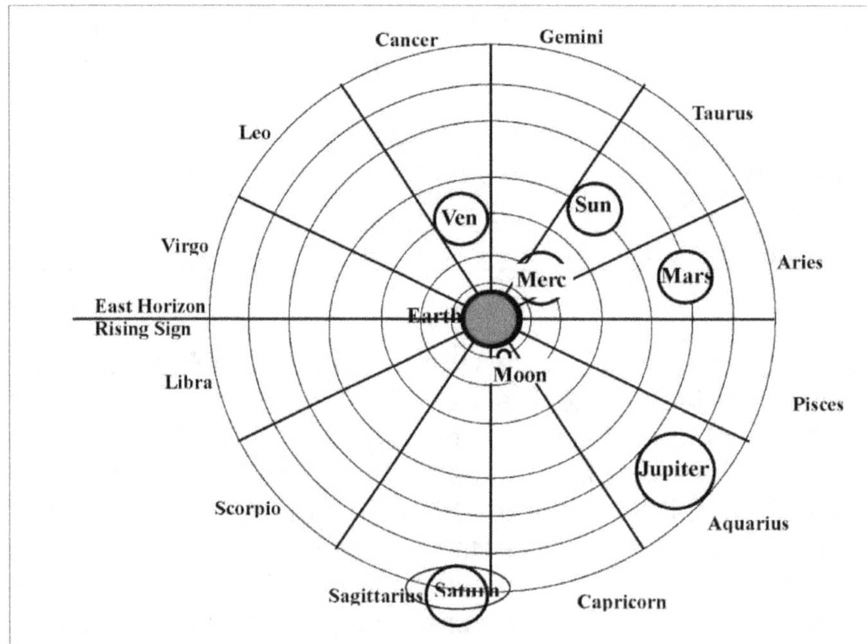

Figure 2: Imaginary Movement of Planets as viewed from the Earth

(i) Stationary Motion (Vikal): Some times, planets apparently seem to slow down or stand still or proceed backwards and then again stop, turn around and go forwards again. There is a point where for a short time, to its cresting on the edge of its orbit in relation to us; it appears to stand still or stationary, when we judge by looking at the sign (Aquarius) behind it in Zodiac. This is called 'Stationary Motion' (see Fgure-3). Except the Sun and the Moon, the other planets, such as, Mars, Mercury, Jupiter, Venus and Saturn change their proper motion in the Zodiac periodically and appear to move backwards for short period and are called "Retrograde" and after some time they resume their direct motion. Most of the time, they move in the "Direct" way, but at some times they fall into a Retrograde cycle. Before retrogression occurs, the planet gets stationary for a certain period of time. The same thing occurs when the planet ends its retrograde cycle. Sun and moon will never retrograde. The planet gives a very strong and steady effect while it is stationary.

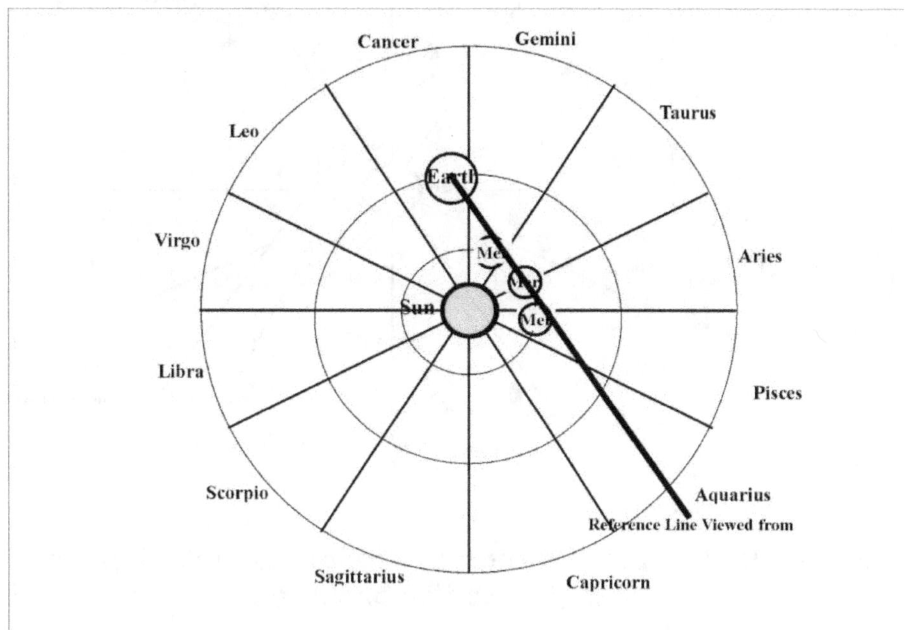

Figure 3: Mercury appears "stationary" or "standstill" as viewed from Earth

ii) Retrograde Motion (Vakra): The planets move within the belt of the zodiac with a different average of speed. When any planet appears moving apparently in the opposite direction to the Sun, as viewed from the earth, that motion is called Vakra or Saktha Avastha or Retrograde Motion. The two planets, i.e. the Sun and the Moon have steady and direct motion. The planet in retrograde is marked in the horoscope with the mark 'R'. A retrograde planet becomes more powerful. The lunar nodes, Rahu and Ketu always move in retrograde direction. In the figure shown below, while Mercury moves from position-I to the position-II (see Figure-4) around the Sun, it cast a shadow of position I and II in opposite direction in the Aquarius Sign, which appears the Mercury moving in Retrograde, i.e. in opposite direction. This is called Retrograde Motion.

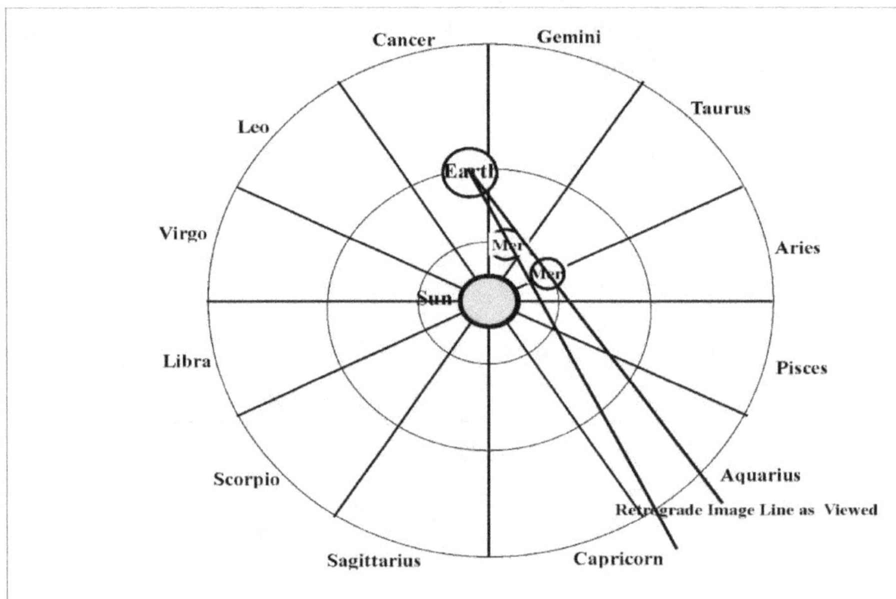

Figure 4: Planet in "Retrograde Motion" against Capricorn Sign (the Zodiac backdrop)

7.3 Planet's Characteristics

The Planet is the most important character in the horoscope. The planets possess some intrinsic properties when they occupy the sign and house. Planets change their behaviour like Favorable/Benefic or Unfavorable/Malefic according to the position they occupy in Sign and House and lordship.
Once the planet's behavior is ascertained because of its placement, predictions are changed with time, depending upon Dasa operating at that time or by Transit.

Table 1: Planet's Characteristics

Planets	Friends	Neutrals	Enemies

Sun (Surya)	Moon, Mars, Jupiter	Mercury	Saturn, Venus,
Moon (Chandra)	Sun, Mercury	Mars, Jupiter, Venus, Saturn	--
Mars (Mangal)	Sun, Moon, Jupiter	Venus, Saturn,	Mercury
Mercury (Buddha)	Sun, Venus,	Mars, Jupiter, Saturn	Moon
Jupiter (Guru)	Sun, Moon, Mars,	Saturn,	Mercury, Venus
Venus (Sukra)	Mercury; Saturn,	Mars, Jupiter	Sun, Moon
Saturn (Sani)	Mercury, Venus,	Jupiter,	Sun, Moon, Mars

Table 2: Plant's Characteristics

Characteri stics	Sun	Moon	Mars	Mercury
Planet's Natural behaviour	Malefic	Benefic	Malefic	Benefic
Karaka	Father, Soul, and 1st house	Mind, Mother and 4th house	Strength, Male's Younger brother, and Female's Husband.	Speech, educatio n, & 3rd house.
Colour	Red-orange	Tawny	Red	Green
Cabinet	King	Queen	General	Prince
Deities	Agni	Varuna	Kartikeya	Maha Vishnu
Sex	Male	Female	Male	Female
Tattwa	Agni (fire)	Jala (water)	Agni (fire)	Bhumi (earth)
Varnas (Deeds)	Kchatriya	Vaishhya (trader)	Kchatriya	Vaishya (trader)
Gunas	Sattva	Sattva	Tamas	Rajas

Abode	Temple	Watery places	Fireplace	Sports ground
Dhatu (Body Part)	Asthi (bones)	Rakta (blood)	Majja (marrow)	Tvak (skin)
Time periods	Ýyana (half year)	Kchana (second)	Vara (day)	Ritu (season)
Taste	Pungent	Saline	Bitter	Mixed
Ritu (Seasons)	--	Varsha (rainy)	Grishma (summer)	Sarad (Winter)
Lord	Lord Siva	Goddess Parvathi	Lord Kartikeya	Maha Vishnu
Dasa Period	6 years	10 years	7 years	17 years

Table 3: Planet's Characteristics

Characteristics	Jupiter	Venus	Saturn
Planet's Nature	Benefic	Benefic	Malefic
Karaka	Knowledge, happiness, Male, 5th house; & 9th house.	Male's wife, Sexual Love, Younger sister, 2nd house, & 7th house.	Misery & Grief, Profession, Elder brother and 10th house
Colour	Tawny	Variegated	Black
Cabinet	Minister	Minister	Servant
Deities	Indra	Sukra	Brahma
Sex	Male	Female	Impotent
Tattwa	Ýaknna (ether)	Jala (water)	Vayu (air)
Varnas (Deeds)	Brahmana (priest)	Brahmana (priest)	Chhudra (worker)
Gunas	Sattva	Rajas	Tamas
Abode	Treasure house	Bedroom	Filthy places
Dhatus Body Part	Vasa (fat)	Virya (semen)	Snayu (muscle)
Time	Masa (month)	Pakcha	Varsha

periods		(fortnight)	(year)
Taste	Sweet	Sour	Astringent
Ritu (Seasons)	Hemanta (Dew)	Vasanta (spring)	Sisir (fall)
Lord	Lord Dakshinamurthi	Maha Lakshmi	Lord Yama
Dasa Period	16 years	20 years	19 years

Note: 1) The Dasa period of Rahu and Ketu are 18 years and 7 years respectively.

2) The Lord of Rahu and Ketu are Goddess Durga and Ganesha respectively.

Table 4: Planet's Characteristics

Planet	Ownership Sign	Exaltation Sign	Debilitation Sign
Sun	Leo	Aries	Libra
Mon	Cancer	Taurus	Scorpio
Mars	Aries/ Scorpio	Capricorn	Cancer
Mercury	Gemini/ Virgo	Virgo	Pisces
Jupiter	Sagittarius/ Pisces	Cancer	Capricorn
Venus	Taurus/ Libra	Pisces	Virgo
Saturn	Capricorn/ Aquarius	Libra	Aries
Rahu	-	Taurus	Scorpio
Ketu	-	Scorpio	Taurus

Table 5: Planet's Characteristics

Planet	Aspects House	Natural Caracter of Planets	Sex of Planets

Sun	7	Malefic	Male
Mon	7	Benefic	Female
Mars	4, 7, 10	Malefic	Male
Mercury	7	Benefic	Female
Jupiter	5, 7, 9	Benefic	Male
Venus	7	Benefic	Female
Saturn	3, 7, 10	Malefic	Impotent
Rahu	5, 7, 9	Malefic	-
Ketu	5, 7, 9	Malefic	-

Table 6: Planet's Characteristics

Name of Planet	Lordship of Signs	Complete Aspect Places	Natural Benefices / Malefic Rating
Sun (Surya)	Leo (Simha)	7th	Malefic*
Moon (Chandra)	Cancer (Karka)	7th	Benefic*
Mars (Mangal)	Aries (Mesh) Scorpio (Vrischika)	4th, 7th , 8th	Malefic**
Mercury (Buddha)	Gemini (Mithuna) Virgo (Kanya)	7th	Benefic**
Jupiter (Guru)	Sagittarius (Dhanu) Pisces (Meena)	5th, 7th, 9th	Benefic****
Venus (Sukra)	Taurus (Vrishabh) Libra (Tula)	7th	Benefic***
Saturn (Sani)	Capricorn (Makara) Aquarius (Kumbha)	3rd, 7th, 10th	Malefic***
Dragons Head	Gemini (Mithuna)	5th, 7th, 9th	Malefic****

(Rahu)			
Dragons Tail (Ketu)	Sagittarius (Dhanu)	5th, 7th, 9th	Benefic****

Table 7: Planet's Characteristics

Name of Graha	Deeptaansh	(Exaltation Sign)	Maximum Exaltation Degree
Sun (Surya)	15	Aries	0-10 0
Moon (Chandra)	12	Taurus	0-3 0
Mars (Mangal)	8	Capricorn	0-28 0
Mercury (Buddha)	7	Virgo	0-15 0
Jupiter (Guru)	9	Cancer	0-5 0
Venus (Sukra)	7	Pisces	0-27 0
Saturn (Sani)	9	Libra	0-20 0
Dragons Head (Rahu)	1	Gemini	0-15 0
Dragons Tail (Ketu)	1	Sagittarius	--

Table 8: Planet's Characteristics

Name of Graha	Deeptaansh	(Debilitation Sign)	Maximum Debilitation Degree
Sun (Surya)	15	Libra	0-10 0
Moon (Chandra)	12	Scorpio	0-3 0
Mars	8	Cancer	0-28 0

(Mangal)			
Mercury (Buddha)	7	Pisces	0-15 0
Jupiter (Guru)	9	Capricorn	0-5 0
Venus (Sukra)	7	Virgo	0-27 0
Saturn (Sani)	9	Aries	0-20 0
Dragons Head (Rahu)	1	Scorpio	--
Dragons Tail (Ketu)	1	Sagittarius Taurus	--

Table 9: Planet's Characteristics

Name of Planet	Karaka of Signs or Bhava	Mula Ansh	Kalaa
Sun (Surya)	Aries ; Thanu Bhava	7	30
Moon (Chandra)	Cancer; Bandhu Bhava	3	16
Mars (Mangal)	Gemini; Sahaj Bhava & Virgo; Ari Bhava	10	6
Mercury (Buddha)	Capricorn; Karma Bhava	6	8
Jupiter (Guru)	Thanu Bhava; Dhan Bhava & Leo; Putra Bhava & Sagittarius; Dharma Bhava & Aquarius; Labh Bhava	9	10
Venus (Sukra)	Libra; Yuvati Bhava	5	11
Saturn (Sani)	Scorpio; Randhra	1	1

Note: (i) Rahu and Ketu do not play any role in these respects.

(ii) The normal motion of a planet. The Sun and Moon are always in Direct motion. All the other planets can move in retrograde motion at times.

Table 10: Planet's Characteristics

Name of Graha	Pindaaya Dhruvaank	Dhruvaank	Badly Affected House during Transit	Best karaka of House
Sun (Surya)	0-10 ansa	19	5	1st , 10th, 11th
Moon (Chandra)	0-3 ansa	25	8	1st, 4th , 7th
Mars (Mangal)	9-28 ansa	15	7	3rd, 6th , 10th
Mercury (Buddha)	5-15 ansa	12	2	2nd, 10th , 11th
Jupiter (Guru)	3-5 ansa	15	3	2nd , 5th, 9th and 11th
Venus (Sukra)	11-27 ansa	21	6	4th, 7th, 12th
Saturn (Sani)	6-20 ansa	20	1	3rd, 6th, 8th , 10th , 12th
Dragons Head (Rahu)	-	-	9	3rd, 6th, 10th
Dragons Tail (Ketu)	-	-	9	2nd, 8th

Table: 11: Planets and Education or Professions

Planets	Educational Field	Semi-Technical Profession	Highly Technical Profession
Sun	Political	Technical,	Physics

	Science	Statistical, Mathematics	
Moon	Music, Fine Arts	Paramedical	Chemistry, Medical
Mars	Logics Related	Mechanic	Engineer
Mercury	Accountancy	Semi-tech Accounts	Higher Accountancy
Jupiter	Vedanta, Classical knowledge, Literature, Preacher	Journalism, Astrology, Philosophy, Psychology	Management
Venus	Music, Dance, Arts, Painting	Hotel Management, Tourism	Computer Animation, Graphics
Saturn	Prachin Vidhya, Labour, Law	Mechanical Work, Geology	Engineering
Rahu	Research, Psycho-analytical work	Pilot, Air-Hostess	Space Engineer
Ketu	Language	Language Specialist	Metrologies, Computer

Table 12: Planet's Characteristics

Graha	Profession for gain of wealth
Sun (Surya)	Sun provides gains of education, wealth through authority; Politicians & political leaders like Ministers, President; professionals like director, Physicians & Doctor, actors, Singer or big Businessman, Government service like I A S officers, Specialist; Social workers, and Philosopher.
Moon (Chandra)	Moon gives gain of finance by profession related to Farmer, nursing, the public health care taker, Advertiser, public relation officer, Journalist, Businessman, marine like sailor, Hawker, restaurants, import/export, Boatman cook, agent, hospitality.
Mars	Mars provides gains of wealth through profession

(Mangal)	like Engineer, Surgeon, Doctor, Dentist, Police & constable, carpenter, mechanic, blacksmith, hunter, military personnel, butchers, barbers, and related to construction, cook, dealing of land, ,spying and Service like, Fire service, Sports, industry, Fire kiln, potter, brick kiln, Instrument manufacturing, stone breaking, granite industries, agriculture and arms industry.
Mercury (Buddha)	Mercury provides gains of wealth through profession like ambassador, lecturer, professor, councillor, teacher, writer, landlord, broker, inspector, publisher, journalist, clerk, audit & accountant, mathematicians, public speaker, imports and exports, Salesman, orator, secretary, traders, poet, scripture, clerk, accountant, Astrologer, Lawyer, editor, mathematician, judge, telephone operator, CBI department officer, legal adviser and businessman.
Jupiter (Guru)	Jupiter gives the status, wealth, gains of finance by profession like Judge, Priest, Lawyer, Manager, Teacher, Chartered Accountant, politics like Ministers, lawyer, bankers, temple workers, Yogasana Teacher, Religious teacher, professor, consultants, spirituality, financial management, mathematics, knowledge of Vedas, philosophy and astrology.
Venus (Sukra)	Venus provides gains of wealth through profession like Artists, treasurer, musician, singer, dancer, dramatist, prostitute, call girl, Minister, Accountant, Auditor, beauty Parlour Saloon, Scented materials business, writer, tourism, sculpture, make-up job, beauty context, money lending, financial organizations, commission agent, cinema theatre & Five-star hotels.
Saturn (Sani)	Saturn provides gains of wealth through Labourer s related to real estate, agriculture, building trades, mining, monk, industry, public dealing, service, and industrial workers and in Government Service and service related to Speech, dealing with labour jobs, policeman, renunciation man, cheater, killer, dacoit and thief.

Dragons Head (Rahu)	Rahu gives gains of wealth by profession like Labourers related to Chemical Industry, Electronic Industry, cigarette factory, wine manufacture, bomb manufacture, CBI department, Secret organization (detective), defence department, foreign trader, smuggler, robber, cheater, pick-pocketed, chain snatcher, corruption maker, underground work, researchers, engineers, physicians, medicine/drugs, speculators and aviation.
Dragons Tail (Ketu)	Ketu provides wealth through profession like Doctor, priest, astrologers, beggars, saints, Lawyer, Tailor, Choir Manufacture, Cable manufacture, weavers, power loom industries, spinning mills, religious teachers, enlightenment (Guru) and metaphysics.
Uranus	Uranus provides wealth through profession like scientists, inventors, computing, astrologers, lab technicians, electronics.
Neptune	Neptune provides wealth through profession like photographers, movies, marine, oil, pharmaceutical, psychics, and poets.
Pluto	Pluto provides wealth through profession like research, insurance, death, longevity-related technology, espionage. Mars, Saturn and Rahu are considered the role of technical planets.

Table 13: Planet's Characteristics

Graha	Education Field	Relation	Diseases & Affected Parts of Body
Sun	Political – Science, Social – Science	Father, Son	Heart attack & disease, Bones, Back, Veins, Right Eye disease, loss of appetite, Fever, Indigestion, skin disease, Hysteria.
Moon	Arts, Psychology, Chemistry, Hotel	Mother, Mother-in-Law, Maternal	Brain, Lungs, chest, Left-eye, Breast, Venereal disease, Skin disease, Cold cough, Phlegm,

	Management	Uncle wife	smallpox, Mental disorder, stomach ache, Sleeplessness, Indigestion, Ajeerna, Arthritis, Jalaghaat, Animal attack.
Mars	Engineering, Physics	Brother, Husband	Nose, Fore Head, intestine, Bone narrow, Accidents, piles, blood pressure, heart disease, constipation, anaemia, Kantha, Rakta-Vikaar, Gathiaa-Bukhaar, Aagajani, poison, blood related, heart attack.
Mercury	Mathematics, Statistics, Drawing	Maternal Uncle, Younger Brother / Sister	Veins, lungs, tongue, arms, Mouth, Hair, Stomach disorder, Leprosy, Childlessness, intestinal complaints, skin diseases, mental disorder, throat, tonsils, leucoderma, Nose, ears, baat, peet, koda, jahar, falling from height, madness, asthma, hakalaanaa & Gungaapan.
Jupiter	Audit and Accounts, Philosophy	Teacher, Self	Lever, digestive organs, Kidney, Hernia, Bronchitis, liver complaints, jaundice Gastric, fever, Plane accident, Liquid from Rt. ears.
Venus (Sukra)	Commerce, Marketing, House-keeping	Wife, Sister, Daughter, Daughter-in-Law	Left ear, glands, sex-organs, sperm, Generative organs, Venereal diseases, uterine disorders, diabetes, diseases of ovaries, Abortion, Koda,

			vaat, peeta, cough, birya-vikaar, eye, urinal, gudaa, ultras-related, pedu's pain.
Saturn	Mining	Elder Brother	Bone joint, Bones, teeth, Knee, cough, Nerves, gas troubles, rheumatism, arthritis, defective speech, indigestion, ulcer, asthma Stomach, leg, gas related disease; falling from tree, Hit from stones.
Rahu (Dragon Head)	Chemical, Nuclear Physics	Paternal Grand Father	Lungs, Lower part of wind pipe, Cataract, leprosy, suicide, rheumatism, sudden death, murder, accidents, sexual perversity, homosexuality, lesbianism, Heart attack, Poison–eating, weakness of heart, Digestive systems.
Ketu (Dragon Tail)	Medical, Theology, Astrology, Law	Maternal Grand Father	Throat, leprosy, suicide, small pox, piles, cancer, sudden death, murder, tumour, Wound, Skin.

Maraka Graha: If Saturn, being the lord of 2nd and 7th houses from the Moon-Sign, is position in the Lagna, then it is a powerful Maraka. The Jupiter being Lagnesh, if positioned in the 6th house, indicate his/her premature death.

Aspect Effects of Planets: Effect of the planet's aspect to the house is more as compared to the planet sitting in the house. The aspect of the planet to any Bhava gives the effect in equal and same amount as it gives by sitting in the house.

Planetary Camps (Friends/Enemies): Sun, Moon, Mars, Jupiter and Ketu are in one friendly Camp. Mercury, Venus,

Saturn and Rahu are in other friendly Camp. The planets positioned in the 2nd, 3rd, 4th, 10th, 11th, and 12th houses from the Planet considered, are immediate (Tatakalik) friends to that planet. But the planets positioned in the 1st, 5th, 6th, 7th, 8th and 9th houses are considered immediate (Tatakalik) enemy.

7.4 Judgement of Benefic and Malefic Planet:

There are certain rules in determining a functional Benefic and Malefic Planet for each lagna in the Kundali. Lordship of certain houses in the chart makes the planets deviate from their natural tendencies of doing good or bad. Example: Jupiter, even though it is naturally the best beneficial planet, but it is Malefic Planet for Kanya (Virgo) Lagna as he, being natural benefic, is the lord of 4th and 7th house (Kendradhipatya Dosha) and hence it does not give good results for Virgo Lagna. Similarly, Jupiter is a neutral planet for Simha (Leo) lagna. Dr. Raman also talks about Kendradhipati (Lordship) Dosha, the blemishing acquired by a natural benefic by virtue of lordship of an angle 4th, 7th or 10th house and thus the planets become functional malefic in the Kundaly. The lord of Lagna does not get affected by this dosha and so Lagnesha is always a befic planet in Kundali.

Lord of 1st, 5th, and 9th House: The lords of Trikona are three and are always considered auspicious or benefic and produce good results always. If exalted planets are placed in Kendra to Trikona, it again causes Raja yoga.

Lord of 4th, 7th, and 10th House: The lords of Kona are are considered auspicious or benefic, if they are natural malefic. The lords of Kona are considered inauspicious or malefic, if they are natural benefic planet. Further, lords of 4, 7, 10 houses, if placed in auspicious houses, produce good results. If it is placed in 6th, 8th or 12th houses, it produces bad results. The conjunction of the Kendra lords or the Kona lords or the Kendra to Kona lords is highly auspicious and produces good results. These are called Raja yoga combinations. The Raja yoga combinations give rise to authority, power, position and wealth.

Lord of 3rd, 6th, and 11th House: Sage Parashara states that whosoever planet even including benefic, if owns 3rd, 6th, and 11th are considered inauspicious or malefic, and will produce sinful effects and are trouble maker, such as, (1) the Dasa of 11th lord will be gainful (Laabh) monetary wise, but it will cause obstacles (Vighna) and diseases (Rogi) to the native simultaneously. During the Dasa of 11th lord, one will get evil news. His co-born will undergo troubles. His offspring will fall ill. The native himself/herself be miserable, be cheated and may suffer from ear diseases; (2) The Dasa of the lord of 3rd (Tri), 6th (Shad) and/or 11th (Labha) and the Dasa or Bhukti of planet joining with the lords of 3rd, 6th, and /or 11th will be inauspicious and produce evil effects; however, benefic occupying these "Upachayas" give rise to so called "Vasumati yoga". Some Seers have views that the lord of 11th house or planet posited therein or the planet associated with the lord of 11th house, is very enigmatic and mysterious during its Dasa or Bhukti, and is capable of improving acquisitions but at the same time, during this Dasa or Bhukti, the same planet gives difficulties, litigations and serious diseases to the native too. The reason is that (a) the 11th house is 2nd from the 10th house, the central point for one's gains of money or wealth through the Karmas of native. (b) The 11th house being 6th from the 6th house, an epicentre of one's loss of money or wealth through diseases, litigations, punishments and debts. 11th lord is better placed in the auspicious houses and produce good results, otherwise produce bad results.

Lords of 12th, 2nd and 8th House: The remaining three houses as 2nd, 8th and 12th, according to Parashara, are not as bad as the 3rd, 6th and 11th. The houses 2nd, 8th and 12th tend to modify their attributes according to relationship; the lords of 12th, 2nd and 8th (strictly to the order of these lords) will prove auspicious, if they are associated with the angular or the trinal lords. But they become deadly inauspicious by their association with the lords of 3rd, 6th, and 11th. While grading these two groups the sage states the intensity of malfeasance will increase in the ascending order i. e. 6th lord is more harmful as compared to 3rd lord while 11th lord is most harmful / inauspicious of the three. The relationships with favourable or unfavourable lords to the lords of 12th, 2nd, and 8th houses (in this very order) give good or bad results as detailed here. Sage Parashara gives some relief to the lord of 8th house but with certain conditions, that the said planet should have simultaneous lordship over a trine i.e. Saturn in case of signs Gemini rising in the lagna, Jupiter in case of Leo lagna and Mercury in case of Capricorn lagna. Here the sage is silent about the Mercury's lordship over 8th house and simultaneously its lordship over a trine viz. 5th house in case of Aquarius lagna and the reason for dropping this planet from the above list can be the exaltation as well as Mooltrikona sign of Mercury falling in the 8th house. There is a saying "If a planet is the ruler of the 12th house from his position, he destroys that Bhava of occupancy in the Dasa and Antar Dasa of such planet. Lords of 6, 8, 12 house, if placed in another 6, 8, or 12 houses, produce good and auspicious results, but produce bad results if placed in the good houses. 2nd lord is better placed in the auspicious houses and produce good results, otherwise produce bad results. The conjunction of the Kendra lords or the Kona lords or the Kendra to Kona lords is highly auspicious and produces good results. These are called Raja yoga combinations. The Raja yoga combinations give rise to authority, power, position and wealth. The Lords of 6, 8 and 12 house, if conjunct or joins Kendra lords or the Kona lords; it spoils even the the Raja yoga formed in the Kundali and produce bad results. If two Dustasthana lords, (6th or 8th or 12th), exchange their houses of lordship, then also it is auspicious. However, if a benefic house lord exchanges houses with a malefic house lord, like 9th lord with the 12th lord, then the exchange is inauspicious.

Functional Benefic Planets in Horoscope: The benefic Planet in the Kundali for a particular Lagna can be determined as follows:

- The planet being the lord of the first house (Lagna), other than moon. The lord of the Ascendant is especially auspicious (benefic) because Ascendant is an Angle (Kendra) as well as Trine. The Lords of 1, 5, 9, Trikona house, (whether Natural benefic or malefic) are the most benefic, even if, they rule a bad house, such as 6, 8 & 12th houses too. The lords of 4, 7 or 10 house (Kendra), if they are natural malefic, are considered Benefic Planets due to Kendradhipati Dosha. The Natural Malefic planet when they become lord of 10th, 7th and 4th house (Kendradhipati

Dosha) only becomes a benefic planet. A planet, even if they are natural malefic, owning a 1, 4, 7 or 10th house (Kendra) first, and then own 2nd, 4th, or 10th (Bhoga houses) next, becomes the Most Benefic Planet (Raja Yoga Karaka Planet) and if owning first Kendra, (1, 4, 7 or 10th house), and then owning one of the 6th, 8th or 12th (Trikas) houses, are ordinary benefic planet. Waxing Chandra (Shukla Paksh Birth) is a benefic. If Chandra is Yuti with a benefic, or receiving a Drishti from a benefic, she turns a benefic, even if in a waning state. Mercury becomes a malefic, if he joins a malefic. If waning Chandra and Buddha are together, both are benefic. The following planets are considered Auspicious/benefic, such as natural benefic, lordship benefic, in friend's sign of Lagna or Navamsa, in Lordship of Benefic's sign, in natural Benefic's sign, in functional friend's star, in own star, in lordship Benefic's star, in natural Benefice's star, functional friend of ascendant lord, in association with functional friend, in association with lordship benefic, in association with natural benefic, aspect by functional friend, aspect by Lordship benefic, aspect by natural benefic, natural malefic if posited in 3rd, 6th, 10th or 11th Bhava, natural benefic but not posited in 12th Bhava, malefic with lordship of 6th, 8th or 12th Bhava, benefic without Lordship of 6th, 8th or 12th Bhava, Vargottama, Yoga karaka and a malefic planet owning a Kendra and simultaneously Lords over a Kona and not by merely owning a Kendra. The Sun and the Mars in the 10th Bhava, the Moon and the Venus in the 4th Bhava, the Mercury and the Jupiter in Lagna and the Saturn in the 7th Bhava are strong and gives excellent result.

- If the lord of 3rd sits in the 3rd, the lord of 6th sitting in the 6th or the lord of 12th sitting in the 12th, then such planets become the benefic planets and give the good result. The lord of the Lagna (Lagnesh) sitting with the Sun gives the good results. The Saturn is said to give good result in the 3rd, 10th and 11th Bhava. If the lord of the Trines and the lord of the Kendra are related to each other, it gives the good result in his/her Antar Dasa. But if not related, then, it gives the bad result in his Antar Dasa. The lords of the two Trines or the lords of the two Kendra, if sitting together in any house, give good result to him/her even though they are malefic. When a planet sitting in the Kendra, have a relation with the lord of the Trines, he is considered an excellent Yoga karaka Graha for him/her, but sitting in the Trines, he is related to the lord of the Kendra, he will be considered an ordinary Yoga karaka Graha. The planet, except the Sun and the Moon, increases the good effects to his second house. The Ketu increases the power of the planet with which it is sitting in the house. If the lord of the Kendra is positioned in the Trikona or the Trikona lord in the Kendra, Parivartana Yoga, the planet gives the excellent result to him/her. Each and every Bhava has its Vishesha Karaka Graha and if that planet occupy or aspect that Bhava, that Bhava gives the excellent result to him/her. When the planet aspects the two Bhava simultaneously, it gives his Dasa effects to the Bhava which is the first from the Lagna in this condition. The lord of the Lagna gives always-excellent result and increases the effect to the Bhava in which it is occupying. The Retrograde Planet is considered stronger and gives excellent result to the Bhava, which it is lord. The Bhava and the sign occupied by the benefic or aspect by the benefic gives only good result. The Debilitated Planets in the 3rd, 6th and 11th house give good result to him/her. When the lord of the 6th is friend of the Lagnesh, the 6th lord is not a Malefic for him/her. The best position of the 8th Lord is the 8th house (Bhava) itself as it will not spoil any Bhava or planet in the Horoscope and at the same time fortify the longevity. A planet, whether benefice or malefic, in exaltation or in his own Sign, will always promote good to the native. If Bhava, Hora and Ghati Lagna, their Drekkana and Navamsa, or the said Lagna and their Navamsa, or the said Lagna and their Drekkana receive a Drishti from a Graha, a Raja Yoga is formed. If two or more inimical planets are occupying a single house, they will not behave as enemies. And each shows his own results of being in that house independently i.e. their conjunction do not affect their individual results. Friendly planets together occupying a house become extremely friendly and help each other in producing good and auspicious results. The 7th house is regarded as the Pucca Ghar (permanent house) of Mercury. If Mercury is placed in the 7th house of a horoscope, even the powerful planets like Sun, Mars, Jupiter and Saturn occupying a Kendra, 2nd or 11th house can not affect the his/her health adversely. All planets in their Pucca Ghar, own house or exalted house give good results. Sun and Moon, even if they are lord of 8th, is not evil. Benefic planets, owning Kendra, will not give benefic effects, while malefic, owning Kendra, will not remain Malefic or inauspicious.

- **Sun:** Sun is benefic as Lord of 1, 4, 7, 10, 5 and 9 houses. This is possible in rising Ascendant as Leo, Taurus, Aquarius, Scorpio, Aries and Sagittarius. **Moon:** Moon is benefic as lord of 1, 5 and 9 houses. This is possible in rising Ascendant as Cancer, Pisces and Scorpio respectively. **Mars:** Mars is benefic as lord of 1 and 8, 2 and 7, 2 and 9, 3 and 10, 4 and 9, 5 and 10, 5 and 12, & 7 and 12 houses. This is possible in rising Ascendant as Aries, Libra, Pisces, Aquarius, Leo, Cancer, Sagittarius, and Taurus. Mars has malefic influence on Gemini Lagna, being the lord of 6th and 11th and for Virgo Lagna, being the lord of 3rd and 8th house. **Mercury:** Mercury is benefic as lord of 2 and 5, 9 and 6 & 9 and 12 house. This is possible in rising Ascendant as Taurus, Capricorn, and Libra. Mercury influences are malefic for Aries Lagna, being the lord of 3rd and 6th house; for Cancer Lagna, being the lord of 3rd and 12th and for Scorpio, being the lord of 8th and 11th and for Aquarius Lagna, Mercury are partial malefic. **Jupiter:** Jupiter is natural benefic but it can give benefic results in limited conditions i.e. as lord of 2 and 5, 6 and 9, & 9 and 12 house. This is possible in rising Ascendant as Scorpio, Cancer, and Aries. Jupiter has partial benefic influence for Leo Lagna, being the lord of 5th and 8th house. Jupiter becomes malefic by position when is the lord of 2 and 11 in case of Aquarius, 3 and 6 in case of Libra rising, 3 and 12 in Capricorn rising, 4 and 7 in Virgo rising, 7 and 10 in Gemini rising, 8 and 11 in Taurus Lagna. **Venus:** Venus is a natural benefic but is not a benefic in all places. It is benefic only in few conditions. Venus is benefic as the lord of 1 and 8, 2 and 9, 4 and 9, 5 and 10, & 5 and 12. This is possible in rising Ascendant as Libra, Virgo, Aquarius, Capricorn, and Gemini. Venus becomes malefic as the lord of 2 and 7 house in Aries rising, 3 and 8 in case of Pisces rising, 3 and 10 in Leo rising, 4 and 11 in Cancer rising, 6 and 11 in Sagittarius rising, 7 and 12 in Scorpio rising. **Saturn:** Saturn is although a natural malefic but can act as benefic as the lord of 1 and 2 house in Capricorn

Lagna, 1 and 12 house in case of Aquarius Lagna, 4 and 5 in Libra Lagna, 9 and 10 in case of Taurus rising. Saturn has malefic influence on Sagittarius, being the lord of 2nd and 3rd house and for Pisces Lagna, being the lord of 11th and 12th. **Rahu & Ketu:** Although, Rahu and Ketu are not considered main planets but they play a major role in a one's life according to astrology. Both, Rahu and Ketu are extremely malefic planets. Rahu is treated similar to Saturn while Ketu to Mars. Rahu and Ketu are considered as both natural as well as functional malefic. We can see from above that the natural benefic are more malefic by position than natural malefic. Thus, the malefic and benefic planets should not be confused with evil and good planets, according to natural classifications.

- **Effects of Functional Auspicious/Benefic Planets:** The Benefic will always bring blessings and produce positive results to the house it occupies. It gives strength, courage, good relations, comfortable life, good health, happiness, peaceful mind, comfort, wealth, good habits, be truthful, upright, beautiful, and splendours. It makes an expert in all deeds, respectful towards the wise and Gods and will be blessed with scents, garlands, robes, ornaments. He will be a king's minister, will be learned, helpful to others, religious and fortunate. If a planet is conjunct with the Moon, or, won in a planetary war, is capable of bestowing complete happiness and kingdom that cannot be seized.

Malefic Planet in Horoscope: The Malefic Planet in the Kundali for a particular Lagna can be determined as follows:

- **Functional Inauspicious/Malefic Planets:** The lord of the 6th, 8th, and 12th are the Malefic Planet in the Kundali and makes the house weak in which he is sitting. If the planet rules a difficult house, such as 6, 8 &12th without ruling any auspicious house, it is more malefic. The 8th Lord is the villain of the Kundali and is the main culprit and capable of shattering the vocational ambitions. The influenced of the 8th Lord to any Bhava or any planet spoil the good effects of that Bhava or planet. Venus as a lord of the 6th house will creates problems for him/her for the house related activities in which he is sitting. If the 6th lord is sitting in the 2nd (income house), he/she will not gain income even though he/she works hard. The lords of the 6th house in 6th, makes him/her mean. The planet does not give the good result to him/her during his Main Dasa and the Antar Dasa. If the planet is sitting in the 6th, 8th or 12th house from his own house of lord ship, the planet becomes weak and gives the adverse effect to the lord ship house as well as his own Karkatva. Example: If the Mercury is the lord of the Lagna and sitting in the 6th, 8th or 12th house from Lagna, the Mercury will be weak and the 1st house (Lagna) will be weak and he/she will face problems related to his education and will face many difficulties in business. Waning Chandra (Krishna Paksh Birth) is a malefic. If the lord of one Trikona is positioned in the other Trikona, it is considered a malefic planet and does not give the good result and is considered bad. Example, the 5th lord sitting in the 9th house, creates problem related to the issues or Santaan. If a planet is the lord of two Kendra simultaneously, it is considered a malefic planet and he does not give good result to him/her.

- All Lords of 3, 6, and 11 irrespective of malefic or benefic is considered a malefic planet. If the lord of 8th is simultaneously is owner of 3rd, 7th, or 11th, he will prove especially harmful. The Natural Benefic planet being the lord of 10th, 7th or 4th (Kendradhipati) house (in that order) only, it is considered a malefic planet (if it is a Natural Benefic planet). This is Kendradhipati Dosha, which gets cancelled only when the aforesaid planet occupies its own sign.
- When a planet, first owns 2nd, 4th or 10th and then owns one of the 3rd, 7th or 11th (Pratipas) next, then it is the Most Malefic Planet, except the 6th lord, if is friendly with the Lagna lord. A benefice planet, if positioned in unfriendly house or inauspicious Sign in the chart, it is considered a malefic planet and is harmful and not beneficial and hence does not give good result to him/her and is called unlucky planets for him/her. If a female planet is placed in any house along with Saturn and both are being aspect by any other planet, then the aspecting planet will become afflicted and affect adversely to him/her. A planet in inimical house or in a house of debilitation gives bad results. If a planet is not placed in his Pucca Ghar and the same is occupied by a planet inimical to such planet, then that planet will destroy the good results of the house in which he stands placed. The planets of the 1st house aspects the planet of the 7th house. If they are mutually inimical, then the bad results

accrue because of his/her foolishness or unworthiness. The planets of the 8th house aspect the planets of the 2nd house and thereby affect their results same as above.

- The lord of the 8th, except Sun and Moon, is especially inauspicious (malefic) as he owns the 12th from the 9th house. When a planet is in yuti with a malefic, it is called Vikala Avastha. Venus presence in Libra, Leo and Gemini is considered dangerous and damaging for happy domestic life. Mars, Mercury, Jupiter and Venus are handicapped because of Kendradhipaya (lordship of Kendra).
- **Effects of Functional Inauspicious/Malefic Planets:**
- Functional Inauspicious/Malefic or afflicted Planets gives mental disorder, orphanage, many face, and expulsion from country, public disgrace of children, wife and friends and he/she will be deprived of his place, or position, dirty-hearted, poor and will wander here and there. He/She will be frightened by his/her enemies. These planets will give bad results during their Main Dasa (periods) or Antar Dasa (sub-periods) or during their transits (Gochara). The aspects of life, which generally receive such bad effects, will be (1) those for which they are Karakas. (2) The aspects represented by the Bhava which are owned by the planets, (3) those represented by the signs, which are resided in or aspect by them.

Neutral Planet in Horoscope: The Neutral Planet in the Kundali for a particular Lagna can be determined as follows:

- Moon being the lord of Lagna. Any planet, irrespective of benefic or malefic, ruling 2nd, 8th and 12th house will become individually neutral planet

Table 14: Planet's Characteristics

Lagna	Beneficial Planets	Planets with evil streak (Malefic)
Aries (Mesha)	Jupiter, Sun, Mars	Saturn, Mercury, Venus
Taurus (Vrishabha)	Venus, Saturn, Mercury, Mars, Sun,	Jupiter, Moon

Gemini (Mithuna)	Venus	Mars, Jupiter, Sun
Cancer (Karka)	Jupiter, Mars	Venus, Mercury
Leo (Simha)	Mars and Sun	Mercury and Venus
Virgo (Kanya)	Venus	Moon, Mars, Jupiter
Libra (Tula)	Saturn, Mercury and Venus	Sun, Jupiter, Moon
Scorpio (Vrishchika)	Moon, Sun, Jupiter	Mercury, Venus
Sagittarius (Dhanu)	Mars, Sun	Venus, Saturn, Mercury
Capricorn (Makara)	Venus, Mercury, Saturn	Mars, Jupiter, Moon
Aquarius (Kumbha)	Venus, Mars, Sun	Jupiter, Moon
Pisces (Meena)	Moon, Mars	Saturn, Sun, Venus, Mercury

7.5 Signification (Karaktva) of Planets

Sun: The Sun is the natural ruler (Karaka) of the fifth house. Sun is Dig Bala in 10th House. Sun gets zero dig bala strength when he is in the 4th house. The Sun provides good results if placed in houses 1 to 5, 8, 10 and 11. Sun provides much better results if he is in conjunction with Mercury. The Sun provides bad results in houses 6th, 7th, 9th and 12th and it gives adverse effects on the things associated with the house in which he is placed. The Mars in the 6th and Ketu in the 1st house make the Sun produce results of an exalted Sun. If the Sun is exalted or placed in an auspicious house of a person's horoscope, he is bound to rise higher and higher in power, position and popularity. If afflictions are created against Sun, that person is bound to meet his doom.

Benefic Sun, if strong and non-afflicted, gives power, position, honour, royalty, and vitality, energy, courage and thin hair. The Sun provides the Government Service and the ability to lead and be forceful and makes the powerful person in any organization. The persons will have a hard time as subordinating and surfaces as leaders by their own force and gives good health, general prosperity and high office, positions, rank and title, government affairs and offices, publicity, popularity, and superiority. Strong Sun brings success, fame and, wealth.

Malefic Sun or if adversely placed, causes separation in married lives and in business partners. Malefic Sun indicates physical, mental and emotional problems, pessimistic attitude, and humiliation at the hands of others and he/she is inclined to be arrogant and extravagant, sick of a feverish inflammatory, eye affliction and heart disorders as well as loss of position and esteem. The Sun is troubled when it is near the Nodes of the Moon, such as Rahu or Ketu. Ketu with the Sun gives insecurities. Sun & Venus in 10th promote fine arts. If the Sun, as a 12th lord, is in the 2nd house in the Nakshatra of Rahu, he/she earns money in foreign Muslim countries. If the Sun is in Leo, his/her immediate co-born will die an untimely death. If by chance, the co-born survives, they will not have any relations with him/her so that they are as good as dead for him/her. The Sun position in Libra, Virgo, Cancer, Aries and Gemini signs is not good for his/her happy domestic life.

Sun represents the self and its expression, personal power, pride, authority; leadership qualities; education, ego, and vitality, the sum of which is named as the "life force". Sun is associated with the heart, circulatory system, and the thymus.

Moon: Moon represents person's emotional make-up, unconscious habits, rhythms, memories, moods, ability to react and adapt to those around them. It is also associated with the mother, maternal instincts or the urge to nurture, the home, the need for security, and the past, especially early childhood. Moon is associated with the digestive system, stomach, breasts, the ovaries, and menstruation (ladies monthly cycle), and the pancreas. Moon is Dig Bala and works best in the 4th house. The Moon gets zero dig bala points when it is in the 10th house (opposite). The Moon provides very good results in 1, 2, 3, 4, 5, 7 and 9 houses. The Moon provides very bad results in 6th, 8th, 10th, 11th and 12th houses. As a fastest moving planet, Moon is responsible for fluctuations in human mood on daily basis. The Moon produces good results for Aries, Cancer, Libra, Scorpio and Pisces Ascendant natives or being the lord of the 4th, 1st, 10th, 9th and 5th house.

Benefic Moon, if strong and free from malefic influence, it cancels all the blemishes and makes the native healthy, cheerful and prosperous. The strength of the Moon boosts up the strength of all other planets. When at birth Moon is weak, conjoined with malefic and without any benefic aspect or if there is no benefic in a Kendra, it shows danger to the life of the child native as Moon primarily governs the childhood period of life. According to Brihat Jataka, "If the Moon with malefic occupies the 5th, 7th, 9th, 12th, 1st or 8th house, and is not aspect by or combined with powerful Venus, Mercury or Jupiter, death to native comes early. If the Moon is too close to the Sun, his/her life can be darkened. If the Moon is close to Saturn there will be depression. When Mercury is too close to the Moon the mind can be overly active. The Moon with Jupiter brings happiness and spiritual knowledge. Jupiter and Moon contacts are the best (Gajakeshari Yoga). The Moon and the Sun give trouble when they are near the Nodes of the Moon (Rahu/Ketu) and brings out fears and obsessions, addictive behaviour and an odd mother. He/She ruled by Moon generally have white complexion and charming eyes and will be plump and short stature. The Moon in its exaltation sign, Taurus and Pisces gives talents and literature. The Moon in Leo gives success in life during Makha or Pubba. The Moon in the 5th and aspect by the Venus gives him/her big gain through lottery. The Moon, afflicted or debilitated, position in the 3rd, 5th, 6th, 8th or the 12th house cause the death of the child or parent or cause handicap from which he/she is not able to overcome for rest of their life. The Moon position in Aquarius or Pisces or Capricorn or Leo signs is considered inauspicious for the happy domestic life. The Moon presence in the 4th and 9th house is found adverse for happy family life. If there is the Moon in the 8th house, there is premature death due to water accident. The Moon in the 12th gives sharp brain, but such a sharp brain harms him/her itself in latter stage. The Moon if aspect by Mercury in the Navamsa makes him/her a good poet and instrumentalist. The Moon in the Navamsa of

Mercury and aspect by Venus makes him/her the best musician.
Chandra in the upachaya bhava improves with age. The native tends to enjoy considerable material success when Chandra occupies a Kendra or a trine. Chandra in 8, may self-destruct via work and partnership-tress. L-2 Chandra in 10, the peak of the pyramid. Uttarabahdrapada is a tremendously orderly and compassionate builder. The scope of his corporate and philanthropic financial (2) responsibilities is entirely global reaching to the full extent of world society. L-8 Chandra in 4, the home-base of family. The scope of her parenting and school-teaching responsibilities is narrower than but relatively more deep. L-8 provides sudden changes. She will probably become a single parent while her children are young in order to affect this destiny. L-10 Chandra in 6, a location of social conflict and war. The scope of his government responsibilities was truly global but the focus of his ordering and regulating behavior was in the area of animosity and accusations leading to fighting. There was so much internal dissension in France for decades after the war that de Gaulle's expertise in warfare remained essential, creating great work for him to do until the moment of his death. L-12 Chandra in 8 works like a pup but he is concerned neither with regulating governments nor with philanthropic finance. His work is to transform society (8) via imaginative visionary (12) insights. The imagination-fueled social transformation must materialize (Shani) into goods and services (Shani, economies). L-4 Chandra in 12 seeks emotional stability and security through cultivating the old ways of one's people (4), Manas (the mind) is tranquil despite exterior fluctuations of material life. It may need to go abroad for schooling or family support and has affinity for the folkways, the place, the homeland, relatives and familiar surroundings.

Mars: Mars is associated with confidence and self assertion, aggression, sexuality, energy, strength, ambition, and impulsiveness. Mars governs sports, competitions and physical activities in general. Mars is the passionate impulse and action, the masculine aspect, discipline, will-power, and stamina. In medicine, Mars presides over the genitals, the muscular system, the gonads, and adrenal glands. It is associated with fever, accidents, trauma, pain, and surgery.

The Mars is Dig Bala and works best in 10th House. Mars gets no dig bala strength in the 4th house (opposite house). Mars acts as a malefic in the 4th and 8th houses. Mars provides very good results in the 2, 3, 6, 9, 10, 11 and 12th houses. Mars provides bad results in the 1st, 5th, and 7th house and is considered damaging for happy married life and causes mass scale destruction through fires, accidents, natural catastrophes. If Mars is badly placed in the birth chart, it creates tensions in mutual relations and it causes the demise of the spouse. If Mars is benefic, strong and un-afflicted, it signifies wealth, power, courage, generosity, ingenuity, rise to authority, martial success, subject to cuts, wounds, accidents, sores and injuries to the face, pains in the head and fevers. If culminating, it indicates martial eminence, success in trade and in occupations. Saturn, if aspects Mars, is a very stressful combination and creates frustration to him/her. Mars with the Sun is very courageous and competitive. Mars with the Moon (Chandra Mangala yoga) gives business success. Mars with Mercury will give power to debate and skill as a lawyer. It is widely known fact statistically that a significant number of sports champions were born just after the planet Mars rises or culminates. Mars with strong placements indicates violent tempers, martial happiness, and influences energy, drive, administrative spirits and leadership abilities, self-confidence, courage, faculty to argue, quick mental activity, extraordinary muscular strength, great organising ability, strong determination and success. If Mars is weak and afflicted, he/she will be rash, will lose temper quickly, and will be fooled and also quarrelsome and will use brute force to settle his/her affairs and may become a drug addict, the worst type of sexuality to satisfy his/her sexual urges. He/She, with Mars as ruling planet, will have white complexion, tall and muscular, round eyes, narrow waists and will have special ability to do business, and become good designers, builders and managers, soldiers and doctor. If he/she has a good Mars, he/she is into enforcing rules, good at following orders and has plenty of energy. A weak Mars makes him/her tired, jittery, angered and reactive. When the Mars occupy the 1st house, there will be irritability between the couple and lack in enjoying the pleasure among them and maintaining the marital happiness. The domestic life will be unhappy. The Mars in 2nd house except Gemini & Virgo Sign is not good. He/She will always be aggressive and demanding the crude form of satisfaction in sex regardless of the feeling of partner. If the Mars is associated with Ketu in 2nd house, his/her spouse will not be useful to him/her, because it will be like a pearl in the monkey hand. Mars occupying the 4th house, he/she will ill-treat his/her spouse. If the Mars occupy the 4th (his own house) in Makara (Capricorn) Lagna, his inimical Sign, he will be exalted in that house and also aspect the 7th house (his marriage house); he will always promote good to him/her. If the Mars is with the Saturn in 4th house or if the Mars is 4th from the Saturn, it will give bilateral and deep attachment in conjugal life, even though it is bickering. But he/she will not be satisfied with his/her spouse fully and he/she will have extra marital relations with outsider and so there will be always quarrelling, misunderstanding among the couple. If the Mars be in a malefic Sign such as Aries, Leo and Aquarius or occupies the 5th, 7th or 9th from the Sun, the Karaka (significator) for father, his/her father will be in the state of forced seclusion or confined in a foreign place or country. Mars in the Sixth House provides good opportunity to work with use of tools, machinery, or sharp instruments and makes you surgeon (Doctor). Mars presence in the 7th house is considered the most unwanted and undesirable. Mars in the 7th House will cause early widowhood. If Venus and Mars exchange their divisions, he/she will go after other sex. If the 7th House falls in a Navamsa of Mars and is aspect by the Moon or Saturn, he/she will have a diseased vagina. If the Mars is occupying own house in 7th or Rahu constellation, the marriage tie will break or snap. If the Mars is in 7th and the Rahu in 2nd, his/her spouse will die due to snakebite within three days of his/her marriage. If the Mars occupies the 7th house along with 7th lord, there will tension and clashes between couple even though he/she is submissive before his/her spouse, but there will not be separation. In Libra Lagna, the Mars is the 2nd (Scorpio) and 7th (Aries)'s lord and if he is in the Kendra (in his Moolatrikona Sign), the Mars will not be evil so much. His/Her spouse will hail from a rich and noble family but the marriage tie will break ultimately due to quarrelling. If Sun and Mercury are placed together in one house, Mars would be positive but if Saturn and Sun are placed in one house Mars becomes negative. If malefic Saturn alone aspect Mars without aspect or association of any benefice planet, then it is said the Kalatra Dasa, which cut short the couple's marriage life duration. If he is male, he/she will be dictatorial and industrious having an upper hand in family. He/She will be proving a virago to his wife and he will certainly loose his partner. The Mars' occupying the 8th is not considered good as it indicate sorrow and sufferings. The Mars in the 8th in Virgo Lagna, it is bad and painful. In this case, bickering, fighting is common and such native seeks pleasures through extra relations from outside partners. When the Mars will ill placed, he/she will be evil minded person, defective in limb and devoid of partner and indulge in sinful acts or may be one of the couple will die shortly after marriage. The Mars occupying the 8th house brings disasters and causes more havoc than the Mars occupying the 7th house. But both positions will certainly cut short or curtails the husband's longevity and cause widowhood. If Mars occupying 7th or 8th house in constellation of benefice or association or aspect by benefice, it will certainly prevent widowhood in woman. If the 12th house falls in Taurus, Gemini, Libra and Scorpio, he/she does not care his/her partner even though his/he partner will be good looking and beautiful and he/she will seek pleasure through untoward ways. If Mars is associated with Ketu in 12th house, his/he partner will suffer from ailments of an inflammatory nature. Mars is definitely a Maraka for Pisces Lagna. Mars position in Libra, Cancer, Leo, and Virgo is considered inauspicious for married life. The first cycle of Mars runs between the ages 28 and 33, the 2nd between 63 and 68 years and the 3rd between 98 and 103 years.

Mercury: Mercury represents the principles of communication, mentality, thinking patterns, rationality and reasoning, and adaptability and variability. Mercury governs schooling and education; the immediate environment of neighbours, siblings and cousins; transport over short distances; messages and forms of communication such as post, email and telephone; newspapers, journalism and writing, information gathering skills, and physical dexterity. In medicine, Mercury is associated with the nervous system, the brain, the respiratory system, the thyroid, and the sense organs.

Mercury is Dig Bala and works best in the 1st houses. Mercury gets no dig bala in the 7th house (opposite). Mercury is considered auspicious in the 1, 2, 4, 5, 6 and the 7th houses. 7th house is the pukka ghar of mercury. Mercury works badly in the 3rd, 8th, 9th and 12th houses and gives bad results. It stands exalted in the 6th house and gets debilitated in the 12th house. If mercury is placed alone in any house he/she keeps running and wasting time here and there. If Mercury and Venus are together in the Ascendant, he/she will be beautiful, fortunate and skilful in arts. Mercury governs good luck, joy, mentality and rewards. If mercury and Rahu both are in their auspicious houses then Mercury causes havoc in his/her houses and produces disastrous result like putting him/her behind the bars or creating troubles of the same sort. Mercury, if strong, he/she is a good trader, mathematician, editor and publisher and has excellent argumentative and analytical power. Exalted Mercury gives a high education, well-known personality and popularity in the society and he/she will be versatile, good in mathematical calculations, engineering, and accounts. Mercury indicates, if rising, honour, intellect, great learning. If culminating, business activity, gain through books and intellectual matters. Mercury with Saturn will give a serious, disciplined mind, and Mercury with Jupiter gives an optimistic, expansive outlook on life. Mercury with Ketu can give profound perception and with Rahu can cause obsessive thoughts. The Moon is not friendly to Mercury and can give an over active mind. Since the Moon rules memory, Mercury with the Moon can give a great memory, but dwells on the past. The weak and affiliated mercury gives rough and irritating voice and he/she will be cunning and mischievous and will turn out to be a great gambler, the worst liar, a conceited showy person, pretend as though he/she knows everything while actually he/she is devoid of real learning. The affliction of Mercury will also cause excess nervous activity both of body and mind. There is also consequent effect on the health detrimental to the operation of bowels and other organs. Mercury is the chief governor for education and uncle. Mercury is the negotiator, so it rules in us the ability to be diplomats, negotiators, deal makers, or good liars. Bad Mercury can work towards making him/her dumb or bad in mental workings. Mercury has been found prominent in Leo, Virgo, Libra, in Lagna, and its two adjoining houses, i.e. 2nd and 12th. The weak and affiliated mercury gives rough and irritating voice to him/her. Exalted Mercury in Virgo sign gives him/her a high education and well-known personality and popularity in the society. If mercury is occupy or aspect to Lagna, it promotes arts, poetry and crafts in him/her. If Mercury heads the 7th House or its Navamsa, the spouse will be skilful and knowledgeable. Mercury & Moon in 7th, it promotes poetry. Mercury & Jupiter in 9th, it promotes craft & sculpture.

Jupiter: Jupiter is associated with the principles of growth, expansion, prosperity, and good fortune. Jupiter governs long distance and foreign travel, higher education, religion, and the law. It is also associated with the urge for freedom and exploration, humanitarian and protecting roles, and with gambling and merrymaking. In medicine, Jupiter is associated with the liver, pituitary gland, and the disposition of fats.
Jupiter is Dig Bala and works best in the 1st houses. Jupiter gets zero dig bala in the 7th house (opposite). Jupiter provides good results if placed in houses 1, 2, 5, 8, 9 and 12. Jupiter provides bad results in 6th, 7th and 10th houses and gives adverse results. Jupiter is Karaka for 2nd, 5th and 9th houses. Jupiter stands exalted in the 4th house and the 10th house is the house of his debilitation. Jupiter gives bad results when Venus or Mercury gets placed in the 10th house of a horoscope. However, Jupiter never gives bad results if placed alone in any house. A malefic Jupiter affects the Ketu (son) very adversely. Jupiter offers malefic results if he is placed with Saturn, Rahu or Ketu in a horoscope.
Jupiter can always correct an erring Saturn. The combination of these two planets, if strong, is no doubt an asset to him/her. All the evil propensities of the Saturn will vanish into air on its association with a strong Jupiter. The function of Jupiter is to enlarge, make fruitful and to bestow a feeling of hope and optimism. Responsible for luck and fortune, Jupiter brings success, fortune and fame, and is an indicator of one's faith, confidence, buoyancy, higher mind, wisdom, optimism and generosity. He governs the blood, liver, veins, arteries and thighs. He rules financial dealings, speculation, shipping business and foreign affairs. If Jupiter is afflicted, he/she will be careless, and in debts, disputes, failure in speculation, gambling, worry through children, and loss of reputation, miss-judgement and miscalculation. A good Jupiter in a chart relates priests, lawyer, higher education, the legal system and the financiers of the world. If Jupiter affects the Lagna, he/she will tend to be large or fatty. When a good Jupiter aspects housing ruling money, 2nd house, then the money will tend to grow.

There will be more Children when Jupiter rules children house. Jupiter is found prominent in Lagna, 9th and 10th houses. Jupiter is not prominent in the 4th, 5th and 12th houses. Jupiter aspect Lagna makes him/her a religious and philosopher and gives intuition. The spouse will be virtuous and be a conqueror of the five senses, if Jupiter rules the 7th House, or it's Navamsa. Jupiter in the 7th house gives a good partner but also the illegitimate affairs. Jupiter in 2nd makes him/her rich. Jupiter in the 9th house takes him/her to all temples and to saint and receive the blessings and give longevity. Jupiter in the 5th house gives him/her daughters. Jupiter in the 4th house gives him/her very bad effects and is detrimental because of Kendradhipatya. Jupiter in the 1st and 5th house for female and in 7th for male found inauspicious for happy family. Jupiter in the 1st, 9th & 11th houses gives him/her a good life and prosperity. Jupiter in the 12th house gives him/her a good finance and wealth. Jupiter in the 6th house makes him/her poor even if he/she is born in an affluent circumstance and rich family. The Jupiter in cancer, Libra, Pisces and Scorpio spoils domestic family life.

Venus: Venus represents harmony, beauty, balance, feelings and affections, the urge to sympathize and unity with others, the desire for pleasure, sensuality, personal possessions, comfort, and ease. It governs romantic relations, marriage and business partnerships, sex, the arts, fashion, and social life. In medicine, Venus is associated with the lumbar region, the veins, parathyroid, throat, and kidneys. Venus is the modern ruler of the second and seventh houses, but traditionally ruled the fifth and twelve houses.

Venus is Dig Bala and works best in the 4th houses. The Venus gets no dig bala points in the 10th house (opposite). Venus offers very good results if posited in the 2nd, 3rd, 4th, 7th and 12th houses. Venus offers very bad results if posited in the 1st, 6th and 9th houses for him/her. If Saturn and Venus both are weak and are posited in 6th, 8th or 12th house or owning these houses or in conjunction with the lords of 6th, 8th and 12th houses, then they give auspicious results, comforts and pleasures in each other's major and sub period. If one of them is the lord of an auspicious house and the other is that of an inauspicious house then also auspicious results can be expected and if both of them are extremely inauspicious then also they produce lot of comforts and happiness. A Navamsa of Venus denotes him/her a fortunate, who will be dear to his/her spouse. If Venus heads the 7th House or its Navamsa, the spouse will be very beautiful and fortunate. Venus stands for the husband for a female and represents the wife in the horoscope of a male. Venus offers good results if placed alone in the birth chart. Affected Venus causes diseases of tooth and nervous system. Venus offers very good results in the houses of Mercury, Saturn and Ketu, whereas evil effects will follow if posited in the houses of Sun, Rahu and Moon. When Rahu aspects Venus or vice versa, or when both are placed together in a house, the good results of Venus will be nullified and he/she will get deprived of money, wealth and family comforts altogether. The eyes of his/her mother will become severely defective if Moon and Venus are placed just opposite to each other. Afflicted Venus causes trouble in the eyes, diseases of the ovaries, gout, anaemia and other complications due to over indulgence in amusements and sex, including gonorrhoea and syphilis. An afflicted Venus may cause vehicular accidents, faithlessness in love and marriage and will deprive him/her of the comforts of vehicles, conveyance. Venus, if rising, indicates good fortune, happiness, gifts, fortune for love and marriage, gain by legacies and inheritance. If culminating, Venus indicates honour and success, dealings with and help through women, success in occupations. Lakshmi yoga is when Venus is in its own sign of ruler ship and it gives riches and wealth Venus, when it is with Mars, it creates a strong sexual attraction. Venus with Rahu can magnify creativity and attraction with Ketu it will be more of an ascetic. Venus with Saturn gives nun-supportive relationships. Venus and Jupiter are arch enemies. The house they are in together gives contradictions due to their opposing forces. Venus Bhukti in the Saturn Dasa gives a sudden down fall to him/her or otherwise harms him. The Venus deals with all matters of the heart and also known as Kalatrakaraka. A badly placed Venus indicates dysfunction of the related parts of the body. The Venus Dasa of twenty years is divided into two parts, namely, the first part, the ascending part that is considered as "gain" part. The second part is the descending part, which is considered loss part. Venus is the chief planet connected with vehicles. Venus is the signification for mother. A good Venus helps in bringing in material wealth, sense gratifications, and all types of material fortunes, and famous, rich and beautiful people. Venus is found prominent in cancer, Virgo, Libra, and Scorpio signs and in Lagna, & in 2nd houses. Libra is own sign but it is interesting to see that Venus is prominent in its Debilitation sign Virgo and in non-reconciling watery signs Cancer and Scorpio. It is surprise to see that the Venus is not prominent in its own sign Taurus and in Exaltation Sign Pisces. The position of Venus is the prime factor of marital happiness. Venus in 2nd, it promotes poetry. Venus in 4th, it promotes proficiency in music. Venus and Sun in 10th promote fine arts. The Venus in the 7th gives a warm and affectionate tie between husband and wife. Well-placed Venus anywhere in chart always assures a medium of marital happiness even if the other affliction is present in marriage house, i.e. the 7th house. For example, if the 7th lord (benefice) is aspect by the Sun and the Mars also aspect the 7th house adversely but the Venus is well placed in the 9th (fortune) house with the Lagna lord, his friend, the Mercury (a friendly house), the couple will be very happy and will enjoy the married life. But if the Venus is in cuspal degree and placed in 6th, 8th and 12th house or 12th from the 7th house, that means, it is not placed well and hence will not give him/her happy married life. It is not that the relation will be bad between the couple, but because of the other problems of the husband such as arrest or litigation, as the case may be. The Venus in the 2nd makes him/her rich man. Mercury in Kendra from the Moon cancels the debilitation of Venus. Venus Bhukti in the Saturn Dasa gives a sudden down fall to him/her or otherwise harms him/her. The Venus presence in Libra, Leo and Gemini is dangerous and damaging for happy domestic life. The Venus in the 5th and 10th is found adverse for happy domestic life. The Venus occupies the 1st and 2nd house for a female; it is found adverse for his/her happy domestic life. The Venus is Karaka for the conveyance (four wheelers). If the Venus is well placed in 4th or is with 4th lord, unaffiliated, he gets good conveyance.

Remedies: According to Lal Kitab, when Venus is in an auspicious position in the birth-chart it brings happiness and love in life. Whereas, inauspicious position of Venus can be painful and he may get drawn towards negative actions and people. Venus is supreme (benefic) in the second, third, fourth, seventh and twelfth house in the birth-chart while in the 1st, 6th and 9th house it is malefic. Venus is exalted in Pisces and debilitated in Virgo. In Gemini, Venus is a Yogakarak planet. If it is present with any other planet in the seventh house, it gives its effects to that planet. If Venus is in conjunction with Saturn and Sun in the seventh house, then Sun becomes malefic and Venus becomes inauspicious. In this situation it causes financial loss and also affects relations with father. Venus when placed in the twelfth house provides higher position and wealth. Venus also causes influence on the body parts and the eyes of the native. Venus is the lord of love, marriage, passion, luxury, singing and dancing. If Venus is auspicious and exalted in the birth-chart of a man, he will be very concerned about his looks and also likes the company of women. When the effects of Venus become malefic, the native may be bad in character, it also creates conflicts and disturbances in the family and household. He may have problems related to skin. Unnecessary pain in the thumb is considered the symptom of malefic Venus.

Saturn: Saturn is associated with the principles of limitation, restrictions, boundaries, practicality and reality, crystallizing, and structures. Saturn governs ambition, career, authority and hierarchy, and conforming social structures. It concerns a person's sense of duty, discipline and responsibility, and their physical and emotional endurance during hardships. In medicine, Saturn is lord of the right ear, the spleen, the bladder, the phlegm, and the bones. Saturn is Dig Bala and works best in 7th house. Saturn is the malefic in 6th, 8th and 12th house and is considered the most dreaded planet. Saturn is considered well in houses 2nd, 3rd and 7th to 12th, whereas 1st, 4th, 5th and 6th houses are bad for Saturn. Saturn gets exalted in 7th house and the 1st house is the house of its debilitation. Saturn gives good result when posited in an Upachaya, such as 3, 6, 10 or 11 house.

Saturn is the Chief Justice in his court and acts as the ultimate teacher and eye-opener. In a Drekana chart, when Saturn aspects Moon (and vice versa) the native acquires the powers to become a saint of high order. Saturn, if malefic by Janma-Lagna, is very harmful but if it is benefic or Raja Yoga Karaka, then a beggar can also become a king. It is a planet whose impact can be known in advance and its maleficience can be reduced substantially by Karma. It is highly difficult to change the malefic impact of other planets like Rahu, Jupiter or Venus.

However, the inauspiciousness of Saturn can be reduced by its worship extensively. Actually, the sub period of Saturn in major period of Venus is very important to judge the effects. This period yields opposite results when Saturn is Yogakaraka but one gets auspicious results if it is malefic. If Saturn transits over Badhaka planets placed in these negative houses (8th or 12th) the malefic impacts reach its climax. One gets death like trouble if the periods are that of Venus in Saturn or Saturn in Venus.

However, Saturn gives the best result when it becomes functional benefic or Raja Yoga Karaka. For example, for Tula (Libra) lagna Saturn gives the best result, and second best for Vrishabha (Taurus) lagna by virtue of becoming a Yogakaraka (Lord of Kendra and Trikona houses). It gives good result for Makar (Capricorn) and Kumbha (Aquarius) lagna as lagna lord. When Saturn is well posited and aspect by benefic for these four Lagna, it produces excellent results during its dasa-bhukti and transit. Saturn gives mixed result for Mithuna (Gemini), Kanya (Virgo), Dhanu (Sagittarius) and Meena (Pisces) Lagna, and proves unfavourable for the remaining Lagna. In addition, Saturn's location in any Kendra, which is its own or exaltation sign, forms Sasa Yoga, one of the Panchmahapurusha Yoga that bestows name, fame, wealth and authority to the native during its dasa and bhukti. Lucky is the person for whom Saturn is favourable and strong in the horoscope. Saturn has special 10th and 3rd full aspects, in addition to the normal 7th aspect. These aspects adversely affect the affairs of the house receiving it, and afflict the planets posited there, but these aspects are favourable when Saturn becomes a functional benefic or Raja Yoga Karaka in the horoscope. Among these aspects, the 10th aspect is the most powerful, and then comes the 7th, and the 3rd aspect is the weakest. Saturn's aspect on Venus spoils the conjugal life of the person in case of functional malefic. Its aspect on Jupiter induces the native to self-abnegation in religious and spiritual pursuits and sometimes causes compromise of moral principles. Saturn's aspect on the Sun makes the native toil hard in life and face separation from, or even death of father in some cases. Saturn's aspect on Moon makes the native simple, calm and calculative and gives distress to mother. Saturn's aspect on Mars makes the native aggressive and cruel, while their mutual aspect portends violent end. Saturn's aspect on Mercury gives crooked and selfish mentality. Its aspect on Rahu makes the native shrewd and causes chronic ailments. The aspect of powerful Jupiter and Venus on Saturn tones down its malefic effect.

The Saturn in the 7th with malefic aspects will make him/her old soon and he/she will be given up by his/her spouse. If the 7th House or its Navamsa be ruled by Saturn, his/her spouse will be old and foolish. Saturn is considered to be very favourable for people born in the signs owned by Venus, whereas Saturn is evil to those born in the signs owned by Mercury. Saturn is described as a serpent, whose head or mouth is Rahu and Ketu is its tail. If Ketu is posited in earlier houses than Saturn, the latter becomes a great benefic for him/her. However, if the position is otherwise, the Saturn throws highest poisonous results on him/her. Venus and Jupiter placed together act like Saturn in that house. Similarly Mars and Mercury placed in a single house act like Saturn in that house. In the former case Saturn behaves like Ketu, while in the latter case it behaves like Rahu. Venus gets destroyed if Saturn is being aspect by the Sun in any horoscope. The aspect of Venus on Saturn causes loss of money and wealth, but the aspect of Saturn on Venus proves highly beneficial. Collision of Saturn and Moon causes operation of his/her eyes. Saturn gives good results if posited in house earlier than sun. Saturn can never give malefic results if posited with Sun or Jupiter in a single house, but highly adverse results would follow if posited with Moon or Mars in any house. Saturn releases its poisonous results on the sign and Mars, if it is posited in 1st house, on Mars only if posited in 3rd house, on moon if posited in 4th house, on sun if posited in 5th house, and on Mars in posited in 3rd house. Saturn in 3rd house deprives him/her of the accumulation of cash money and kills his/her children when posited in 5th house and 10th house is not empty. It becomes highly benefic in 12th house if friendly planets are posited in 2nd house. Saturn provides very good results if placed in houses 1 to 7 on the condition that 10th house is empty. Saturn in 1st house and sun in 7th, 10th or 11th houses causes all sorts of troubles for his/her wife. Combination of Mars and Saturn gives adverse results all through.

Saturn, if rising, indicates disgrace, ruin, calamity, grave, melancholy, liable to disgrace, too much anxiety. When the

Saturn is debilitated or un-favourably positioned, he gives him/her a pessimistic outlook, a lot of misery, makes him/her sluggish, malevolent, ruthless, unhappy and irritable. When the Planet is weaker, it gives to him/her greater injury. When the Saturn is strong, well positioned and well dignified, he makes him/her a man of drive, energy, courage and conviction and gives him/her a great happiness, position, pomp and glory. When the Planet is strong, it gives to him/her greater benefits without any restraint. If it is in 7th house afflicting Moon or ruler of 7th, he/she gets a slovenly wife. If Saturn is culminating, he/she gets trouble through old people, disgraces, trade losses, deceitful associates, rise followed by fall. With the luminaries, it indicates lean and infirm and many miseries. If afflicting Mercury, deafness. Saturn gives the experiences of delays, depression, restrictions, setbacks, destruction, disease and death. Saturn with the Sun will inhibit and lower a person's self-confidence. Saturn with Jupiter will bring frustration. The combination of these two planets, if strong, is no doubt an asset to him/her. All the evil propensities of the Saturn will vanish into air on its association with a strong Jupiter and the native will have more spiritual gains. The combination of the Saturn and strong Venus, a bosom friend, is also a happy combination from the point of view of material gains, since Venus stands for art and luxury and embodies the wise old man. If the Saturn be in a malefic Sign such as Aries, Leo, Aquarius and occupies the 5th, 7th or 9th from the Sun, the Karaka (significator) for father, his/her father will be in the state of forced seclusion or confined in a foreign place or country. If Saturn is well placed or well aspect, he/she is brave, profound, prudent, and cautious and an excellent organiser and will have executive ability. If Saturn is afflicted, he/she is apt to be bigoted acquisitive irritable, discontented and complaining. Saturn when benefice makes him/her true, reliable, honest, sincere, faithful and chaste. He aids concentration, meditation prayers. Saturn is the chief governor for longevity, called Ayush Karaka. If he occupies the house of longevity, namely, the 8th house in a horoscope, he/she gets a long span of life. Saturn is final, being the last one and rules all types of final things or things which are insurmountable and death. Saturn has the ability to oppress anything. When it aspects the Moon, or Mind, then he/she feels the effects of Saturn in his/her mind, which basically leads to depression. Saturn in 11th, it promotes craft & sculpture. Moon and Jupiter in 7th, it promotes fine arts. The Saturn aspecting the 7th house leads to become a Sanyas. The Saturn in 8th or aspecting the 8th house gives longevity. The Saturn in 5th or aspecting the 5th house gives trouble to him/her through children. The Saturn in 2nd or aspecting 2nd house leads to heavy expense. The Saturn is inauspicious, if he is the 11th or 12th lord. The Saturn presence in Libra, Scorpio, Sagittarius, Capricorn and Virgo is bad for happy marriage life. Saturn in the 10th house is good for him/her, but bad for life/business partners and they will be unhappy. This makes the position of Saturn in any horoscope decisive and it is prudent to properly assess the disposition and strength of Saturn before making any prediction. Saturn is a deterministic planet and acts for good or bad without any restraint.

Saturn has two special features, such as Sadesati (7½ years' period). If Moon in a horoscope is already afflicted by Saturn, Mars, Rahu or Ketu individually or jointly, or Moon is weak and located in a trika (6/8/12) house, the effect of Sadesati is severe. When Sadesati occurs during the dasa of Saturn, Sun, Moon, Mars or Rahu, then it causes considerable hardship in respect of the houses occupied and aspect by Saturn in the

Remedial Measures for Saturn: Saturn is considered dreadful and is said to be 'malefic of malefic'. It is considered to be a reason of troubles, difficulties and hardships. Saturn expresses darkness and gloom. He is known as the servant in the planetary cabinet. Saturn comes last in the order of natural strength of planets. He is said to possess the least Naisargik bala.

Vedic Mantra of Saturn: Chanting the following Mantra by Maharishi VedVyasa pacifies Saturn. "Visanjana sama busam avputhram yama grajam, chaya marthanda sambhootham tham namami Saneescharam". The following Mantra is also considered very well for planet Saturn. You may choose to recite one of the Mantra which is easy to remember.

Beeja Mantra: 'Om pram preem praum saha Sanaischaraya namaha'.

Gayathri Mantra: "Om Sanaischaraya vidmahe soorya puthraya dhimahi tanno manda prachodayat".

Recite all these Mantra 108 times on Saturdays. Recite the Mantra 12 times with devotion and dedication if you don't get time.

Other Remedies: Lord Saturn is believed to be an aggressive deity. It is believed to bring good fortune, success in everything and protection when is pacified. Donate any one of the following things that you can afford on Saturdays. Black cloth, black gram, sesame seeds, mustard oil etc. The ill effects of Saturn are reduced by donating these things on Saturdays. We now know that Saturn is a deep acting planet affecting a human being adversely when is afflicted and bestowing prosperity when is pacified.

Rahu (North lunar Node): Rahu is a legendary master of deception who signifies cheaters, pleasure seekers, and operators in foreign lands, drug and poison dealers, insincere & immoral acts. Rahu can cause suffering to the house He is in. It is the significator of an irreligious person, an outcast, harsh speech, falsehoods, dirtiness, abdominal ulcers, bones, and transmigration. Rahu is instrumental in strengthening one's power and converting even an enemy into a friend.

Rahu in Vedic texts represents the chief, the advisor of the demons, the minister of the demons, ever-angry, the tormentor, bitter enemy of the luminaries, lord of illusions, one who frightens the Sun, the one who makes the Moon lustreless, the peacemaker, the immortal (having drunk the divine nectar), bestowal of prosperity and wealth and ultimate knowledge. Rahu is friends with Ketu, Saturn, Mercury and Venus. It is the enemy of Sun, Moon, Mars, and Jupiter. It is important to note that Rahu is more inimical to the Sun. When Rahu forms close conjunction/aspect with natal planets, it causes affliction and destroys the significations of these planets. The harm is less when the afflicted planets are strong and more when they are weak. When the planets are weak, badly placed and closely afflicted by Rahu (North lunar Node), it is indicative of tragic. Rahu is Dig Bala in 10th House. The Rahu works best in 3rd & 6th and 10th houses and will give success, strength & power and can convert enemies into friends & boost for long life. In other houses it can be a bit too much causing problems. Rahu gets exalted in houses 3 and 6, whereas he gets debilitated in houses 8, 9 and 11. Rahu gives bad results in 1, 5, 7, 8 and 11th houses. 12th house is his 'Pakka Ghar' and he proves highly auspicious in houses 3, 4 and 6. As a node of Moon, Rahu does not provide adverse results so long as 4th house or Moon is not afflicted. He gives good results when Mars occupies houses 3 and 12, or when Sun and Mercury are in house 3, or when he himself is posited in 4th house. Rahu further provides good results if placed together with Mercury or aspect by him. Rahu offers highly beneficial effects if placed in houses earlier than Saturn. But if it is otherwise, Saturn becomes stronger and Rahu acts as his agent. Sun provides very good results when Rahu is aspect by Saturn, but Rahu gives the effects of a debilitated planet when Saturn is aspect by Rahu. If Sun and Venus are placed together in a horoscope, Rahu will generally provide adverse results. Similarly, Rahu will provide bad results if Saturn and Sun are also combined in a horoscope. Mars will also become negative, if Ketu is placed in houses earlier than Rahu. Rahu will provide adverse results, whereas Ketu effect would be zeroed. If there is Rahu in the Fifth (5th) house and has a relation with the sun, such girl does not have Menses.

There is a proverb that, no planet can bless like Rahu and no planet can curse like Ketu. The Dasa of Ketu spoils and pulls him/her down. Some body's father lost his/her fortune in Ketu Dasa period and his/her family was ruined whereas, the Dasa period of Rahu helped the noted actor and chief minister of Tamil Nadu, Sri M.G. Ramachandran. Rahu indicates fame, extremes, foreigners and foreign lands, fulfilment of worldly desires, status, prestige, power, worldly success and outer success with inner turmoil. Rahu becomes strong in a Kendra from Mercury or the ascendant. Our most intense desires are granted under Rahu. It grants us fame and fortune. Rahu is the material world, and gives all the desires. Rahu deals with fear, obsessive and compulsive behaviour. Rahu with Venus or Jupiter can bring wealth. Rahu with Saturn can cause suffering to the house they are in.

Rahu is instrumental in strengthening the power and can convert enemies into friends. It is popularly called the planet of success. Rahu does not affect the materialistic pursuits adversely during the Transit, as he is a very fond of sensual pleasures. So during Rahu Transit over the planets holding the charge of the Bhava/House other than those of the 6th, 8th and 12th is, generally, beneficial. The Rahu energize the planets during his transit over the planets. But while crossing over the lords of the Dustasthana planets, such as 6th, 8th and 12th house, insures to create bad results such as diseases, disgrace, impediment and heavy and wasteful expenditure.

Remedies of Rahu:

Worship: There is a dedicated temple to Rahu at Naganatha Temple at Thirunageswaram, Tamil Nadu. There is a milk abhishekam everyday during Rahu Kaalam to appease Rahu. The milk turns light blue when it flows down after touching the statue of Rahu. This practice has been followed for over 1,500 years.

Mantra: Beej Mantra: Chanting of Rahu mantra "Om Bhram Bhreem Bhroum Sah Rahave Namah", 18000 times in 40 days. **Ved Vyaas navagraha Mantra:** "Ardha kaayam maha veryam chandraditya vimardshanam, simhika garba sambootham tam rahum pranamamyaham". **Gayathri Mantra:** "Om nagadwajaya vidmahe padma hastaya dheemahi thanno rahu prachodayat". Reciting above Mantras with faith, devotion and sincerity will help to ward off the evils of Rahu during the Dasha and Antardasha of Rahu. One can also worship lord Shiva for this.

Chanting of Rahu Stotram with Meaning, such as, (1) Rahur dhanavamanthri cha simhika chitha nandana, Ardha kaya, sada krodhi, chandradhithya vimardhana. (Rahu, Minister of Rakshasa, one who makes Simhika happy, Half bodied one, one who is always angry, Tormentor who troubles Sun & Moon). (2) Roudhro rudhra priyo daithya swar bhanur, bhanur bheethidha, Graha raja sudhapayee rakadhithyabhilashtaka. (Angry one, Devotee of Rudhra, Ogre, One who is near the Sun, one who terrifies the sun, King of planets, one who got nectar , One who desires the moon and the sun). (3) Kala drushti kala roopa, Sri Kanta hrudayashraya, Vidhunthudha saimhikeya, ghora roopo, maha bala. (One who has death inflicting sight, one who likes death, One who lives in the heart of Shiva, one who made moon dim, One who is the son of ogress Simhika, One who has terrifying form, One who is very strong). (4) Graha peeda karo damshtri raktha nethro mahodhara, Panchavimsathi namani sthuthwa rahum sada nara, ya paden mahathi peeda thasya nasyathi kevalam. (One who torments planets, one who has big teeth, One who has red eyes and one who has a big paunch, If a man recites these twenty five names and prays to Rahu And as soon as he reads big tormenting troubles vanish immediately). (5) Aarogyam putham athulam sriyam dhanyam pasum sthadha, Dadhathi rahu sthasmai thu ya padeth sthothramuthamam. (Health, incomparable sons, wealth, cereals and animals would be given to him by Rahu, to the one who reads this great prayer). (6) Sathatham padathe yasthu jeeveth, varsha satham nara. (The man who reads this regularly would live for one hundred years)

Other Remedies: Feeding the ants is considered one way to propitiate Rahu. In some parts of India feeding ants is considered one of the ways of propitiating Rahu. Rahu Dan (donation like mustard, radishes, blankets, sesame, lead, saffron, satnaja (a mixture of seven grains), and coal on Sunday morning is a remedial measure in astrology. These articles are to be donated by a person facing the evil effects of Rahu, or if Rahu is not in a good position in one's horoscope.
Gems: The gem stone related to Rahu is hessonite and it rules the number 4.

Ketu (North lunar Node): The Ketu is Dig Bala in 12th House and works best in the 2nd & 8th and 12th houses and will give spiritual influences, spiritually speaking, highly renounced, separations, cuts, loses and deprivations. Ketu gives its exalted effect when in 5th, 9th or 12th house and its debilitated effect in 6th and 8th house. Ketu represents son, grandson, ear and spine. 6th house is considered to be its 'Pucca Ghar.' Dawn is its time and it represents Sunday. Venus and Rahu are its friends, whereas Moon and Mars are its enemies. Forty two years is the age of Ketu. Ketu is also considered to be the bed. So the bed given by in-laws after marriage is considered to be auspicious for the birth of a son and as long as that bed is in the house, the effect of Ketu can never be inauspicious. Ketu has tail but no head and represents broken relationships, cleverness, changing events, accidents. Ketu with the Moon can give psychic abilities. The placement of the Ketu in the 4th house causes mother in debt. He promotes ill will and is the trouble-makers of the zodiac. They attack in a highly clandestine manner. The native pines for happiness and becomes a recluse, always cursing his/her fate. Those ruled by Ketu, are secret adversaries and relates to people in pharmaceuticals industries, doctors and astrologers. Ketu is the most spiritual planet, with Saturn being next in line, and Jupiter takes third place. Sun-Ketu combination may make him/her a spiritual leader, Moon-Ketu may give good ability to separate emotions from logic, and Jupiter-Ketu may give deep intellectual or philosophical abilities. If Ketu is not strong to give spiritual benefits, then he would rather give disappointment in material things, and may even prove harmful, giving injuries, sudden changes. Then with the Sun, he will harm status, with the Moon, the mind and with Jupiter, his/her wisdom or religion, similarly as Rahu does. Ketu usually makes him/her unfortunate, i.e., causes loss or lack of opportunity. When Ketu is in the 10th house and aspect by Saturn, Mars and Rahu form Gemini ruled by Mercury, confer him/her the internationally famous mathematician. The Ketu in 10th house gives him/her a fertile brain, happiness, religious and pilgrimages to sacred rivers. When the Ketu is in 10th and in Gemini, it gives him/her good health, happiness, elevation to responsible and exalted positions, good food and at the old age venereal troubles, death of near relatives and loss of reputation and self respect in its own Dasa. If the Ketu is in 10th house in Sagittarius sign, it makes him/her quite disorganized though he/she is very sharp mentally. He/She will say and do the things in such a way the guru (teacher) will fail to understand him/her. If Ketu is in the 10th house, which is aspect by the Saturn without any beneficial conjunction, he/she will earn the money from government service or municipal office.

Remedial Measures for Ketu: Ketu is signified as 'Mathamaha' (father's mother). Ketu is a headless half planet not as malefic as Rahu. It is considered Mokshakaraka (cause of liberation from the cycle of birth and death). It gives spiritual knowledge and non attachment to material pleasures. Ketu and Rahu are always in retrograde motion. Ketu gives us wisdom, healing power, tantric healing, psychic abilities and healing powers possessed by evil spirits. Ketu influences pharmaceutical industry, medicine and astrology. There is a common misconception that Ketu is a dreaded planet. But actually, Ketu bestows supreme form of spiritual enlightenment which is similar to a great boon for all the natives.

Worship: Worship lord Ganesha.

Mantras: (1) Beeja Mantra: 'Om shraam shreem shraum saha ketave namaha'. **(2) Ved Vyasa navagraha Mantra:** "palasa pushpa sankaram taraka graha mustakam, raudram, raudrathmakam goram tam ketum pranamamyaham", **(3) Gayathri Mantra:** "Om aswathwajaya vidmahe shoola hastaya dheemahi,thanno ketu prachodayat". Recite the above mantras at least 108 times during the Ketu Dasha or Antardhasa. If time does not permit, chant it 12 times daily with absolute faith and devotion.

Other Remedies: Donate black things like mustard on Sundays.

Gem: Ketu's gemstone is cat's eye and he rules number 7 in Indian numerology.

Uranus: Uranus represents the principles of genius, individuality, new and unconventional ideas, discoveries, electricity, inventions, and the beginnings of the industrial revolution. Uranus governs societies, clubs, and any group dedicated to humanitarian or progressive ideals. Uranus, the planet of sudden and unexpected changes, rules freedom and originality. In society, it rules radical ideas and people, as well as revolutionary events that upset established structures. In art and literature, the discovery of Uranus coincided with the Romantic Movement, which emphasized individuality and freedom of expression. In medicine, Uranus is believed to be particularly associated with the sympathetic nervous system, mental disorders, breakdowns and hysteria, spasms, and cramps. Uranus is considered by modern astrologers to be co-ruler of the eleventh house alongside Saturn
It gives unexpected gains, restlessness, electricity, aviation, antiquities and abrupt. If rising, he is eccentricity. If it is culminating, he is learned, eminent in arts and sciences, mechanical and inventive ability. Uranus indicates change, erratic, eccentric, air planes, astrology, sudden and unexpected gain, lightening, rebellious and computers

Neptune: Neptune represents idealism and compassion, but also with illusion, confusion, and deception. Neptune governs hospitals, prisons, mental institutions, and any other place, such as a monastery, that involves a retreat from society. Its appearance coincided with the discovery of anaesthetics and hypnotism. In medicine, Neptune is seen to be particularly associated with the thalamus, the spinal canal, and severe or mysterious illnesses and neuroses. It gives compassion, dreams, fashion, alcohol, drugs, intoxication, illusions, deception, denial, romantic, confusion, glamour, spirituality, higher consciousness, devotion, the ocean, liquids, oil and gas, sensitivity, psychic, art, music, and dancing and altered States. It gives us inspiration and vision, imagination to dream. Many poets, psychics, astrologers and writers have strong Neptune placements in their natal chart. It is Ruler of Pisces. Neptune is a lot like Ketu. It is the illusions of the world, spirituality, and psychic powers. Neptune is the higher octave of Venus.

Pluto: The dwarf planet Pluto is associated with extreme power and corruption. Pluto represents renewal through bringing buried, but intense needs and drives to the surface, and expressing them, even at the expense of the existing order. A commonly used keyword for Pluto is "transformation. It is associated with power and personal mastery, and the need to co-operate and share with another, if each is not to be destroyed. Pluto governs big business and wealth, mining, surgery and detective work, and any enterprise that involves digging under the surface to bring the truth to light.
Pluto is the god of the underworld and death. Pluto represents the dark side of life and fear, and shadow side. The Pluto is the Karaka of transformation and elimination, power, empowerment and truth. Pluto deals with unconscious drives and desires. It also deals with metamorphosis, which can be mentally or physically. Pluto controls darker and subliminal drives. It is ruler of Scorpio. Pluto resembles Rahu. Pluto is the higher octave of Mars. The Pluto represents explosive, power, controlling, manipulative, birth and death, surrender, transformation, under world, Mafia, secrets, obsessions, compulsions, big money, sex, atomic energy, revenge, big government, corruption, healing.

Functional Malefic Saturn as per Tajik Astrology

(i) Saturn, in the Ascendant, give immense physical sufferings and loss of wealth. Donate mustard oil on Saturday and tilak with milk or curd to calm Saturn in first house. (ii) When Saturn is in the second house, he faces severe problems due to the govt. and even gets punishment from the govt. Feed fishes with flour and pour un-boiled milk in a well on Saturday to calm Saturn in second house. (iii) Saturn in the third house provides a person financial gains and stability. (iv) If Saturn occupies the fourth house, he may lack happiness and peace and financial problems and health challenges. Pour wine in a river and avoid taking milk during night time. (v) Saturn in fifth house creates fear of theft in home, and struggle to acquire wealth and problem related to children. Adopt a black dog and take aniseed in a new piece of a cloth, honey, silver and copper, and after tying them together keep it in the darkest corner of your room. (vi) Saturn in the sixth house is very good as it and brings wealth and keeps away from enemies. (vii) Saturn in seventh indicates problems in married life. Avoid eating non-vegetarian foods and take care of a black cow. (viii) Saturn in the eighth brings physical and mental problems and creates challenges in the career. Avoid meat and pour milk in river on Monday. (ix) Saturn in ninth house gives pain to the siblings. Don't store unnecessary things on the terrace and keep fast and worship Jupiter on Thursdays. (x) Saturn in tenth creates problem in animal husbandry related profession and challenges in the financial sphere. Worship Lord Ganesha and help blind people. (xi) Saturn is good in the eleventh house and strengthens financial stability. Keep fast and worship Saturn on Saturday. A silver brick in your house will be helpful to restrict the negative impact of Saturn. (xii) Saturn in 12th house, spends money excessively but it provides him success. Eat vegetarian food, fast properly and avoid making windows or doors at the backside of house.

8

Lords in Twelve Houses

8.0 Predictions by Combust Lord Effects:

We will discuss the result of a combust lord of a house in the kundli-Birth Chart.

Combust lord of 1st house: If the lord of the 1st house is combust then the native may have a life of a prisoner. He or she may have to face fear, disease and tension very often and this may result in lowered lifestyle, money crisis and obstacles in the way of success.

Combust lord of 2nd house: If the lord of the 2nd house is combust then the native may behave improperly. She may even do some inconceivable acts. This combination can also cause eye pain, stammering and give a squandering nature. She may have a problematic life.

Combust lord of 3rd house: If the lord of the 3rd house is combust then this is harmful for the native's brothers. He or she might get a bad advisor and enemy in guise. She may feel mental stress and her self-esteem may be at a strike.

Combust lord of 4th house: Combust lord of the 4th house means the native's mother may be in pain. The native's land asset or pet may be in danger and he or she may be at risk from water and vehicle. Native's personal happiness is hampered in this situation.

Combust lord of 5th house: Combust lord in this house may cause problems for children, failure of plans, problems in legs and unnecessary loss of money.
Combust lord of 6th house: If a native's kundli has combust lord in 6th house then she or he may face defeat, conspiracy and may have to work under others in unfavourable conditions. He may be hurt, afflicted by theft and may not be able to take intelligent decisions.
Combust lord of 7th house: Combust lord in this house indicate separation from life partner. He may get in trouble from the opposite sex and develop illegitimate relationships. He might suffer from hidden diseases.
Combust lord of 8th house: Weakness, failure, despair, hunger, disease and death are created by combust lord in the 8th house of a native's kundali Janam Patrika.
Combust lord of 9th house: Ill fate, poverty and foolish behaviour are caused by combust lord in this house. The natives who have this combination in their kundli may be cursed by elder people and teachers.
Combust lord of 10th house: If the lord of the 10th house of a kundali is combust then a native may fail at work, get demotion at the job, and may lead a difficult life where unpleasant incidents may happen very frequently.
Combust lord of 11th house: Combust lord in this house means danger for elder brother of the native, bad news, loss of wealth, ear related diseases and separation from friends.
Combust lord of 12th house: A native if has this situation in kundali may suffer various types of diseases, and may face a situation in which they are either imprisoned or in threat of imprisonment.

8.1 Predictions by Lagna Lord (Lagnesha) in 12 Bhava:

Lagna (First House) represents fame, complexion, body, appearance, features, wealth, profession, first marriage, social personality, social placement or social identity. Any planet in Lagna is a rising planet and is always benefic and gains strength and dignity by occupying the lagna. Lagnesh become exceptionally strong when located in lagna. Chandra Rashi lord in Chandra Lagna will give his/her mother a leadership, public responsibility, and dignity. If the Ascendant lord is exalted or aspect by a benefic, is in own house or friend's house, posited in quadrants or trines, he/she will be blessed with wealth, happiness and the eightfold Siddhis or paranormal powers. He/She will rise to a high position, will be happy, having good health and will attain gradual prosperity. There will be physical well-being, gains, and happiness. There will be good environments, good mood, fame, gain of maternal grandfathers' wealth, residence abroad, and pride during Maha Dasha of Lord of the Ascendant. If on the other hand, the Ascendant lord is weak, debilitated, associated with malefic planets, is in inimical Amsa and/ or aspect by malefic, he/she will be unhealthy, poor and will be indulging in self-destruction. If the North Node (Rahu) is posited in the Ascendant with malefics, he will be insulted by people. He will be tormented by enemies and crooks. If Ascendant Lord is badly placed or afflicted, the health of native will suffer during his Lagna Lord Maha Dasha. He/She will suffers with mental anxiety, accuse of personality, general retardation of success and failure in attempts leading to frustration.

8.1.1 Predictions by Lagna Lord (Planetwise) positioned in Lagna

Followings are the specific effects of Lagna Lord Position in different house:

Lagna Lord (Surya) in Bhava-1: If Surya is strong, full of vitality and well disposed in bhava-1 (in his own sign in Kendra from Ascendant for Leo Lagna), he is excellent proprietor or entrepreneur, the creative director, the politician, a gifted actor; a brilliant theatrical director; emperor or monarch, self-employed, dancer, athlete, warrior, center-stage performer, iconic representative of a social movement, exhilarated in politics and political theatre. He is a magnificent stage performer or military general. He marries an elder, experienced, practical and socially networked spouse. He is admired for creative leadership in his profession and has prosperity. If Kuja is favourable, pioneering, he gets success in politics and military professions. He is invincible in his approach to life, mathematical genius and a superb philosopher and is famous. He will be King or queen of one's own body, brilliant light of confidence emanates from the body, excellent selfishness, a perfect autocrat and very selective.

Lagna Lord (Chandra) + Guru in Bhava-1: If Moon is strong, full of vitality and well disposed in bhava-1 (in his own sign in Kendra from Ascendant for Cancer Lagna), he is a gifted actor, a brilliant theatrical director, gregarious; cunning, short-statured, valorous, happy, likes milk-based foods, interested in astrology, famous in the later part of life, subject to perils, and can suffer from somnambulism, or sleepwalking or daydreaming in life, and is handsome but fickle minded, scared of water. He can travel much in his 15th year.

Lagna Lord (Mangala) in Bhava-1: If mangal is strong, full of vitality and well disposed in bhava-1 (in his own sign in Kendra from Ascendant for Aries and Scorpio Lagna), it forms a 'Ruchaka Yoga' and is identified by king or an equal to a King, is famous, wealthy, long-lived and leader of an army. He is martial, a leader, a great Commander, and an aggressive but patriotic ruler or an equal. He is vitally athletic, super-competitive, warrior & success in battle, endurance walker, and independence-winner, innovative, energetic, athletic, and in-charge and succeeds in a physical and muscular-movement environment. Mars-1 is unfortunate for father, typically accidents to him. He will have good health, wealth and get higher position in defense or police. He is victor. He earns lands and houses, millions of worth riches and money. His wife and children prosper well and enjoy good health, happiness and wealth. His brothers prosper and help him and cause him happiness. He gains huge profits by agriculture, horticulture too. He feels most natural and comfortable in youthful environments; wins in physical athletic events, in sporting and in combat or hunting competitions like Mr. Federar, and is in the center of attention for one's expertise or specialty,

Lagna Lord (Budha) in Bhava-1: If Budha is strong, full of vitality and well disposed in bhava-1 (in his own sign in Kendra from Ascendant for Bemini and Virgo Lagna), it forms a 'Bhadra Yoga' and he is excellent, strong, intellectual, learned and rich, and becomes good in commerce and communication and will help relatives and will live up to a good old age. He is identified for his communication skills, talking, gesturing, reporting, writing, painting, tool-using, and technology-applying, arguing, criticizing, explaining, instructing, interacting with others, competition, and winning as political figure. He will have dispute with brother in the teen years. He is expert communicator, easy and friendly, personal styler. He is socially identified with mental abilities, science, adolescence, planning groups and associations. He prefers the company of younger people and has advertising and explaining skills, and travels far and wide; interested in the higher branches of learning, polite, kind and conciliatory and pilgrimage and great gain in 27th year. He is blessed with spouse and children; wealth, and truthful.

Lagna Lord (Guru) in Bhava-1: If Guru is strong, full of vitality and well disposed in bhava-1 (in his own sign in Kendra from Ascendant for Sagittarius and Pisces Lagna), it formed a 'Hamsa Yoga' and his legs will have the markings of a conch, lotus, fish and ankusha. He will be pure in mind, spiritual, selfless, fortunate, and has the voice of a swan. He gets a beautiful wife and possesses all comfort. He is religiously inclined and favourably disposed towards spiritual studies. He may be a writer, religious doctrine, dramatist model, a prosperous and high-profile career in modeling or drama, a mystical fiction writer or a multi-cultural teacher or an anthropologist-teacher. He *has* accomplishments in language, literature and culture. He is well liked by the public. He has good fortune; longevity, power and dignity to lead the social and commercial world; internal power through moral and religious self. He may hold a high position; such as, bankers, judges, doctors, lawyers, professors, preachers, government officers, shipping or large scale wholesaler. He is recognized by people for his proficiency, affluent, graceful.

Lagna Lord (Shukra) in Bhava-1: If Sukra is strong, full of vitality and well disposed in bhava-1 (in his own sign in Kendra from Ascendant for Taurus and Libra Lagna), it forms 'Malavya Yoga' and he is fortunate, renowned, learned and has fame and name, immensely richness and wealth, happy children and wife, conveyances, sensual pleasures. He is millionaire or a beauty queen, wealthy, good marriage and strong sense of justice and has a great deal of sexual enjoyment having good comforts in life. He always involves him self in good deeds, pleasant speech and skilled work. He is skilled in all trade, adviser, preceptor, home bound, powerful, fond of sour and salty food, mathematician, fond of beautiful garments, meritorious, social, artistic and has love for art, music, drama, poetry, singing-all that is beautiful, fair, generous and refined; admired by the opposite sex. He is very attractive, and has an important position in politics.

Lagna Lord (Shani) in Bhava-1: If Shani is strong, full of vitality and well disposed in bhava-1 (in his own sign in Kendra from Ascendant for Capricorn and Aquarius Lagna), it forms 'Sashya Yoga' and makes a philosopher of political economy and may be a president of a country along with Guru and he will command many good servants, but his character will be questionable. He has competitive attitude, good life, wealth, intelligence and physical health at all times. He feels most natural and comfortable in youthful environments, physical athletic competitions, physical performance, in sporting, combat or hunting and surrounded by followers. His character will be questionable. He will be head of a village or a town or even a King, will covet other's riches and will be wicked in disposition. He no doubt becomes famous and happy but his sexual outlook would be perverse. He would be sporting with other men's wives and he would employ every unscrupulous means to gain other's money. Most corrupt and powerful or Mafia person would perhaps be having Sasa Yoga; otherwise they could not have minted millions at the cost of the poor man. In interpreting the Sasa Yoga, due consideration should be bestowed on the disposition of the Moon. If this luminary is free from affliction, the person having 'Sasa Yoga' will not covet other's wealth nor will he be unscrupulous. Where the

Moon is not afflicted, the evil results attributed to Sasa Yoga can have an only 'restricted play.

8.1.2 Predictions by Lagna Lord in Various Bhava:

1. Lagna Lord in Thanu: If Lagnesha is strong, full of vitality and well disposed in bhava-1, **he** will be famous; endowed with material happiness, fame & wealth. He/She will visit foreign or faraway lands. He/She will have a magnetic personality & majestic appearance and respect in his circle. His identification in the society is done by one of the Pancha Maha Purusha Yoga and feels most natural and comfortable as an iconic image, physical athletic competitions, physical performance, in sporting, combat or hunting, in the center of attention receiving recognition for his expertise, working as an icon of the athletic in sports industry; or as a warrior and has high vitality, distinctive appearance, warrior-energy, high movement in body, self-determination and enthusiasm for life. He is often photographed. He has personality, physical movement, and athletic prowess as “athlete" originally meant “mobility". He/She will be intelligent, fickle-minded, will have two wives and will unite with other females. He is socially identified strongly with the characteristics of the graha (Saturn) occupying Lagna.

2. Lagna Lord in Dhan: If Lagnesha is strong, full of vitality and well disposed in the Second House, he/she will be graced by scholarship and has many outstanding qualities. He/She will be blessed by the articulate speech and will be sweet- tongued and graceful in speech. The Goddess of Speech will reside in him indicating poetic & communication abilities. He/she is socially identified with musical and artistic abilities, dramatist, industrialist, journalist, stored wealth, knowledge, face, voice, speech and song. He/ She will focus on his family and accumulating wealth and is coming from a strong family background. He/She will be gainful, scholarly, happy, religious, and honourable, and has good qualities, and has many wives. He has treasuries of wealth, storage of bank savings, scholarship, and collector, professor, song-singers; speech-makers on topics of historic values; writer; philosopher; mathematician; song-singers; speech-makers; writer; philosopher and mathematician. The attachment to family is strong. This is a common position for actors who look attractive, have nice teeth, strong memorization skills, and favorable speech or song. He/ She is physically attractive as Lagnesh, the lord of the body is in Shukra natural Bhava-2.

3. Lagna Lord is in Sahaj: If Lagnesha is strong, full of vitality and well disposed in the Third House, his courage will be indomitable and has unorthodox and unconventional action and will not let anything mar his reputation. He/she has the potential to become a great musician. He/She can become great mathematicians. He is socially identified with teamwork, office work, business, government administration, commerce, writing of all kinds, auditing the works of others, announcements, short-term travel, planning, scheduling, meetings, and small group process, favouring neighbour, commercial, and regional communications, pragmatic writing and media work, and sibling/team-mate/close-friend relationships, frequent business travel, messaging, business activities which generate self-made wealth. He is most natural and comfortable with siblings. He is connected with his younger brothers and sisters, will acquire college degrees and spiritual initiations. He/She will be equal to a lion in valour, be endowed with all kinds of wealth, be honourable, and be intelligent and happy. If lagnesha is strong such as uchcha Kuja, he is powerfully identified with corporate travel.

4. Lagna Lord in Bandhu: If Lagnesha is strong, full of vitality and well disposed, in the Fourth House, he/she will have good physique and impressive personality and happiness from mother. He/She will be blessed by a house and conveyances and a lot of friends and becomes a leader amidst friends & relatives. His uncles may be favourable to him and will have a good house with all comforts. He is socially identified by childhood education, patriotic and parenting, ownership of the land and house properties, home, building and property management, vehicles, schooling, very parental, great teacher, a natural educator, school-teaching, good education and financial security and provides better security and lifestyle for his people. He/She will be concerned with mother, home, heart and happiness. He/She will spend a lot of time at home, is concerned with the career of the spouse and will work for his/her spouse or work with her in his/her own home in perhaps a home-based business. He/She will be endowed with paternal and maternal happiness; will have many brothers, be lustful, virtuous and charming. He does not prefer to advertise or promote himself, but prefers to be sought out. He is associated with landed property. If the Graha is inauspicious, there may be disruption of ties to the land.

5. Lagna Lord in Putra: If Lagnesha is strong, full of vitality and well disposed, he/she may gain the good will of the political parties in power. He/she will benefit from commercial & diplomatic services and will shine in Govt Service and will get the support of officials and seniors. He/she will be lucky in speculation and investments. He is socially identified as lucky person with speculation, gambling, authoring, performance arts, politics, and theatrical stage or political campaign. He has good life wealth, intelligence, physical health at all times, politics, well-known for theatrical celebrity (5), remarkably "lucky" in life, a natural celebrity with performance in the musical or theatrical stage and is dear to king. Children are fortunate with good intelligence. He has too much romance in life, body-based performance such as acting, modeling, dancer, may be a brilliant, performance arts, speculation, gambling, and creative environment. Mesha lagna, Simha lagna, Thula lagna, Vrischika lagna, Dhanau lagna, Makara lagna, Meena lagna have the best results. He receives opportunities, much praise and applause for performances and privileges in creative activity, giving audiences, directing, acting, orating, playing games, being on stage. L- 1 (Surya) in Bhava-5 makes a great religious Guru like Guru Paramahamsa Yogananda, Saint Josemaria and is karaka for authorship (writer), esp. biography (writing about one's own self) like Rudyard Kipling.

6. Lagna Lord in Ari Bhava: If Lagnesha is strong, full of vitality and well disposed in the Sixth House, he/she will be courageous and distinguished and financially good. He/she may join the armed forces or in the health sector and reach a high position and will be helped by brothers and sisters. He/she will have enemies in plenty and have to be careful about their machinations, but will be successful generally. He is physically and socially identified with medicines, drugs, poisons, illness, poverty, finance, loan, social work and family conflict, accusations and blaming, criminals and crime, victimization, low class, exploitation and animosity. He may be a professional in conflict management such as physician, attorney, police officer, or judge. He is a gifted advocate for victims in the practice of law and medicine, a conflict management professional such as physician, criminal or divorce attorney, police officer, social worker, or judge. If Lagnesha is not strong, and not well disposed,he/she is identified with illness, poverty, conflict, unemployment, war, enemies, toxicity, pollution, debt, disease, exploitation, medicines, drugs, poisons, finance & loan-making, family conflict, accusations and blaming, criminals and crime, victimization, outsiders & low class, exploitation and animosity, and inability to achieve agreement. He/She will be devoid of physical happiness and will be troubled by enemies, if there is no benefic Drishti to 6th house. The affects of deaths, diseases and enemies overwhelm his/her life.

7. Lagna Lord in Yuvati Bhava: If Lagnesha is strong, full of vitality and well disposed in the Seventh House, he/she to be careful about the health of their partner and he may develop vairagya and turn to asceticism during the latter part of his life and wander without gaining momentum. He/she experience vicissitudes and moods always alternate between dejection & elation. He/she may end up in a foreign country. He/she is identified with negotiations, brokering, deals, partnerships, agreements, trades, match-making, terms of the contract, an agent of making deals within the law courts, in professional advising and counseling activities, executing barters, broker, match-maker, trader and negotiations of all kinds, such as, justice, equity, fairness, arrangement. He/She is identified with negotiations, brokering, deals, partnerships, agreements, trades, match-making, terms of the contract, an agent of making deals within the law courts, in professional advising and counseling activities, executing barters, broker, match-maker, trader and negotiations of all kinds, such as, justice, equity, fairness, arrangement. His/Her wife will not live long. It is a benefic; he/she will wander aimlessly, face penury and be dejected. He/She will alternatively become a king (if that Graha is strong). He/She is heavily focused on his/her partner, spouse, in life. If Lagnesh is Budha, he is a natural advocate. Attorneys with this placement are professionally successful in law courts and legal offices, mediation and arbitration tables, counseling rooms, and discussion with partners including spouses and is typical wealthy due to commercial acumen (unless the Rashi is unsuitable for either Budha or Shukra). If it is Mangala (Karaka for two unions), he is best match-maker for quick-in, quick-out of partnerships. It is favouable for a stable and respectable alliance and a strong factor for marital longevity and one is motivated to persist in the union even when unstable conditions exist. He will reside in foreign countries or in the house of his father-in-law. He may have multiple intimate relations and will be given to pleasures and will adorn himself, self with scents, and other beauty care products. If L-2 is in Yuvati Bhava, family interests and concerns are expressed through the marriage. L-3 in Yuvati Bhava, the spouse is administratively competent and inclined toward communicative aspects of business. L-4 in Yuvati Bhava, it is favorable for a stable and respectable alliance and a strong factor for marital longevity.

8. Lagna Lord in Randhra Bhava: If Lagnesha is strong, full of vitality and well disposed in the Eighth House, he/she will be a scholar with excellent academic records and has health. He/she should avoid gambling & speculation. Unfortunate events may invade his life. He/she may try to escape tension by taking to drugs, alcohol & day dreams instead of meditation & prayer. He/she will be an accomplished scholar, be sickly, thievish, be given too much anger, be a gambler and will join others' wives. He/She will face serious physical harm, unforeseen difficulties and serious problems in lifetime. There will be sudden changes, shock, surgery, evolution, transformation, mysteries, and discovery of treasures, rebirth, secret dealings, treasures and piracy. Effects of any Graha in 8 are eruptive, secretive, occult, mysterious, hidden, magical, and changing-changing-changing. Lagnesh in Randhra or join a malefic is considered a "Deha-Kashta Yoga" meaning" bad , ill , evil, wrong, painful, grievous, severe, miserable, difficult, troublesome for the body". He will be devoid of bodily comforts. Most of the manual workers belong to this category. The Yoga is said to become defunct or Deha-Kashta yoga does not apply, if a benefic aspects or conjoins the ascendant lord in Randhra. Lagnesh in Randhra carries strong circumstances of sudden death. If the lord of the 1, who represents the body in the chart, conjoins a malefic planet or is in 8, and is not aspect or conjoins by a benefic, the simple pleasures of life, such as peace and physical comfort, are difficult to achieve throughout the life. If Lagnesh is a strong malefic or associated with a malefic (esp Rahu), one may be employed in emergency services, such as first-responder in fire, police, military, or medical intervention services and is at home in secret meetings, handling confidential or volatile materials, in tantric practice, in secret discovery modes and wants to be in domains where hidden matters are revealed, where healings occur as secrets are brought to the surface like Tantric, the psychiatrist's office, psychic reading room, stock market trading desk, oil drilling and mining, treasure hunter diving ship, police detective, archeological dig are all appropriate to his identification with concealed matters. Lagnesh in 8 is identified with secret transactions, hidden sources of wealth, including private investments, confidential therapeutic privilege, and classified diplomatic information and is often a secretive person

9. Lagna Lord in Dharma Bhava: If Lagnesha is strong, full of vitality and well disposed in the Ninth House, he/she will be fortunate both material & spiritual and has good oratorical skills & communication ability. He/she will be a silver tongued orator and well appreciated by the opposite sex. He/she is likely to inherit a paternal legacy and has domestic happiness. He/she will be deeply spiritual and religious and devoted to the preceptors and God. He/she may take to asceticism during the latter part of their life.

he is socially identified with wisdom, preacher & teaching, fortunate, dear to people, devotee of Shri Vishnu, skillful, eloquent in speech, endowed with wife, sons, and wealth, sacred texts-knowledge, world travel, life philosophy, worship, and a fortunate person who can speak well, who has a good

spouse and children, acquisition of wealth will be through her father and elders and who has a good future life as well.
He/She will be fortunate, dear to people, be a devotee of Sri Vishnu, be skilful, eloquent in speech and be endowed with wife, sons and wealth. He/She is an overall fortunate and has all general good things in lifetime. L- 1 (Shani) in Bhava-9 produced Swami Vivekananda, Rush Limbaugh and Elizabeth Kubler Ross as a preacher. He is a preacher-teacher with philosophical beliefs, religious doctrine, ritual priesthood, the culture of liturgy and worship, or dharma teachings who feels most natural and comfortable in the university, the temple, delivering sermons (in person, by writing, or by communications media) . The perceived life-defining task is transmission of higher knowledge.

10. Lagna Lord in Karma Bhava: If Lagnesha is strong, full of vitality and well disposed, he is strongly identified with leadership, lawful environments, symbolic social roles, public responsibilities and institutional governance roles, positioned near the top of some hierarchy, well-known for dignified leadership roles such as parenting, business director, and governance of organizations, kinetic energy of the physical body movement and fleshly appearance and dynamically into high positions for public viewing and socially respected regardless of the job function. He/she will have professional enhancement & reputation, fame and patronage in the professional sphere and has discipline and workaholic and will be appreciated by the seniors He/she will be a self made man with self made wealth and may have income from many sources.
There will be easy success from positive recognition. He/She will be endowed with paternal happiness, royal honour, fame among men and will doubtlessly have self-earned wealth. He/She is heavily focused on the attainment of career, status position and success. If Lagnesh is fallen, there will be self-esteem issues can sabotage career performance. He is a dignified parent with the family's social standing, often going back several generations living long, devoted to his well-being, help to raise his children, and are persons of respect and reputation themselves. He is most natural and comfortable in the corporate or government in executive office, leading mass movements, and being a public icon. He is an exceptionally

public person, quite famous and identified with prominent social position and recognized for lifetime public service, hierarchical leadership roles, making decisions which affect others, setting cultural standards for their group, attracted to domains of world government and policy, dignified public assemblies, regal ceremonies of state, the throne-chair in the corporate board room; enjoys roles such as village headman, president of trade guild, and has fame among men and will doubtlessly have self-earned wealth like USA Pres-41 George H. W. Bush "Sr." (with Surya), UK-Prince Charles & Meryl Streep (with Chandra), Mother Teresa of Kolkata & UK-Princess Diana Spencer (with Mangal), Albert Einstein (with Budha), Catherine Zeta-Jones & Larry King (with Guru), UK-Princess Anne Laurence & Tom Cruise (with Sukra) and poet and patriot Jiddu Krishnamurti & William Butler Yeats (with Sani & Rahu).

11. Lagna Lord in Labh Bhava: If Lagnesha is strong, full of vitality and well disposed in the Eleventh House, he/she will be inclined to beauty and harmony in music, art and romance and associated freely with the opposite sex. He/she has to guard against marital infidelity. He/she will be subject to delays or difficulties in marriage. He/she will not experience financial difficulties and will succeed in business with the help of elder brother. Elder sisters will also be favourable to him. He is identified with environments that feature active networking and gain via links of profitable association in commercial and community organizations, in large social groups and assemblies, as "fund-raisers", and in the marketplace of goods and ideas, the principled and well-connected social progressive activist, social net worker, economies, distributive and associative networks, social movements, marketplace income, large group assemblies, gains and profits and is happy when moving from large event to large event. He/She will always be endowed with gains, good qualities, and fame. He/She is heavily focused upon the achievement of desires and certain gainful things in life and attached to friend and oldest siblings. If L-1 in 11 is Surya, he is engaged in political or entertainment career focused upon economics (11) or community development (11), global politics, theatre of philosophical explication. If L-1 in 11 is Mangala, he is engaged in muscular movement, career focused upon active pursuit of social and economic gain. If L-1 in 11 is Budha, he is engaged in large-assembly communications, rallies, meetings, planning, and advertising. If L-1 in 11 is Shukra, he is engaged in large-scale negotiation, deals, bargaining, brokering; also, can be decorating, or matching-making profession. If L-1 in 11 is Shani, he is engaged in extensive social networking necessary to establish effective governance. He/She is recognized for ability to coordinate and distribute resources on a large and well-ordered scale, like Elvis Presley who has the huge fan-base assemblies which crowded his concerts and the "mela" scene in Las Vegas and USA President-39, Jimmy Carter who concerned to distribute system-participatory opportunity via networks of charitable and political association just as easily mixing with the crowd.

12. Lagna Lord in Vyaya Bhava: If Lagnesha is strong, full of vitality and well disposed in the Twelfth House, he/she is likely to be a spendthrift. He/she has a dual or polar character and should beware of gambling. He/she is well in education. He/she derives more pleasure in serving than being served and from giving than receiving and are likely to become altruists consecrating his life for the welfare of his fellowmen. He is socially identified with his "bed pleasures", visionary imagination, prayer and spiritual guidance, private bedroom, hospital and hotel, meditative spiritual retreat, divination, an internationalized individual who feels equally comfortable in the homeland and abroad, trait emerging most vividly in foreign lands during the Vimshottari Mahadasha period of the Lagnesh. He/She will be busy in charity, donations, and may live for some other purpose other than own life. He wants to be in an enclosed, sanctuary space like, bedroom, dormitory, attic, a spa, ashram, hospital, monastery, prison, peaceful courtyard, the ordered space of the meditation hall, the interior compound, the super-private meeting room, concentration camp, survivalist compound and any walled enclosure with very high levels of privacy in all undertakings, both personal and professional. If associated with other Graha, he will have high profile public life; creative ideas, decision-making skills, core relationships in a very private space and his appearance on the world stage. If Lagnesh is strong along with Surya, he can be a dramatic articulator of the collective imagination, like Madonna and Jack Welch. If Shukra is involved, may be a seductive tempter/temptress into a world of forbidden, private delights like Madonna and wealthy.

8.1.3 Predictions by Ascendant Lord Combinations (Yoga):

Sareera Soukhya Yoga: If the Lord of Lagna, Jupiter or Venus, are in Kendra, the Yoga is highly powerful and native is endowed with long life, wealth and political favours.

Deha-pushti Yoga: If the Ascendant lord is in a movable sign and aspect by a benefic gives rise to this Yoga. The native will be happy, will possess a well-developed body, will become rich and will enjoy life.

Deha-kashta Yoga: If the Ascendant lord join a malefic or occupy the 8th house, he will be devoid of bodily comforts. Most of the manual workers belong to this category. The Yoga is said to become defunct if a benefic aspects the ascendant lord.

Roga-grastha Yoga: (a) If the Lord of Lagna occupies the ascendant in conjunction with the lord of the 6th, 8th or 12th; or (b) if the weak Lord of Lagna (i. e. Shadbala Pinda (sum total of strength) falls short of the requisite quantity) joins a Trine or a Kendra, he will not have a healthy constitution. It will lack the requisite power of resistance so that he falls an easy prey to disease. He will possess a weak constitution and is sickly and will suffer from disease, infirmity, sickness, possessed by a demon; tormented; eclipsed; slurred; and has inarticulate pronunciation of the vowels

Krisanga Yoga: If the Ascendant lords occupies a dry sign (Aries, Taurus, Gemini, Leo, Virgo and Sagittarius) or the sign owned by a 'dry' planet (Sun, Mars, Mercury and Saturn), or the Navamsha Lagna is owned by a 'dry' planet and malefic joins the Lagna, he will be lean, emaciated, thin, spare, weak, feeble, small, insignificant and will have an emaciated or lean body and will suffer from bodily pains. The same result can be judged if the Lagna and the lord have acquired a large number of dry Varga or Divisions, such as Rashi, Hora, etc.

Deha-Sthoulya Yoga: If Lord of Lagna and the planet, in who's Navamsa the Lord of Lagna is placed, occupy watery sign (Cancer, Aquarius, Capricorn, Pisces, Scorpio and Libra) or the Lagna is occupied or and aspect by Jupiter from a watery sign or the ascendant must fall in a watery sign in conjunction with benefic or the ascendant lord must be a watery planet (Moon, Mercury and Jupiter and Venus), the native will have corpulent appearance, stoutness of the body and has well-built or strong physical appearance. He will have bulky, big, huge, coarse, rough; dull, stolid, doltish, stupid, grown fat body.

Sada Sanchara Yoga: If the lord of either Lagna or the sign occupied by Lagna lord is in a movable sign or both the Lagna and the Navamsha Lagna are in movable signs, the native will always or most of the time, walking about , wandering, roaming , driving or riding, and he will almost always be a wanderer. This combination is very common in the horoscopes of travelling agents, diplomats and globe-trotters.

Bahudra-vyarjana Yoga: If the Lord of the Lagna in the 2nd, lord of the 2nd in the 11th and the lord of the 11th in Lagna or if a point of contact is established between the Lagna, 2nd and 11th houses and the lords of these three houses interchanging their respective positions, the native will earn lot of money and will amass a good fortune. Here again, the real value of the Yoga depends upon the strength of the lords concerned and how they are disposed in regard to the general scheme of the horoscope.

Swa-veerya-ddhana Yoga: If Lord of Lagna, being the strongest planet, occupies a Kendra in conjunction with Jupiter and the 2nd lord should join Vaiseshikamsa (a Graha which occupies a favorable sign in 13 out of 16 varga) or the lord of the sign, in which the lord of the Navamsha occupied by the Ascendant lord is, is strong, and join a Kendra or a trine from the 2nd lord or occupies his own or exaltation sign, or the 2nd lord occupies a Kendra or trine from the 1st lord or the 2nd lord being a benefic is either in deep exaltation or in conjunction with an exalted planet, then the native will earn money by his own efforts and exertions as a self-earned money and will have inherited wealth.

Daridra Yoga: If the lords of the 12th and Lagna exchange their positions and is conjoined; or is aspect by the lord of the 7th; or the lords of the 6th and Lagna interchange their positions and the Moon is aspect by the 2nd or 7th lord; or Ketu and the Moon are in Lagna; or the Lord of Lagna is in the 8th aspect by or in conjunction with the 2nd or 7th lord; or the Lord of Lagna joins the 6th, 8th or 12th with a malefic aspect by or combined with the 2nd or 7th lord; or Lord of Lagna is associated with the 6th, 8th or 12th lord and subjected to malefic aspects; or the lord of the 5th joins the lord of the 6th, 8th or 12th without beneficial aspects or conjunctions; or the lord of the 5th is in the 6th or 10th aspect by lords of the 2nd, 6th, 7th, 8th or 12th; or Natural malefic, who do not own the 9th or 10th, occupies Lagna and associate with or is aspect by the maraka lords; or the lords of the Lagna and Navamsha Lagna occupy the 6th, 8th or 12th and have the aspect or conjunction of the lords of the 2nd and 7th; then, this yoga indicates poverty, financial straits, wretchedness and miseries and the native is Daridra and roving, strolling, poor, needy, deprived of and a beggar, living a life of a horrible specter more grim like than even death, wretched life. He is compelled to live in squalid and unhealthy houses. His children have no facilities for education, health and decent existence and are shunned everywhere.

8.1.4 Predictions by Lord of a Kendra in Good Divisions [Amsha's]:

(i) Effects of the Divisional Dignities of the Lords of Kendras: If the Lord of a Kendra is in Parijata-amsha, the native will be liberal; in Uttama-amsha, will be highly liberal; in Gopura-amsha, will be endowed with prowess; in Simhasana-amsha, will be honourable; in Paravata-amsha, will be valorous; in Devaloka-amsha, will be head of an assembly; in Brahmaloka-amsha, will be a sage; and in Iravata-amsha, he will be delighted and be celebrated in all quarters.

8.1.5 Predictions by Lords of Kendra and Lords of Kona in Good Divisions [amsha's]:

If the Lord of a Kendra and the Lord of a Kona are yuti or having Drishti or have a relationship in Parijata-amsha, he will be king and will protect men; in Uttama-amsha, he will be an excellent king, endowed with elephants, horses, chariots; in Gopura-amsha, he will be a tiger of kings, honored by other kings; in Simhasana-amsha, he will be an emperor, ruling over the entire earth; in Paravata-amsha, he will be like as Manu was born; in Devaloka-amsha, he will be like as the Incarnations of Lord Vishnu took place; in Brahmaloka-amsha, he will be like as Lord Brahma was born; and in Iravata-amsha, he will be like as the Swayambhu Manu was born.

8.2 Predictions by Dhan Lord in Various Bhava:

If second Lord is well placed or un-afflicted, he/she will be blessed with wealth, additions to the family, success due to speech and good food. There will be overall happy financial position, inflow of finance, wealth possessions, education, grains, family, happy domestic life, Sanyas, inheritance, law suits and domestic comforts during Maha Dasha of Lord of the Second. If second Lord is afflicted or badly placed, he/she has to incur heavy expenses on family affairs, eye trouble and fear from government due to financial irregularities. There will be bad food, suffering due to his/her bad speech or loose talking. There will be distress and possibility of death during Maha Dasha of Lord of the Second. As it is the lord of Maraka Bhava, so, if second Lord is associated with 6, 7, 8th house, the troubles will multiply. Followings are the specific effects of Dhan Lord in different house:

The second house is the significator of wealth and speech and Domestic happiness. If the second house is a benefic sign and if the second house is tenanted by benefics and if the second lord is in a quadrant or a trine, he will become immensely rich with oratorical prowess. If the Ascendant lord is weak and if the second lord is weak with malefic posited in 6,8,12 houses, he will be poor and helpless. If the second lord is exalted in a quadrant or trine, or in benefic houses, he will have tremendous wealth and will protect a lot of people. If the second lord is aspect by benefic and is posited in a quadrant or trine, he will be handsome. If not, he will be ugly and wicked. If the second lord is aspect by malefics and is with malefic, he will stammer and stutter. If the second lord is with malefic, is posited in the tenth with an afflicted Sun, and if Mars and the Sun are posited in the second house, his speech will be afflicted and he will be rejected by others.

The Effects of the Second Lord in the 12 Houses:

1. **Second Lord in the 1st House:** He/she will be endowed with sons and wealth, be inimical to his/her family, lustful, hard-hearted and will do others' jobs. He/she will be wealthy, but will have strained relationships with other members of his/her family. He/she wants to be away from home and he/she will seek for pleasures outside his/her home and family. He has to be careful not to get involved in fraudulent transactions. He will be subjected to vicissitudes as financial ups and downs mark in his career. He does not care much for his manners & his manners may be frowned down by society. Domestic happiness becomes a problem. He will be sweet tongued but sometimes she/he will be prone to anger and will use angry words.

2. **Second Lord in the 2nd House:** He/she will be wealthy, proud, will have two or more wives and be bereft of progeny. L-2 Surya in bhava-2 makes him king of the Bankers. Has a regally entitled way of handling the money, but Bhava-2 is Shukra-sthana, so there is bound to be trouble with autocratic finance and sensual tastes. He/she will be financially sound. He/she is most likely to have a small family and will be sweet tongued and will be known for his gift of the gab. His wealth will be above medium and above want.

3. **Second Lord in the 3rd House:** He/she will be blessed with the qualities of valour, wisdom and economic prudence. His sister will help him. He/she is interested in music and fine arts and this will help him professionally. He/she has business, but looked down upon by people as a miser. Dhan Lord is in Sahaj Bhava, he/she will be valorous, wise, virtuous, lustful and miserly; all these, when related to a benefic. If related to a malefic, he/she will be a heterodox.

4. **Second Lord in the 4th House:** Dhan Lord is in Bandhu Bhava, he/she will acquire all kinds of wealth. If Dhan Lord is exalted and is yuti with Guru, he/she will be equal to a king. He/she will be blessed by house and conveyances and a pious

and religious mother. He/she will earn well from land & automobiles as 4th house represent vehicles and land. He/she may be helped by maternal relatives, his mother and her brothers & sisters.

5. **Second Lord in the 5th House:** Dhan Lord is in Putra Bhava, he/she will be wealthy. Not only he/she, but also his/her sons will be intent on earning wealth. He/she will be lucky with investments which will pave the way for prosperity even for the next generation. He/she may become a victim of abandonment & lack of sympathy shocking him to the extreme and so he becomes unkind to people. He/she may gain unexpectedly in the form of lotteries likely. He/she can try his luck in lotteries & the stock market.

6. **Second Lord in the 6th House:** Dhan Lord is in Ari Bhava along with a benefic, he/she will gain wealth through his enemies; if Dhan Lord is yuti with a malefic, there will be loss through enemies apart from mutilation of shanks. He/she gains from and through enemy's power, privilege and wealth. He/she will cut anyone to size that go against his wishes & aspirations. He/she always achieves what he really wants even though people say he employs means that are not above board. His wealth and power may be attributed to black marketing, black mailing and deceit. He/she may be assailed by health hazards in the latter part of life. He/she is likely to be subject to persecution from enemies. He/she will emerge victorious in the end.

7. **Second Lord in the 7th House:** Dhan Lord is in Yuvati Bhava, he/she will be addicted to others' wives and he/she will be a doctor. If a malefic is related to the said placement by yuti with Dhan Lord, or by Drishti, his/her spouse will be of questionable character. The medical profession may attract him with its treasures. He/she always helps others who are lacking in luck. He/she may waste much money for the gratification of the senses. He/she may have many sources of income, one of them foreign. His partner may be religious and pious. His partner will have a good aristocratic background. His wealth will be subject to fluctuations as alternate ups and downs mar his career.

8. **Second Lord in the 8th House:** Dhan Lord is in Randhra Bhava, he/she will be endowed with abundant land and wealth. But he/she will have limited marital felicity and be bereft of happiness from his/her elder brother. It will be difficult for him to retain wealth & assets. He/she has misunderstandings with members of family likely. His hypersensitivity is so high that even a hint of emotional abandonment can bring about unreasonable panic, if not terror. He is likely to lose wealth. Elder brother may misunderstand him. His relatives will turn enemies. Domestic happiness becomes a problem. Beware of anger and angry words.

9. **Second Lord in the 9th House:** Dhan Lord is in Dharma Bhava, he/she will be wealthy, diligent, skilful, and sick during childhood and will later on be happy and will visit shrines, observing religious code. He/She will earn finance or wealth by fortune and will earn money in the lottery and spend in religious work. He/she will attain to wealth by his professional expertise during the latter part of the career. During the early period of his life there will be considerable suffering, both physical & mental. He/she may secretly resent the loss of childhood innocence & pleasures. He/she may get legacies & benefits will come from many a source.

10. **Second Lord in the l0th House:** Dhan Lord is in Karma Bhava, he/she will be libidinous, honourable and learned; He/She will have many wives and much wealth, but he/she will be bereft of filial happiness. He/she will have professional expertise in whatever profession he has chosen. He/she may not get much happiness from progeny. He/she will be engaged in many a vocation, such as, business or agriculture He/she will also indulge in philosophical dissertations or lectures. As a result he may get fame, name & largesse.

11. **Second Lord in the 11th House:** Dhan Lord is in Labh Bhava, he/she will have all kinds of wealth, be ever diligent, honourable and famous. He/she will attain fame by virtue of his/her diligence. During childhood he may have health problems, which will progressively improve. He/she can earn money by banking and finance. His self development stuns people and growth generates envy, as both 2nd and 11th represents wealth He/she will be sufficiently well-off and

above want. As the 11th represents the fulfillment of all desires, the second lord posited therein indicates that all his desires will be fulfilled.

12. **Second Lord in the l2th House:** Dhan Lord is in Vyaya Bhava; he/she will be adventurous, be devoid of wealth and be interested in other's wealth, while his/her eldest child will not keep him happy. He/she will be devoid of immense wealth. It is quite likely that he may be interested in other's material possessions. He/she may not get much happiness from the elder brother. He/she may be cheated by people. There may be violation of confidentiality as the trust he places in people may be violated. The native should curb the tendency to criticize others. He/she is quite likely to be involved with the government.

8.3 Predictions by Sahaj Lord in Various Bhava:

If third Lord is well placed or un-afflicted, there will be cooperation and help from co-born, rise in career, rise in self-earning, profitable short journeys and success in educational attainments during Maha Dasha of the planet. If Third Lord is afflicted or badly placed or afflicted, there will be quarrel and misunderstanding with brother/cousin/neighbour, mind will be in tension, activities will be performed at a low spirit; troublesome journeys, mental anxiety and may have to suffer due to his wettings/agreements. There will be unfavourable effects. There will be breaking of relation with younger siblings or younger brothers & sisters; and loss of vitality, personality and power to face the struggles, loss in business, and journey (short trips), parent's death, and diseases like cough, respiratory system during Maha Dasha of the planet. Followings are the specific effects of Lagna Lord Position in different house:

The Effect of the Third Lord in the 12 Houses:

1. Sahaj Lord is in Thanu Bhava, he/she will have self-made wealth, be disposed to worship, be valorous and be intelligent, although devoid of learning. He/she is a self sufficient and self made person. His intelligence & know ledge will be impressive despite his/her qualifications. He/she has to control anger which can be his/her greatest enemy. He/she tends to be attracted to fine arts & will be drawn to acting, music & dancing

2. Sahaj Lord is in Dhan Bhava; he/she will be corpulent, devoid of valour, will not make many efforts, be not happy and will have an eye on others' wives and others' wealth. The younger siblings will take the wealth his/her wealth. It also indicates loss to the younger siblings, but of course, all these lies in the nature of the planet concerned. He/she may be lazy or lethargic and is attracted to his neighbour's assets which are considered to be mischievous by many. His image may be spoiled by his headstrong behaviour and his

3. Sahaj Lord is in Sahaj, he/she will be endowed with happiness through co-born and have wealth and sons, be cheerful and extremely happy. He/She will have company of brothers and sisters. L-3 Simha Surya in bhava-3 makes him a king of the messengers, who lords it over the siblings and the work group. He wants to be admired for publishing and writes autocratically.He is not interested in the opinions of others. He has announcement by royal decree. He is very selective - even elitist - in communications style. He/she will have the company of brothers & sisters. He/she cannot don the deceptive role. Optimism is deep rooted in his heart and he never cries over split milk. He/she does not bother about the past and believe that today is his eternity. He/she may not enjoy good relationship with younger co-born

4. Sahaj Lord is in Bandhu Bhava, he/she will be happy, wealthy and intelligent, but will acquire a wicked spouse. He/she will have conveyances, houses & everything. He/she

will be intelligent and wealthy. His spouse will be shrewd and intelligent and unorthodox in many ways. He/she may have stepbrothers. They may not have much happiness from his mother. Some of his friends will turn away from him and his uncles and aunties may turn hostil.

5. Sahaj Lord is in Putra Bhava; he/she will have sons and be virtuous. If in the process Sahaj Lord be yuti with, or receives a Drishti from a malefic, he/she will have a formidable spouse. He/she will be virtuous and chivalrous. Her/his married life will turn into a paradise as both of them are ready to forget & forgive even after the most vicious encounters. His brothers will definitely help him in his hour of need. He/she knows when to sow and when to reap. He/she may not have much happiness from his children. His neighbours also may turn hostile towards him.

6. Sahaj Lord is in Ari Bhava, he/she will be inimical to his/her co-born, be affluent, will not be well disposed to his maternal uncle and be dear to his/her maternal aunt. He/she will not maintain good relationships with brothers, sisters and uncle. He/she will have honesty and sincerity in financial dealings. He/she will be extroverts by nature. One of his brothers may opt for the Army. Another one may join the medical profession. His will be troubled by enemies.

7. Sahaj Lord is in Yuvati Bhava, he/she will be interested in serving the king. He will not be happy during boyhood, but the end of his/her life he/she will be happy. He/she is an employee than a businessman. His income will be regular, steady & not subject to vicissitudes. He/she will excel as subordinates than as commanding officers. He/she will always be in the good book of the boss. He/she is careful not to incur the wrath of the Law

8. Sahaj Lord is in Randhra Bhava; he/she will be a thief, will derive his/her livelihood serving others and will die at the gate of the royal palace. He/she will be basically altruistic & may even be ready to die for his love. He/she has honesty and sincerity and good intentions and is deeply hurt by innuendos on his honesty and having a heart of gold. He/she may miss the company of his younger brother or sister.

9. Sahaj Lord is in Dharma Bhava; he/she will lack paternal bliss, will make fortunes through wife and will enjoy progeny and other pleasures. He/she may not get much from the parental side and his spouse may bring everything for him. They may be much bothered by father's reputation and attitude which may cause despondency and alarm. His relationship with father will be tainted by misunderstanding.

10. Sahaj Lord is in Karma Bhava; he/she will have all lands of happiness and self-made wealth and be interested in nurturing wicked females. He/she will be self sufficient and self made and has all sorts of comforts for him, even the forbidden. He/she has pleasing personalities and sincere approach and has a sympathy wave amidst the public. His profession may be connected by travelling and is the gainers. Almost all the brothers become successful and he will help them in many ways. He/she will attain to professional reputation

11. Sahaj Lord is in Labh Bhava; he/she will always gain in trading, be intelligent, although not literate, be adventurous and will serve others. He/she will have business knack. He/she is selfish and can become vindictive. He/she depends on his brothers and is ready to sacrifice for them. He/she may be helped by his younger co-born. He/she will have gains of a high order and his desires will be fulfilled in time

12. Sahaj Lord is in Vyaya; he/she will spend on evil deeds, will have a wicked father and will be fortunate through a female. He/she earns wealth via member of the opposite sex, most probably his/her partner. His relationship with father is likely to be tainted by personality clashes. He/she is quite domineering and wants his wishes to be granted and his orders should be obeyed by all. They are He/she is rebel and terrible antagonists.

8.4 Predictions by Bandhu Lord in Various Bhava:

The fourth house represents happiness, mother, friends, uncle, house properties & conveyances. All these significations get a boost if the 4th lord or 4th house becomes strong. If the fourth house is vitiated by malefic, it will make happiness a hasty retreat. Relatives turn hostile and there will be difficulty in acquiring houses or conveyances. If fourth Lord is well placed or un-afflicted, he/she will be helpful to family; will gain good relations from mother; acquire land, house, property, agricultural land; acquire vehicle and overall a happy period. There will be acquisition of house and land, family affairs, happiness, possession of vehicles, higher education such as master's degrees, or doctorates, short journeys and pleasure trips, gain of fame and wealth, pleasures, luxury, increase in savings, reputation/popularity, and pets during Maha Dasha of the planet. If fourth Lord is afflicted or badly placed or afflicted, t there will be Distress due to mother; Disputes in family leading to loss of property; litigation; discomfort of conveyance; danger from water; no peace of mind during Maha Dasha of the planet. Followings are the specific effects of Lagna Lord Position in different house:

The Effect of the Fourth Lord in 12 Houses:

1. If Bandhu Lord is in Thanu Bhava, he/she will be endowed with learning, virtues, ornaments, lands, conveyances and maternal happiness. He/she will have all sorts of domestic comforts, houses & conveyances. He/she is outspoken & independent, clever and intelligent. His mother is gentle and tender. He/she has an academic mind and his qualities will be appreciated in the field of education. He/she will have the help of many friends and uncles. They He/she will have a well decorated house.

2. If Bandhu Lord is in Dhan Bhava, he/she will enjoy pleasures, all kinds of wealth, family life and honour and be adventurous. He/She will be cunning in disposition. He/she will inherit much from his mother or maternal relatives and his mother in turn must have received much from her own sisters & brothers. He/she is cunning and quite clever when occasion demands. He/she will have house & conveyances.

3. If Bandhu Lord is in Sahaj Bhava, he/she will be valorous, will have servants, be liberal, virtuous and charitable and will possess self-earned wealth. He/She will be free from diseases. Because of his educational background he may get the much needed help from office levels. Subordinates love working for him as he leads by example. He/she will retain self earned wealth. Uncles may turn hostile. He/she may have trouble regarding house & conveyances.

4. If Bandhu Lord is in Bandhu, he/she will be a minister and will possess all kinds of wealth. He/She will be skilful, virtuous, honourable, learned, and happy and be well disposed to his spouse. L-4 Simha Surya in bhava-4 makes him a king of one's own castle with royal dominion over one's own home. He has splendid house, vehicles, and furnishings. He is impatient with schooling and might not go far in the educational world. He has self-centered parent, but wishes to be admired for his parenting. He is very selective - even elitist - in regard to the home and schooling. He/she will have a good house and conveyances. He/she may be connected with powerful, political people. He/she has the knowledge to manipulate ideas, men and things with dignity and honour. He/she maintains a spotless character.

5. If Bandhu Lord is in Putra Bhava, he/she will be happy and be liked by all. He will be devoted to Sri Vishnu, be virtuous, honourable and will have self-earned wealth. He/she will have vehicles & conveyances. He/she maintains happy relations with people. He/she will enjoy wealth. He/she will have a comfortable life with riches. His children will prosper well & may get happiness from them. He/she may gain in speculation with whirlwind profits if he speculates.

6. If Bandhu Lord is in Ari Bhava, he/she will be devoid of maternal happiness, be given to anger, be a thief and a conjurer, be independent in action and be indisposed. He/she will have problems with respect from house & conveyances. He/she may receive affection from somebody other than his mother during childhood. His mother's health may be imperilled. Some of his friends, his uncles and aunts may turn hostile. Uncles and aunts may not get much happiness from mother & conveyances.

7. If Bandhu Lord is in Yuvati Bhava, he/she will be endowed with a high degree of education, will sacrifice his/her patrimony and be akin to the dumb in an assembly. He/she will shine well in the field of education. He/she will sacrifice & relinquish his rights on property at the slightest request from their mother. He/she acquires vast areas of land and houses. His public relations are good. He/she will have good friends.

8. If Bandhu Lord is in Randhra Bhava, he/she will be devoid of domestic and other comforts, will not enjoy much parental happiness and be equal to a neuter. He/she will have problems in education and will face difficulties in childhood. He/she may not fulfil mother wishes and this makes her dejected. He/she may be separated from father quite early in life. He/she may have to encounter difficulties arising out of litigation. Some of his friends, uncles and aunts may turn hostile. He/she has problems regarding house & conveyances.

9. If Bandhu Lord is in Dharma Bhava, he/she will be dear to one and all, be devoted to God, be virtuous, honourable and endowed with every land of happiness. He/she will be blessed by a loving and compassionate mother. He/she may inherit paternal legacy. This is a fortunate combination with regard to father and properties. He/she will have a good house & conveyances. He/she will be blessed by many good friends who come to him help in his hour of crisis.

10. If Bandhu Lord is in Karma Bhava, he/she will enjoy royal honours, is an alchemist, be extremely pleased, will enjoy pleasures and will conquer his/her five senses. He/she will

have professional enhancement and reputation professional expertise. He/she will have political success and knowledge to handle any situation. He/she will be good at handling chemicals. He/she will be quite domineering and have a knack of making his presence felt and will vanquish his enemies. He/she will be blessed by house & conveyances. He/she will have good friends who help him in their hour of crisis.

11. If Bandhu Lord is in Labh Bhava, he/she will have fear of secret disease; he will be liberal, virtuous, charitable and helpful to others. He/she will be and will have a lot of friends who will be help him. He/she have lots of gains and the fulfilment of all desires. He/she will be blessed by conveyances and good well decorated house. He will have lot of mental tensions also as 11th is 8th to the fourth. Lack of mental peace and bliss can result.

12. If Bandhu Lord is in Vyaya Bhava he/she will be devoid of domestic and other comforts, will have vices and be foolish and indolent. The mother will be strong in his/her religion and it will cause a great impact him. So, he/she will have strong faith, belief in spirituality or religion. He/she will have to face many ills & unhappy situations in life and loss of happiness (Sukha). He/she may have to encounter litigation and many problems regarding house and mother. Uncles and aunts and friends will turn hostile. He/she may have to spend much money on house and conveyances. He/she may have to face losses in speculation.

8.5 Predictions by Putra Lord in Various Bhava:

The Fifth House represents progeny and divine merit. I Q is ruled by the fifth house. The placement of the Fifth lord in a benefic Sign makes the native intelligent and meritorious. The reverse may happen if the fifth lord is weak or is placed in malefic Signs. If fifth Lord is well placed or un-afflicted, there will be attainment of education; passing a competitive examination; birth of a child (according to age); gain and promotion in career; political success; sudden gain of money; sudden rise in attraction to women. There will be progress in education, child birth and happiness from the children. There will be love affairs & relationships, gain by speculation, lotteries, luck, and pregnancy during Maha Dasha of the planet. If fifth Lord is afflicted or badly placed or afflicted, there will be sickness of child; worst affliction may lead to death; worries on account of children like failure in examination; poor performance in studies; victim of fraud and deceit and a life full of distress during Maha Dasha of the planet. Followings are the specific effects of Lagna Lord Position in different house:

The Effect of the Fifth Lord in 12 Houses:

1. Putra Lord is in Lagna; he/she will be scholarly, be endowed with progeny happiness, be a miser, be crooked and will steal others' wealth. He/she will be lucky regarding investments and children and getting happiness from children. He/she will be a scholar in whatever profession he chooses. He/she will command a number of servants & will have punitive powers and will be successes in speculative ventures

2. Putra Lord is in Dhan Bhava, he/she will have many sons and wealth, be patter families, be honourable, be attached to his/he spouse and be famous in the world. He/she gains from speculation and will attain to fame & status at the international level based on his/her background. He/she will hit headlines of the media one day. He/she feels proud of the achievements of his children. He/she will have a beautiful partner and well behaved kids. He/she has leadership qualities and initiative. His sons may shine in their fields & they get the benefit thereby.

3. Putra Lord is in Sahaj, he/she will be attached to his/her co-born, be a tale bearer and a miser and be always interested in his/her own work. He/she will have tremendous communication ability. He/she is miser.

4. Putra Lord is in Bandhu, he/she will be happy, endowed with maternal happiness, wealth and intelligence and be a king, or a minister, or a preceptor. He/She will be money earner right from boyhood. He/She will live in a luxurious building with beautiful surroundings for his/her children to enjoy. His/Her mother will have good longevity. He/She may have more daughters than sons. He will become rich due to investments and speculations. He/she will be a money earner right from boyhood. He/she will be living in a luxurious building with beautiful surroundings for his children to enjoy. Mother will have good longevity. He/she may have more daughters than sons. He/she become rich due to investments & speculation.

5. Putra Lord is in Putra Bhava, he/she will have progeny, if related to a benefic; there will be no issues, if malefic is related to Putra Lord, placed in Putra Bhava. Putra Lord in Putra Bhava will, however, make one virtuous and dear to friends. L-5 Simha Surya in bhava-5 makes him a king of the royal court, queen of the theatrical stage; a highly effective and credible political position with creative charismatic performance. He is very selective - even elitist - in regard to lovers, children, presentation style. He/she will have happiness from children. He/she lives dangerously and likes thrills and excitement. He/she will have children who achieve greatness in their professional spheres. He/she gains from children even economically and his sons take care of his every need.

6. Putra Lord is in Ari Bhava, he/she will obtain such sons, who will be equal to his/her enemies, or will lose them, or will acquire an adopted or purchased son. He/she may have problems relating to one of the sons likely. Children pose problems. Losses can accrue if they indulge in speculation. He/she may have to face problems in investments since the 5th rules investments.

7. Putra Lord is in Yuvati Bhava; he/she will be honourable, very religious, endowed with progeny happiness and be helpful to others. He/she will be known for his/her public relations. His son will go abroad & attain fame and wealth. His children will have everything even though it is beyond their means. They are careful not to be misunderstood by others, particularly by members of family. They will have a spouse who is cordially

disposed. Since 5th lord is the owner of a trine in a quadrant, this is a powerful position for both gains from spouse and children.

8. Putra Lord is in Randhra Bhava; he/she will not have much progeny happiness, be troubled by cough and pulmonary disorders, be given to anger and be devoid of happiness. The happiness from progeny will be below expected standards. They may be beset with cough and lung disorders. Speculations can be disastrous. He/she should be careful about jumping into investments and should stay away from gambling race courses & stock markets, eschew evil & sinful acts. Children pose problems

9. Putra Lord is in Dharma Bhava, he/she will be a prince, or equal to him, will author treatises, be famous and will shine in his/her race. He/she will be recognised as a new star that zoomed on the family firmament and he/she will be treated like a prince. He/she will have innate abilities to be an author. He/she enjoys being with children and it will inspire them to reach dizzy heights. One of their children will reach a high state as an author or
orator. They will be blessed with fortunate children.

10. Putra Lord is in Karma Bhava; he/she will enjoy a Raja Yoga and various pleasures and be very famous. The 5th lord in the10th itself is a Raja Yoga and he/she will be a powerful individual in many respects. He/she will have fortunate children. One of their sons will be renowned and will be the cynosure of all eyes. One of his sons may join the investigative department.

11. Putra Lord is in Labh Bhava, he/she will be learned, dear to people, be an author of treatises, be very skilful and be endowed with many sons and wealth. He/she will be a renowned author He/she will be learned and will be the darling of the crowd. He/she becomes known and famous in academic circles. He/she is capable of building a secure and luxurious future for his children. He/she will feel proud of his children later on. Later on his children will bring him all that he/she wants as they become successful.

12. Putra Lord is in Vyaya Bhava, he/she will be bereft of happiness from his/her own sons, will have an adopted, or purchased son. He/she may derive unhappiness from children. He/she should be prepared to face problems associated with children. He/she is at times stubborn & purblind & impetuous. Eventually he/she will lead a life of detachment and attain to Self Actualisation.

Effects of the Divisional Dignities of Putra's Lord:

If Putra's Lord is in Parijatamsha, the native will take to the branch of learning, befitting his race; in Uttamamsha, will have excellent learning; in Gopuramsha, will receive world-wide honors; in Simhasanamsha, will become a minister; in also Paravatamsha, will be endowed with Vedic Knowledge; in Devalokamsha, will be a Karma Yogi (performer of actions, worldly and religious rites); in Brahmalokamsha, will be devoted to the Lord; and if in Iravatamsha, will be pious.

8.6 Predictions by Ari Lord in Various Bhava

The Sixth House represents enemies and wounds. This is an important house as enemies can destroy our mental peace. Even though philosophically the adversary is an active collaborator and is the builder of the strength in man, practically he can make us cry with his machinations! Hence a powerful Sixth lord is called for; if the sixth house is weak or aspect by malefic, we are bound to suffer. The best Yoga for the destruction of enemies is Jupiter in the Ascendant & the North Node in the sixth. This is also known as Ashta Lakshmi Yoga. This Yoga will wipe out enemies.

If sixth Lord is well placed or un-afflicted, then he/she will have victory over enemies - success all of a sudden; rise in personal efforts, good health; redemption of debt during Maha Dasha of the planet. If sixth Lord is afflicted or badly placed or afflicted, he/she will suffer loss through enemies/colleagues/business opponents, reversals in life; constant troubles due to labour

unrest, shortage of raw material; sudden troubles from income Tax/Excise. If Sun and Jupiter, being the lord of Sixth house, are badly afflicted, there ill be humiliation in family circles; loss of vehicles; loss of job due to negligence. There will be danger from enemies and ill health. There will be debts, law suites, challenges from competitors or thieves and robberies, accidents, acute illnesses, obstacles in life, mental worries, calamity, fighting, imprisonment, misery, success over enemies, loss of moneys, cheatings and calamities (troubles) through women during Maha Dasha of the planet.
Followings are the specific effects of Lagna Lord Position in different house:

The Effect of the Sixth Lord in 12 Houses:

1. Ari Lord is in Thanu Bhava; he/she will be sickly, famous, and inimical to his/her own men, rich, honourable, adventurous and virtuous. He/she will be rash & adventurous and inimical to own people. He/she may join Defence or may be a jail superintendent. He/she may be plagued by enemies. He/she join a criminal group.

2. Ari Lord is in Dhan Bhava; he/she will be adventurous, famous among his/her people, will live in alien countries, be happy, be a skilful speaker and be always interested in his/her own work. He/she will be enterprising and will live in a place where he/she are more surrounded by enemies than friends. He/she is workaholics. He/she is always shaky about scarce economic resources & financial conditions. He/she has a good health.

3. Ari Lord is in Sahaj; he/she will be given to anger, be bereft of courage, inimical to all of his co-born and will have disobedient servants. He/She has to face tremendous enemies and non co-operation from neighbours. It makes him/her angry and loses confidence. Enmity to brother is also predicted. He/she may have to face tremendous enmity & non-cooperation from brothers, sisters and neighbours. It will be full scale war if he clashes with someone he loves. Neighbours become hostile & he may have to face their wrath.

4. Ari Lord is in Bandhu Bhava, he/she will be devoid of maternal happiness, be intelligent, be a tale bearer, and be jealous, evil-minded and very rich. may not have much happiness from his/her mother. Education gets disrupted. Troubles through servants indicated.

5. Ari Lord is in Putra Bhava, he/she will have fluctuating finances. He/She will incur enmity with his/her sons and friends. He/She will be happy, selfish and kind. There is a draining away of the negative things in life such as death, diseases and enemies. He/she will be subject to dire vicissitudes. The atmosphere at home may not be smooth enough. His maternal uncle may help them. Due to stress intense he/she may shatter all ties with relatives and fall into a sad neurosis. They are humorous & strong willed.

6. Ari Lord is in Ari Bhava; he/she will have enmity with the group of his/her kinsmen, but be friendly to others and will enjoy mediocre happiness in matters, like wealth. L-6 Simha Surya in bhava-6 makes him a king of the Servants. He wants to be admired for work in medicine and human services. He is very selective - even elitist - in regard to medical, military, and legal services. He/she attains strength to fight his enemies. Their inner conscience conflicts against He/she will be blessed by conveyances and will have good longevity. Enemies pose problems but ultimately he triumphs. He/she will recover from illnesses and debts due to the sixth lord's strength.

7. Ari Lord is in Yuvati Bhava, he/she will enjoy happiness through wedlock. He/She will be famous, virtuous, honourable, adventurous and wealthy. He/She will be lucky with respect to 12 house of final emancipation. His/Her expenditure will be curtailed and there will not be any incarnation. He/she will be surprised by the difference between actuality & the marital life which he/she dreamt. He/she will marry from within his family. His maternal uncle may live abroad. He/she may have clashes with life partner.

8. Ari Lord is in Randhra Bhava; he/she will be sickly, inimical, will desire others' wealth, be interested in others' wives and be impure. Enmity increases from his associates. The adverse position of the 6th lord can give diseases and debts. They may

be subject to the machinations of enemies. Enemies may go in for a campaign of character assassination to destroy his image.

9. Ari Lord is in Dharma Bhava; he/she will trade in wood and stones, 'Pashan' also mean poison, and will have fluctuating professional fortunes. His professional fortunes will be subject to severe vicissitudes. Ultimately he triumphs over his enemies with the touch of poetic justice. His father will be quite renowned. His enemies will be fooled & their friends will benefit from their association with them.

10. Ari Lord is in Karma Bhava; he/she will be well known among his/her men, will not be respectfully disposed to his/her father and will be happy in foreign countries. He/She will be a gifted speaker. He/she is bound to be successful in foreign countries or faraway lands. He/she will have articulate speech. His ancestral properties which he inherited will be subject to litigation and disputes

11. Ari Lord is in Labh Bhava, he/she will gain wealth through his/her enemies, be virtuous, adventurous and will be somewhat bereft of progeny happiness. He/she will gain a lot from enemies and acquire wealth thereby. His spouse is in the habit of spending more than he earns.

12. Ari Lord is in Vyaya Bhava; he/she will always spend on vices, be hostile to learned people and will torture living beings. He/she becomes a sadist taking pleasure in torturing other beings. He/she will be subject to considerable persecution from their enemies.

8.7 Predictions by Yuvati Lord in Various Bhava:

The Seventh House represents the Life-Partner & love-life. Malefic afflicting the Seventh House indicate many affairs and divorce. Benefic posited therein means a smooth love life. If seventh Lord is well placed or un-afflicted, he/she will marry during this period; can successfully enter into partnerships; chances of foreign tours; if already married, then happiness, good relations with wife, pleasure and enjoyment from wife or opposite sex; acquisition of luxury household items, and marriage of self or someone in family. There will be travel, trade, marriage, partnership in business, conjugal happiness, honour and reputation in foreign country, residence in foreign lands, and journeys to distant places during Maha Dasha of the planet. If Seventh Lord is afflicted or badly placed or afflicted, he/she may suffer due to wife and partnership, and extent of sufferings depends on the association of malefic planets with 7th Lord. There will be bad relations with in-laws; separation from wife, contacts with ill-reputed women during Maha Dasha of the planet. As it is Marakasthan, the Lord of 7th house, if it is in 2nd, 8th and 12th house or associated with 2nd, 8th and 12th lord, will give physical disturbance/ailments. There will be distress to spouse and the possibility of the death, litigations and courts cases, divorce, and increase in expenditure. Followings are the specific effects of Lagna Lord Position in different house:

The Effect of the Seventh Lord in 12 Houses:

1. Yuvati Lord is in Thanu Bhava, he/she will go to others' wives, be wicked, skilful, devoid of courage and afflicted by windy diseases. He/she will marry an acquaintance. He/she is of wavering minds and does not stand firm on his words & actions. He/she is quite flexible and not at all domineering. He/she will get a cordially disposed spouse as the seventh lord is well placed. His public relations will be good.
2. Yuvati Lord is in Dhan Bhava; he/she will have many wives, will gain wealth through his/her wife and be procrastinating in nature. He/she will find inordinate wealth & prosperity after marriage. He/she may be subject to ordeals by fire & will have to face Govt enquiries and litigation, but ultimately he emerges graceful & successful.
3. Yuvati Lord is in Sahaj, he/she will face loss of children and scarcely there will be a living son. He/She may have a daughter. He/She will be devoted to his/her wife. His/Her in-laws will help him/her whenever he/she is in crisis. He/She will

have immense literary talent and will enjoy fine arts, ornaments and dress. He/She will have good public relations which advance his/her development. He/she may have to bear losses on account of the upbringing of children. His daughter will bring him better luck. His partner is loving and endearing personality.

4. Yuvati Lord is in Bandhu Bhava, his/her wife will not be under his/her control. He/She will be fond of truth, intelligent and religious. He/She will suffer from dental diseases. He/she will have a lucky partner who gives a satisfactory married life with children & comforts. He/she is noble and is good friends always & remains. If difference of opinion surfaces he clears it out and maintain harmony.

5. Yuvati Lord is in Putra Bhava, he/she will be honourable, endowed with all (i.e. seven principal) virtues, always delighted and endowed with all kinds of wealth. He/she has a rich partner. He/she will achieve respectability and a majestic position in life. He/she is born salesmen & shine in marketing. He/she perform well in jobs where the gift of the gab and travelling are required.

6. Yuvati Lord is in Ari Bhava, he/she will beget a sickly wife and he/she will be inimical to him/her. He/She will be given to anger and will be devoid of happiness. His marital happiness is under threat. Their partner's constitution may not be at par with theirs. His spouse becomes too sickly & jealous when he demands too much

7. Yuvati Lord is in Yuvati Bhava, he/she will be endowed with happiness through spouse, be courageous, skilful and intelligent, but only afflicted by windy diseases. L-7 Simha Surya in bhava-7 makes him self-centered. Surya is less beneficial in bhava of pair-matched Shukra. Partner may be King of the marriage, often solipsistic in relationships and radiantly confident in one's entitlement to hold the center of attention within a marriage. He/she will have indomitable courage, skill & high I Q. Opposite sex will be attracted to him for temporary / lasting relationships.

8. Yuvati Lord is in Randhra Bhava, he/she will be deprived of marital happiness. His/Her spouse will be troubled by diseases, be devoid of good disposition and will not obey him/her. As the 7th lord is in 8th, marriage will be with someone known to them earlier. Their luck is after marriage. He/she will find it difficult to fool his spouse for long as the partner happens to be too clever.

9. Yuvati Lord is in Dharma Bhava; he/she will have union with many women, be well disposed to his/her own wife and will have many undertakings. He/She and his spouse will have faith in religion. He/she will do well with their spouse. He/she will be unorthodox and will be known for versatility & concentration.
10. Yuvati Lord is in Karma Bhava; he/she will beget a disobedient spouse, will be religious and endowed with wealth, sons. He/She works with his/her spouse. He/she gets a devoted and chaste partner who will be a friend philosopher and guide to him contributing immensely to his progress & advancement. He/she will be pious & enjoys all the comforts of life. His spouse is actually shrewd & can read between the lines. He/she will be successes abroad.
11. Yuvati Lord is in Labh Bhava; he/she will gain wealth through his/her spouse, be endowed with less happiness from sons etc. and will have daughters. He/she will gain immensely through marriage. His spouse may be rich and religious. He/she has to adjust with his sons. His spouse may gain a lot of property or may be usefully employed. He/she gains wealth and the fulfilment of all desires via his spouse.
12. Yuvati Lord is in Vyaya Bhava, he/she will incur penury, be a miser and his/her livelihood will be related to clothes. His/Her spouse will be a spendthrift. His partner may spend more than what he/she budgets personally and uses all his tactics to manage his spouse. His income may be related to textiles. He/she may lose largesse due to his partner's imprudence.

8.8 Predictions by Randhra Lord in Various Bhava

The 8th house is very important in VA. It represents Death and the causes of Death. If the 8th lord is weak, longevity of the native is threatened. If on the other hand, the 8th lord is powerful, then high longevity is decreed, provided there is no malefic aspect on the 8th house.If eighth Lord is well placed or un-afflicted, there will be elevation in rank, benefit through court judgments, many quarrels will be settled; and acquisition of immovable or permanent assets, inheritances or money from others such as wills and insurance policies, monetary gains from partner, during Maha Dasha of the planet. If eighth Lord is afflicted or badly placed or afflicted, he/she will have ill health, distress and sorrows. There will be rivalry and opposition in many fields. The Lord of 8th house, the strength will be more during its Maha Dasha, if associated with 2nd and 7th lord, is a potential killer and so there will be death of self or in family and loss of reputation during Maha Dasha of the planet. There will be the possibility of death and financial losses. There will be accidents, worries, inheritance, mental pain, obstacles, and gain from in laws, imprisonment, investigation, scandals, bankruptcy, and trouble from enemies, lost property and unions with others. Followings are the specific effects of Lagna Lord Position in different house:

The Effect of the 8th lord in 12 houses

1. Randhra Lord is in Thanu Bhava; he/she will be devoid of physical felicity and will suffer from wounds. He/She will be hostile to gods and Brahmins. He/she will scoff at religious rituals & practices. He/she may be afflicted by physical ailments right from childhood. He/she may have to suffer bodily complaints from diseases or disfiguration. He/she may have a physical constitution which appears weak. He becomes the victim of governmental displeasure.

2. Randhra Lord is in Dhan Bhava; he/she will be devoid of bodily vigour, will enjoy a little wealth and will not regain lost wealth. He/she is not robust physically. He/she has to consult an ENT doctor. He/she may have to put up with food of inferior quality. It is better that he preserves and protects his wealth as regaining it becomes difficult.

3. Randhra Lord is in Sahaj Bhava, he/she will be devoid of fraternal happiness, be indolent and devoid of servants and strength. Ear problems are to be expected. He/she does not like social occasions & will withdraw himself within. He/she may suffer from hallucination. He/she may have a monetary windfall from literary work.

4. Randhra Lord is in Bandhu Bhava, he/she will be deprived of its mother. He/She will be devoid of a house, lands and happiness and will doubtlessly betray his friends. He/she misses maternal proximity and is away from home. His mental peace is far from satisfactory. Fiscal problems are accompanied by domestic bickering. He/she may be worried on account of parent's health. Reverses in profession and the displeasure of seniors may have to be faced with true fortitude.

5. Randhra Lord is in Putra Bhava, he/she will be dull witted, will have limited number of children, be long-lived and wealthy. He/She will be subject to dire attitude. His good actions and altruistic behaviour may go unnoticed which gives him/her the creeps. He/she will be subject to dire vicissitudes. He/she is extremists by nature and is intense about everything like intensely ruthless, intensely loyal, intensely compassionate and intensely cool. Child mortality and sickness of children create worries for him. He/she have to achieve mental equanimity in order to avoid nervous debility.

6. Randhra Lord is in Ari Bhava; he/she will win over his/her enemies, be afflicted by diseases and during childhood will incur danger through snakes and water. He/she will experience Vipareetha Raja Yoga results like affluence and the fulfilment of all desires. As the 6th rules diseases he may be attacked by ill health at times. There is phobia in them about future health hazards. It is quite possible that he may lose money due to theft and litigation. Maternal uncle's health causes concern. He/she ultimately win over enemies.

7. Randhra Lord is in Yuvati Bhava, he/she will have two wives. If Randhra Lord is yuti with a malefic in Yuvati Bhava, there will surely be downfall in his business. He/she has disharmonies in married life. He/she and his spouse are different in motivation & personality and they may suffer in health.

8. Randhra Lord is in Randhra Bhava, he/she will be long-lived. If the said Graha is weak, being in Randhra Bhava, the longevity will be medium, while he/she will be a thief, be blameworthy and will blame others as well. L-8 Simha Surya in

bhava-8 makes him King of the Mysteries. One may be an important figure in a secret society or in a tantrik setting but the public does not see this bright light. He will be shrouded in mystery, often in secret government functions. He is very selective - even elitist - in regard to hidden liaisons, secret wealth, and confidential information. He/she has tremendous longevity. He/she may have to face scandals & even criticism. He/she will have the active support of his partner. In early life his father will have to pass through a crisis.

9. Randhra Lord is in Dharma Bhava, he/she will betray his/her religion, is a heterodox, will beget a wicked spouse and will steal others' wealth. He/she appears to be always youthful and charming. Relations with father may be strained.

10. Randhra Lord is in Karma Bhava; he/she will be devoid of paternal bliss, is a talebearer and be bereft of livelihood. If there is a Drishti in the process from a benefic, then these evils will not mature. He/she tends to be with his parents. He/she may have to face stiff competition from a cunning subordinate. They are capable of coping with any problem situation and despite the obstacles he faces he will achieve progress in his career.

11. Randhra Lord along with a malefic is in Labh Bhava; he/she will be devoid of wealth and will be miserable in boyhood, but happy later on. If Randhra Lord is yuti with a benefic and be in Labh Bhava; he/she will be long-lived. He/she will have miseries during early childhood. They will be hailed as accomplished raconteurs. His partner becomes a challenge to his managerial skills. Relations with elder brother may be strained.

12. Randhra Lord is in Vyaya Bhava; he/she will spend on evil deeds and will incur a short life. More so, if there is additionally a malefic in the said Bhava. He/she will be seized with an abnormal desire to spend on unwanted and tremendous appeal will be generated by them which few members of the opposite sex can resist. He/she may get caught with bad health and deteriorating bank balance. He/she may have to face many troubles & turmoil as life becomes plagued by miseries.

8.9 Predictions by Dharma Lord in Various Bhava

The 9th House is the most important house as it deals with fortune. It is said that a Wise One prayed that she should have fortunate children, and not scholars or heroes. Fortune is an invisible goddess which no wealth can court. How can we define fortune? It is said that a man needs health, wealth & wisdom. All these three in equilibrium is Fortune. If ninth Lord is well placed or un-afflicted, he/she will have prosperous time with family; gain in wealth and rank; success in higher education/may go to other place for higher education; doing religious and charitable acts; pilgrimage and fruitful journey; develop power of intuition; blessings of preceptors. There will be educational gain, religious gain and unexpected gains of wealth. There will be fortunes, foreign- travels, higher education, writing books, prosperity, pilgrimages or journeys to gain spiritual knowledge, gain from lottery and Diksha (spiritual initiation) during Maha Dasha of the planet. If Ninth Lord is afflicted or badly placed or afflicted, there will be decline in prosperity; domestic unhappiness; weak financial status; fall in respect and reputation during Maha Dasha of the planet. Followings are the specific effects of Lagna Lord Position in different house:

The Effects of the 9th Lord in the 12 Houses

1. Dharma Lord is in Lagna, he/she will be fortunate, will be honoured by the king, be virtuous, charming, learned and honoured by the public. He/she will be a self made person. He/she will definitely get public honour and acknowledgment and acceptance & encouragement will be given by seniors. He/she will reach the top. They are endowed

2. Dharma Lord is in Dhan Bhava, he/she will be a scholar, be dear to all, wealthy, sensuous and endowed with happiness from wife, sons etc. He/she will be the son of a rich and influential man. He/she will inherit paternal property. He/she gains knowledge on subjects selected by him and will become scholars in his specialised fields. He/she cannot concentrate on one subject for long.

3. Dharma Lord is in Sahaj Bhava, he/she will be endowed with fraternal bliss, be wealthy, virtuous and charming. He/she may come up via writing. He/she will have pleasing personalities and charming manners. As the 9th lord is in the 7th from the 9th, his father will be a respectable man in society. He/she becomes suspicious and mysterious.

4. Dharma Lord is in Bandhu Bhava; he/she will enjoy houses, conveyances and happiness, will have all kinds of wealth and be devoted to his/her mother. He/She will be fortunate to have a luxurious house and be healthy. He/she will have beautiful houses & conveyances and is deeply attached to mother who is also a fortunate individual. He/she may inherit father's immovable properties.

5. Dharma Lord is in Putra Bhava, he/she will be endowed with sons and prosperity, devoted to elders, bold, charitable and learned. He/she will have a famous & prosperous father. As the 9th lord is in the 9th from the 9th, father becomes fortunate and successful. He/she will be renowned for his learning and will be of charitable disposition. His character will be spotless. He/she will be prosperous and his domestic life will be satisfactory.

6. Dharma Lord is in Ari; he will enjoy meagre prosperity, be devoid of happiness from maternal relatives and be always troubled by enemies. His/Her father's health may be affected and may have clashes with his/her mother. Wealth will be inherited from his/her father. As the 9th lord is in the 10th from the 9th, his father becomes successful professionally and may have to face health problems, which may develop into a chronic disease. He/she is a cruel revengeful person if he feels that he has been cheated. He/she will gain wealth as a result of successful termination of father's legal problems.

Money compensation comes to him automatically. His father may have to face litigation and other problems

7. Dharma Lord is in Yuvati Bhava, he/she beget happiness after marriage, be virtuous and famous. He/she gets luck after marriage. Father goes abroad and prospers. He/she also will his fortune in foreign lands. He/she will be blessed with a noble and lucky spouse. He/she wants to be with members of his family. As the 9th lord is in the 11th from the 9th, it means fulfilment of fortune & all desires.

8. If Dharma Lord is in Randhra Bhava, he/she will not be prosperous and will not enjoy happiness from his/her elder brother. He/She will lose his/her father early in life and will be raised without a father. As the 9th lord is in the 12th from the 9th, he may have to face separation from his father. Problems manifest for his elder brothers and sisters on account of them. He/she

is undoubtedly very good salesmen. As the 8th is 12th from the 9th, he may have to suffer reverses in luck and constant bickering in married life are to be expected. Children may also pose problems.

9. Dharma Lord is in Dharma Bhava; he/she will be endowed with abundant fortunes, virtues and beauty and will enjoy much happiness from co-born. L-9 Simha Surya in bhava-9 makes King of the Temple, King of the Philosophers, and regal Father-figure to the priests. He is very selective - even elitist - in regard to philosophical doctrine and tenets of belief. As the 9th lord is in the 9th he will have a long-lived & prosperous father. He/she is respected in society and known to be very lucky enjoying paternal legacy. He/she will be extolled as exemplars and his qualities will be a source of inspiration to others. Even his relatives gain immensely from his luck. He/she earns largesse from his foreign visits. He/she will have a cordially disposed spouse and good children.

10. Dharma Lord is in Karma Bhava, he/she will be a king, or equal to him/here, or be a minister, or an Army chief, be virtuous and dear to all. He/she will become famous & powerful. This is said to be a powerful RajaYoga, a

combination for political power. He/she will be blessed with a royal status in Govt or in defence.

11. Dharma Lord is in Labh Bhava, he/she will enjoy financial gains day by day, be devoted to elders, virtuous and meritorious in acts. As the 9th lord is in the 11th and as this is a Dhana Yoga, he will be above want. He/she will have Wealth beyond the dreams of avarice. He/she will be blessed with influential friends. His father is also renowned and well off. His capability is enormous in the sense he will leave his footprints in the sands of time by way of great achievements. He/she is masterminds when it comes to execution of any plan with malice aforethought.

12. Dharma Lord is in Vyaya Bhava; he/she will incur loss of fortunes, will always spend on auspicious acts, and will become poor on account of entertaining guests. As the 9th lord is in the 12th, luck does not come to him. They may have to work very hard in life. Even then success does not come to him. His father may leave him penniless. 9th lord in the 12th does not indicate a rich background. He/she may be faced with financial difficulties due to festivals and celebrations.

Effects of the Divisional Dignities of Dharma's Lord:

If Dharma's Lord is in Parijatamsha, the native will visit holy places; in Uttamamsha, has been visiting holy places in the past births and he will do the same within this life-time; in Gopuramsha, will perform sacrificial rites; in Simhasanamsha, will be mighty and truthful, conquerer of his senses and will concentrate only on the Brahman, giving up all religions; in Paravatamsha will be the greatest of ascetics; in Devalokamsha, will be an ascetic, holding a cudgel (Lagudi), or he will be a religious mendicant, that has renounced all mundane attachments and carrying three long staves, tied together, in his right hand (Tridandin); in Brahmalokamsha, will perform Aswamedh Yagya (Horse Sacrifice) and will attain the state of Lord Indra; and in Iravatamsha, he will be a synonym of Dharmaa, or virtues just, as Lord Ram and Yudhishtira.

8.10 Predictions by Karma Lord in Various Bhava

The 10th House or MC is considered the most important house after the Ascendant. The calculation of the longitudes of the houses starts with the calculation of the longitudes of the Ascendant and the MC. The Tenth House deals with Profession, the important dimension in one's life. One becomes a success in the professional sphere, depending on the strength of the 10th lord and the 10th house. If the 10th lord is debilitated or weak, the professional life of the native will suffer. If he be powerful, professional success is indicated. If tenth Lord is well placed or un-afflicted, he/she will be successful in his/her attempt to improve his sources of income; his/ her promotion or professional elevation; gain of authority and status; good financial gain and political success. There will be recognition and awards from the Government. There will be gain in career, profession, business, honour & respect, popularity, reputation, status in life, power and authority, government favour, livelihood, living in foreign lands and Government employment or authority. There will be gain of wealth of the father, earned money, position, and place of settle in life, self-fulfilment in carrier and promotion during Maha Dasha of the planet. If tenth Lord is afflicted or badly placed or afflicted, there will be fall in position; financial losses; dispute and disgrace in profession; set-back in career; involved in misdeeds during Maha Dasha of the planet. Followings are the specific effects of Lagna Lord Position in different house:

The Effects of the 10th lord in 12 Houses

1. Karma Lord is in Thanu Bhava, he/she will be scholarly, famous, be a poet, will incur diseases in boyhood, and be happy later on. His/Her wealth will increase day by day. Since the 10th lord is in the First, he will be a workaholic and will come up the hard way to the top. He/she will be self employed

and will have independent professions. He/she is talented. He/she has health problems during childhood. The progress he achieves will be steady and slow. He/she will have relations with powerful people, people who are related to politics. This combination is conducive to success in politics. You will be in high ranks with the government or something akin to that. You possess an intelligence about business transactions, which works cooperatively with your artistic spirit. A poet, pioneer and determined worker, you gain success, prosperity and status. You gain a reputation as one of knowing authority. Social projects, the building of institutions and charitable deeds all lay on your path. If the 10th lord is either conjunct or aspected by the lagna lord in the first house it gives great fame as an inventor or innovator.

2. Karma Lord is in Dhan Bhava, he/she will be wealthy, virtuous, honoured by the king, charitable, and will enjoy happiness from father and others. Since the 10th lord is in the 2nd, he is lucky in profession. Since the 2nd is 5th from the 10th, he will be successful in his professional spheres. He/she may develop his family business. A big patrimony may be inherited. You may have the food service industry, specifically restaurants and catering, offer successful business prospects. You possess an ability to generate significant earnings constantly increasing your fiscal well being. And you are driven by the urge to consistently improve your money. Under your beck and call, workers and assistants flock near you. It is indicated that your power and authority become unlimited. Cordial relations exist in your life and especially the relationship with your father. You may absorb the family business, which contains its own trials and tribulations. Some losses might exist in stabilizing this business. Various troubles with such losses face you on occasion, especially if afflicted.

3. Karma Lord is in Sahaj Bhava, he/she will enjoy happiness from brothers and servants, be valorous, virtuous, eloquent and truthful. Since 10th lord is in the 3rd, part of his career will be spent in travelling. He/she may shine as speakers or writers. Brothers will be instrumental in his progress in the professional sphere. He/she may be beset by problems in profession as the 3rd is the 6th from the 10th. He/she will overcome these problems in time. The capacity to work hard

and undertake even strenuous work is your assets. Speaking and writing top the lists of your professional pursuits. Acclaim comes to you as a bold writer. Successes occur swiftly; the climb up the career ladder happens easily. Agency work, newspapers, publishing houses, writing and contracts all provide you with significant gain. You possess musical inclination and talent, possibly becoming famous in this field. Many short journeys fill your travel dockets. Others see you as courageous and truthful. Your siblings and servants surround you with harmonious relations. Should the planet/house be afflicted, successes would be slow and the path ridden with obstacles. Restlessness and aimless wandering would follow, haunting you and your desire for happiness. Rivalry with a brother could lead to career setbacks.

4. Karma Lord is in Bandhu Bhava, he/she will be happy, be always interested in his/her mother's welfare, and will Lord over conveyances, lands and houses, be virtuous and wealthySince the 10th lord is in the 4th, he will be a versatile person with knowledge in various subjects. He/she will be renowned for his learning and generosity. He/she may shine in real estate deals and agricultural pursuits. He/she will wield political power and will be known as good mediators. They will have powerful friends who will help them in his hour of crisis. They will have a well decorated house & conveyances. Because of his wealth and leadership qualities, he will have followers and juniors who admire them. This is a powerful position for public life. Native Land provides you with the location and wherewithal to work within agriculture and to set up industries or business establishments. These ventures proceed successfully bringing you renown in your native land. You gain several vehicles in your acquisitions as well. Authority comes to you politically (especially if the 9th lord also joins the 10th lord in the 4th thus leading to a raja yoga of a high order). You wield positions of great responsibility such as the head of an organization or governmental position. Respect and fame for your significant learning and demonstrated generosity follow you wherever you go. The bond you have with your mother is one of great attachment and concern for her welfare (remember that the 10th lord is the 7th lord from the 4th making him a maraka for the mother). Under aspects of duress you may enter into a life of servitude. You could be inclined

toward poor decisions regarding properties. This would cause economic suffering and impact your reputation.

5. Karma Lord is in Putra Bhava, he/she will be endowed with all kinds of learning, and he/she will be always delighted, and will be wealthy and endowed with sons. He/she will excel in real estate deals and in speculation. He/she leads a simple life with prayer and meditation. He/she is interested in learning from early childhood and will be blessed with all the comforts of life. He/she will have powerful friends. As the 5th is 8th from the 10th, he will have reverses in profession and may be subjected to vicissitudes. They will have powerful enemies also who will try to block his progress and development. You could possess the rank of ruler or minister in your ascension, or be elected to office (10th lord in the 5th leads to a raja yoga). The arts appeal to you. You earn through music, drama or cinema. Indications support you becoming a famous artist in your own right. Heading an orphanage or a house of remand also remains likelihood. Wealth becomes yours to claim. You are blessed with children, more than likely sons. Highly learned, you maintain a cheerful disposition. Disagreements with your boss or higher ups could cause your concern (5th house being the 8th from the 10th)

6. Karma Lord is in Ari Bhava, he/she will be bereft of paternal bliss. Although he may be skilful, he/she will be bereft of wealth and be troubled by enemies. Since the 10th lord is in the 6th, he will shine in occupations which are connected with the judiciary, hospital or prison. He/she will hold responsible posts. He/she will be known as impartial men and will be held in high esteem. There may be transfers and changes in his environment. He/she will subject to trouble through enemies. As the 6th is 9th to the 10th, he will have professional luck and people will recognise them as professionals. He/she will wield political power and have wealth beyond the dreams of avarice. Lucky breaks come to him automatically. Service oriented and work minded you love your job. Naturally, this opens the way to your success typically through quick promotions. Your work relates to hospitals, prisons and the judiciary. In a judicial career, you rise to a high court, Supreme Court or receive a position of importance in this governmental branch. You possess the ability to be a skilled surgeon or physician.

Respect and honor goes with your work. Under stressful aspects your enemies cause you ongoing contention. Disgrace comes to your career, commonly because of exposure to criminal activity or litigations. Incarceration would be possible, especially if afflicted. Despite your skills, money does not materialize in the way you expect. Your parents would be of little or no help.

7. Karma Lord is in Yuvati Bhava, he/she will be endowed with happiness through wife, be intelligent, virtuous, eloquent, truthful, and religious. Since the 10th lord is in the 7th, this is a powerful position for professional life. His IQ will be above the average & he will be renowned for his communication skills. He/she will be blessed with a partner who becomes a cause for his career development. He/she may travel abroad for business. His managerial abilities are well known as he fills his targets in time. He/she believes in people and in delegation. Hence all ventures initiated by them prove to be successful. As the 7th is 10th from the 10th, their professional fame will surpass all boundaries. He is strong in business skills, social and financial success naturally come his way. Perseverance and highly honed oratorical skills add to his success profile. Work in conveyances, railways and automobile companies provide him with professional profit. HIs patterns indicate an aptitude for becoming a pilot. Partnerships and cooperative ventures augment his economic accomplishments. Travel abroad wins him distinction and recognition. Diplomatic missions could occur within your travels. He is religiously inclined and truthful. The opposite sex is comfortable for him to be around. He attracts a mature mate who assists in his work (10th lord in the 7th indicates a career oriented spouse). Conjugal bliss goes with him and his spouse. If the 9th lord also joins the 10th lord in the 7th the father may be rich and successful, because the 7th house is the 10th from the 10th (10th lord could signify the wealth/riches of the father, being 2nd from the 9th).

8. Karma Lord is in Randhra Bhava, he/she will be devoid of acts, long-lived and intent on blaming others. His/Her career is not good and has great difficulty in establishing a career or has some sudden or serious troubles, which greatly harm the progress in his/her career. Since the 10th lord is in the

8th, changes or breaks in career are to be expected. Anyway he will have a regal status in his profession. He/she may become mystics & choose the path celestial. They will be blessed with good longevity. He/she is noble-minded & high principled and upholds lofty principles. He/she will be well appreciated by his juniors & associates. As the 8th is 11th from the 10th, they will have high gains via profession. Their brothers also ascend the ladder of success. His career path provides you with a source of constant challenge. Caution and attention can preclude some difficulties. However, it appears he will rise up to positions of authority, be suspended, and then later reinstated. He may change careers and even pursue a hobby seriously that it may become a source of income one day (8th being the 11th from the 10th). The workplace consists of talebearers seeking to do him harm. Behind the scenes, there will be meetings and back biting conversations become commonplace. He may be forced to resign from some professional positions. Some of his stressful professional situations muddy his mind. The decisions he makes work against him and debilitates his career. Humiliation can be somewhat avoided by shunning all criminal or dubious activities, especially under malefic aspects. He chooses lowly positions like undertaker, graveyard work or burning ghats. Alternately his career could be connected to leather products, insurance, mining etc. If the 10th lord is a benefic or has benefic influences, he may attain high ranks for short periods in his career path. But the 8th house brings breaks in career, ups and downs and dramatic changes at times as other dusthanas too do. Accepting responsibility and avoiding placing blame on others for his life conditions would be helpful to him. He possesses the ability to become a spiritual teacher or mystic within his indicated long life span, especially with Jupiter's aspect or influence.

9. Karma Lord is in Dharma Bhava, he/she born of royal scion will become a king, whereas he/she being an ordinary person will be equal to a king. This placement will confer wealth and progeny happiness. He/she will become a sage & a mystic and guides to those who walk the path celestial. Fortune will favour him generally and he will be well off. A hereditary profession will be taken up by him such as a teacher, preacher or healer. His father will play a dominant role in his development and he

will prove dutiful and will shine as psychological counsellors. He/she will have a regal status and bearing and will be respected for their talents. He will be a spiritual stalwart and a beacon of light for spiritual seekers, his religious inclinations, spiritual understanding and embarkation upon a "spiritual sadhana" (exercise to attain perfection) light his way. He could work as a government officer, auditor, banker, medical professional. Private consulting services fall within his nature. He has a dutiful child, youand he derives pleasures from his own offspring. Astrological afflictions, if applicable, would bring a lowly professional status and lack of success.

10. Karma Lord is in Karma Bhava, he/she will be skilful in all jobs, be valorous, truthful and devoted to elders. L-10 Simha Surya in bhava-10 seeks high-visibility positions; leadership style is radiant, regal and entitled. Surya is less effective in domains of Shani (10, 11) and he is autocratic, not always successful unless Shani is also well placed. Downside: He is arrogant, not necessarily aware of the little people at the bottom of the pyramid. Upside: He is glorious to behold. He is very selective - even elitist - in regard to professional placements. He/she will shine in their profession as this position is conducive to professional brilliance. He/she can wield immense political power & always have contact with those in the Government. A powerfully posited 10th lord confers professional enhancement & reputation. People will look up to them for guidance. Distinction, respect and honor come your way after a hard, but swift climb upward. He shows respect to elders. Business success occurs based upon his use of skill and courage. His influence extends widely. The workers have espect for him, and his word is held as law. He sets up institutions relating to his field. If afflicted, he lacks self-respect, and has vacillating mind due to the afflicted planets in the 10th influencing the 4th by full aspect. If too many planets aspect or occupy the 10th he might opt for the life of an ascetic renouncing worldly affairs.

11. Karma Lord is in Labh Bhava, he/she will be endowed with wealth, happiness, and sons, virtuous, truthful and always delighted. He/She can earn merit and reputation along with money. He/She will display a happy exterior always and show bonhomie and geniality. This earns him/her good reputation

and goodwill among public. He/She will have fame and reputation. He/she will earn merit & reputation along with money. He/she displays a happy exterior always and this earns him good reputation and goodwill among the public. He/she will be in a position to give employment opportunities to many a people. And this makes him the most sought after individuals with many friends. As the 11th is 2nd from the 10th, profession will fetch them immense largesse. Fame and reputation will be his professional enhancement rewards. Fortune in all respects is indicated. He provides employment to hundreds of persons, receiving honor, success and respect for this and other meritorious deeds. Business enterprises include: trade in sea products, milk and restaurants woolen factories, match industries, gold market or chemicals. Becoming a newspaper magnate or prosperous publisher also fits his disposition. Cheerful, truthful, intellectual and a thinking individual, his relations tend to be cordial with others. If afflicted some friends may be hidden enemies and not hold his best interests at heart. Their efforts could bring him hardship and worry. Even under the most difficult aspects he can achieve moderate success and wealth.

12. Karma Lord is in Vyaya Bhava; he/she will spend through royal abodes, will have fear from enemies and will be worried in spite of being skilful. He/she is likely to reside abroad and will be beset with many problems and obstacles. He/she should be cautious in matters of tax or when dealing with government organisations. Beware of involvement with politicians which will result only in major loss for him. He/she may have a vocation linked to rituals and religion. Income may be from ecclesiastical sources. He/she may have many enmity & problems in profession. They are advised to turn to remedial measures for problem solution. Foreign lands or secluded places figure into his money making efforts. His prosperity comes from distant sources. Job changes dot his professional path (10th lord in any dusthana does this). Despite his skills, some professional adversity must be overcome. A fearless approach works best. He passes through the royal abodes of rulers. Within his indicated job changes he likely loses his job as the cause. Temptations in unscrupulous business affairs taint his reputation. While the 10th lord in the 12th indicates changes and ups and downs in one's career path, it does not

necessarily signify an unsuccessful or bad career life. On a positive note, it could indicate links with foreign countries with respect to career, or even a career related to hospitals, prisons, investigative or research organizations etc. But if the tenth lord is with the 12th lord, it may even ruin one's career in extreme cases.

8.11 Predictions by Labh Lord in Various Bhava

The Eleventh House is the House for the fulfilment of all desires. Labha or Gains is its main signification. The profit one makes in Life is dependent on the strength of the Eleventh House. Only if the eleventh lord is powerful, can one achieve success in any venture. If the eleventh lord is too powerful, anarchic qualities may manifest. The eleventh lord has much to do with gains and achievement of one's desires and it's placement in the various twelve houses tells us a lot about achievement of desires in life and where his/her desires lay. For example, if the 11th lord is in the ninth house, then fortune flows to the achievement of his/he desires. If the eleventh lord is in the eighth house, then great trouble comes in the achievement of his/her desires. In this way one of the twelve houses exerts itself upon the eleventh lord and he controls the things. If eleventh Lord is well placed or un-afflicted, there will be continuous flow of money, happiness and prosperity; in family due to improved financial status; increase in bank balance; assets; good position in society and most sought after. There will be gains of wealth and the possibility of diseases. There will be gain in accumulated wealth, fulfilment of desire, children marriage, gains from profession, elder sibling, love affairs and girl friends, envision of new and different possibilities during Maha Dasha of the planet. If Tenth Lord is afflicted or badly placed or afflicted, there will be problem with brothers; loss and disrespects, ear trouble; chances of being cheated financially and problem of money invested in risky ventures during Maha

Dasha of the planet. Followings are the specific effects of Lagna Lord Position in different house:

The Effects of the Eleventh Lord in 12 Houses

1. Labh Lord is in Thanu Bhava, he/she will be genuine in disposition, be rich, happy, and even-sighted, is a poet, be eloquent in speech and be always endowed with gains. As the 11th lord is in the First, he can have immense wealth. He/she may not have an elder brother. As the ascendant is the 3rd from the 11th, he will have the help of younger co-born in the battle of life. He/she will have a good economic background with fulfilment of all desires by this combination.

2. Labh Lord is in Dhan Bhava, he/she will be endowed with all kinds of wealth and all kinds of accomplishments, charitable, religious and always happy. He/she will have the help & guidance of his/her elder brothers. Harmonious relationship will be established with friends and elder co-born. Businesses partner friends will always bring good profits. Friends & elder co-born help them throughout life. This is a powerful Dhana Yoga, a combination for immense wealth. The 2nd is the 4th from the 11th hence he will be blessed by conveyances & a good house.

3. Labh Lord is in Sahaj Bhava; he/she will be skilful in all jobs, wealthy, endowed with fraternal bliss and may sometimes incur gout pains. His main source of income will be music & poetry. Brothers help him throughout life and he will have lot of friends. He/she has undoubted skills and he is likely to enjoy fraternal bliss. As the 3rd is the 5th from the 11th, he will have immense gains of a high order. If he speculates properly he can earn immense wealth. The 5th house rules investments and speculation. Elder brothers and sisters give not only advice but also financial help.

4. Labh Lord is in Bandhu Bhava; he/she will gain from maternal relatives, will undertake visits to shrines and will possess happiness of house and lands. As the 11th lord is in the 4th, he accumulates via estates, products of the soil and rentals. Mother will be highly cultured and possessing an exemplary charac ter. He/she will be renowned for his academic abilities.

He/she will be blessed by a loving and charming partner. As the 4th is the 6th from the 11th, many a problem will have to be faced regarding business. Rivalry & enmity in the professional sphere can be expected. During the periods of business cycles, possibility of loss has to be countenanced. Real estate deals & other ventures are subject to vicissitudes.

5. Labh Lord is in Putra Bhava, he/she will be happy, educated and virtuous. He will be religious and happy. As the 11th lord is in the 5th & aspects the 11th, he will have immense gains and general prosperity. Children come up well. If he indulges in speculation he will get windfall profits as the 5th rules speculation. His sons & daughters will be dutiful & meritorious. They will climb the ladder of success and fame as Labha lord is in the 5th. Elder co-born-s will be cordially disposed.

6. Labh Lord is in Ari Bhava, he/she will be afflicted by diseases, be cruel, living in foreign places and troubled by enemies. As the 11th lord is in the 6th, money will be gained through litigation & running nursing homes. Happiness will be achieved away from the land of birth. He/she excels in service rather than independent business. As the 6th is 8th from the 11th, some problems should be expected in the area of profession. Rivalry and enmity from rivals and their machinations have to be faced.

7. Labh Lord is in Yuvati Bhava; he/she will always gain through his/her wife's relatives, be liberal, virtuous, and sensuous and will remain at the command of his spouse. His spouse's people will help him/her considerably. Luck favours him in the professional sphere. He/she may go abroad and be on a world tour. His spouse bosses over him.

8. Labh Lord is in Randhra Bhava; he/she will incur reversals in his/her undertakings and will live long, while his/her spouse will predecease him/her. He/she may be blessed with better longevity than his/her spouse. His career will be subject to dire vicissitudes and he should be careful about cheats & swindlers who approach him in confidence. As the 8th is the 10th from the 10th, they will be renowned for his professional work. They will be sought after persons having expertise. Losses of a high magnitude may have to be countenanced.

9. If Labh Lord is in Dharma Bhava, he/she will be fortunate, skilful, and truthful, honoured by the king and be affluent. As

the 11th lord is in the 9th, he will inherit a large paternal fortune paving the way for his/her success. Many houses & conveyances come under his/her control. He/she is basically altruists and will set up charitable institutions. The political powers will honour him. This is a powerful Dhana Yoga & can bestow immense wealth.

10. Labh Lord is in Karma Bhava, he/she will be honoured by the king, be virtuous, attached to his/her religion, intelligent, truthful and will subdue his senses. He/she will excel in business & earn fame and reputation. Elder co-born will help him in. As the 10th is 12th from the 11th, he will have to face losses and undue expenditure connected with business. Rivalry & enmity increases. He/she will have to fight immense adversity in the realm of profession. He/she will have immense educational gains and gets prizes or awards. He/she will have good friends who will always help him.

11. Labh Lord is in Labh, he/she will gain in all undertakings, while his/her learning and happiness will be on the increase day by day. He/She will have a comfortable life with his/her partner, children and riches. It makes a powerful Dharma yoga. Immense gains will accrue to him/her. L-11 Simha Surya in bhava-11 (creative networking) makes dramatic center of a network that has no center. Surya is less effective in domains of Shani (10, 11) but still the Simha-Surya is well-connected socially and earns by cult of personality. He is often the figurehead or political icon of a community but does not hold political office. He/she will live a comfortable life with his/her partner, children & riches. He/she will have powerful elder brothers and friends who help him throughout life. This is a fortunate combination. Labha lord in labhasthana is a powerful Dhana Yoga. Immense gains will accrue to him. They will be above want. Elder brothers and sisters attain high status. They gain as a result of their benevolence.

12. Labh Lord is in Vyaya Bhava; he/she will always depend on good deeds, be sensuous, will have many wives and will befriend barbarians. He/she is likely to lose wealth because of his/ her elder co-born. He/she may have desire to spend on things which are not essential in life. Money may be spent on gratification of the senses. He/she will associate with unknown groups and foreigners. He/she may have to spend much

8.12 Predictions by Vyaya Lord in Various Bhava

The Twelfth House refers to Loss, incarceration, expenditure & Final Emancipation. Hidden enemies are also indicated by the 12th. If malefic tenant the 12th house, necessary expenditure, ill health and sorrow are indicated. If the 12th house is tenanted by benefic, expenditure will be under control and there won't be any incarceration. If twelfth Lord is well placed or un-afflicted, he/she will spend money in good causes; have a pleasurable life, good deeds and respect during Maha Dasha of the planet. If twelfth Lord is afflicted or badly placed or afflicted, then there will be ill health; expenditure on treatment; increase in secret enemies; chances of imprisonment and if 6th lord is also weak, then there will be rise of debts during Maha Dasha of the planet. There will be distress and danger from diseases. There will be loss or expenses, death, bed comforts and couch pleasure, miseries, sufferings, troubles, betrayals, law suits, imprisonments, hospitalisation, conjugal relations with opposite sex other than wife, sorrows, debts and loss of goods. There will be mental agony, disputes, the foreign trips, number of the foreign trips, benefits or loss due to the foreign trips and settlement in the foreign lands. Followings are the specific effects of Lagna Lord Position in different house:

The Effects of the 12th lord in 12 houses

1. Vyaya Lord is in Thanu Bhava; he/she will be a spendthrift, be weak in constitution, will suffer from phlegmatic disorders and be devoid of wealth and learning. He/she will be handsome. He/she may be feeble minded and may have a weak constitution. He/she may have breathing problems may be experienced. He/she will suffer from imaginary ills and unnecessary fear of death. Health problems manifest as 12th lord is not desirable in 1st.

2. Vyaya Lord is in Dhan Bhava; he/she will always spend on inauspicious deeds, be religious, will speak sweetly and will be endowed with virtues and happiness. He/she will be subject

to dire financial problems. Irregular food habits take his toll. Lack of harmony at home will be experienced & his eyesight will generally be poor. As the 2nd is 3rd from the 12th, his expenditure will rise considerably. He/she will helped by his younger co-born. He/she will take to the dolorous divine way.

3. Vyaya Lord is in Sahaj Bhava; he/she will be devoid of fraternal bliss, will hate others and will promote self-nourishment. He/she will be shy and diffident He/she may lose one of the brothers. Much of his money will be spent on younger co-born. Since the 3rd is the 4th from the 12th, he will have the comfort and bliss of spiritual freedom. He/she will have spiritual guidance. His younger co-born may turn hostile against them.

4. Vyaya Lord is in Bandhu Bhava; he/she will be devoid of maternal happiness and will day by day accrue losses with respect to lands, conveyances and houses. He/she will be mentally restless and unnecessary worries are created by his minds. Relatives turn hostile and he may have to live in a faraway place. The landlord may torment him most of the time. The maintenance expenditure rises regarding vehicles & equipment. Loss may have to be countenanced regarding vehicles. Loss of comforts indicated. As the 4th is 5th to the 12th, spiritual progress can be expected

5. Vyaya Lord is in Putra Bhava, he/she will be bereft of sons and learning. He will spend, as well as visit shrines in order to beget a son. He/she will be mentally restless and unnecessary worries are created by his minds. Relatives turn hostile and he may have to live in a faraway place. The landlord may torment him most of the time. The maintenance expenditure rises regarding vehicles & equipment. Loss may have to be countenanced regarding vehicles. Loss of comforts indicated. As the 4th is 5th to the 12th, spiritual progress can be expected

6. Vyaya Lord is in Ari Bhava; he/she will incur enmity with his/her own men, be given to anger, be sinful, miserable and will go to others' wives. The 12th lord is in the 6th is a Vipareeta Raja Yoga which gives wealth, fame and all sorts of comforts. The 12th lord in the 6th is a Vipareetha Raja Yoga which gives wealth, fame and all sorts of comforts. As the 6th happens to be the 7th from the 12th and is an angle from it, he will be lucky with respect to 12th house significations His enemies will be vanquished.

7. Vyaya Lord is in Yuvati Bhava; he/she will incur expenditure on account of his/her spouse, will not enjoy conjugal bliss and will be bereft of learning and strength. A socially inferior person may become his life partner. It will be highly difficult to continue the relationship. Later on he may embrace asceticism as separation can happen at any time. His mind and even his learning will be afflicted. This is due to the fact that 7th is 8th from the 12th and damage is caused not only to 7th house significations but also to the 12th house significations. Expenditure rises spirally and he feels that he is bound to terra firma.
8. Vyaya Lord is in Randhra Bhava; he/she will always gain, will speak affably, will enjoy a medium span of life and be endowed with all good qualities. His mind and even his learning will be above want & famous in his circle. He/she will spend a lot for a life full of luxury and he will have many subordinates and legacy from a prominent person. He/she will be renowned for his righteousness and will be recognised as persons with the gift of the gab. Since the 8th is 9th to the 12th, he will be lucky with regard to the 12th house significations.
9. Vyaya Lord is in Dharma Bhava; he/she will dishonour his elders, be inimical even to his/her friends and be always intent on achieving his/her own ends. The religion, spirituality, dharma and god and misfortune (bad luck) will have a great affect upon the losses in his/her life. It means that losses will come to him/her because of his/her bad luck and his/her faith, spirituality, or acts of god. He/she will find it difficult to maintain good relationship even with his/her own people. It is quite likely that he will live abroad and prosper there. He/she is noble hearted, honest & generous to the core. Family life turns out to be unpleasant as he thinks only about profit and loss.
10. Vyaya Lord is in Karma Bhava; he/she will incur expenditure through royal persons and will enjoy only moderate paternal bliss. Expenditure becomes unbearable as his social contacts are with people is on a higher socio-economic pedestal. His sons may turn hostile against him. As the 10th is 11th to the 12th, he will gain with respect to the 12th house significations. Spiritual progress indicated. There won't be any incarceration. He/she will have the contacts of the high & the mighty and the powers that rule help him when he is

in deep trouble. He/she advised to turn to Bhakthi Yoga & Jnana Yoga for deliverance.

11. Vyaya Lord is in Labh Bhava; he/she will incur losses, be brought up by others and will sometimes gain through others. The 12th lord in the 11th is detrimental to his business, profits & expenditure. Because the 11th is 12th to the 12th, many enemies are cultivated and only a few friends. Losses at a critical time are inevitable. Regarding progeny there may be difficulties and delays. Unexpected losses may result as the Vyaya lord takes an adverse stance. Expenditure spirals as a result and spiritual progress is hard to come by. They may be subject to persecution by enemies.

12. Vyaya Lord is in Vyaya Bhava; he/she will only face heavy expenditure, will not have physical felicity, and be irritable and spiteful. L-12 Simha Surya in bhava-12 provides creative imagination and one lives like charismatic royalty at the center of attention in one's own mind. He is often a bit of a "bedroom diva"and plays the role of foreign royalty quite successfully; is not quite so well admired at home. He is a legend in one's own mind. He is very selective - even elitist - in regard to bedroom, hotel, private arrangements, and foreign assignments. He/she will have to cater to heavy expenditure for which he/she is the cause. He/she will spend lavishly for the pleasures of the flesh as the 12th rules Sayana Sukha or the pleasures of the couch. Spiritual progress can be expected after he become blase. He/she is advised to turn to prayer & meditation in order to overcome the lusts of the body. Bhakti & Jnana YogaA can give him final emancipation.

9

Predictions by Planet (Graha) in Sign

9.1 Special Effects of Planets in Sign during his Dasha Period:

Mars (7 Years Dasha):

Mars in Aries: Acquisition of wealth, increase in reputation within the company, much of gain in the business. **Mars in Taurus:** Destruction of all enemies, success in all struggles, respect from rulers and elders, good production from lands. **Mars in Gemini:** Access to a great wealth, good business related to precious stones and Metals, success in agriculture business, Overall a successful periods. **Mars in Cancer:** Danger from competitors, honour and reputation at stake, disharmony within the company or place of work. **Mars in Leo:** Happy and successful through out the period, develop a taste in fine arts and get success in business related to it. **Mars in Virgo:** Successful pursuits, beneficial results, happiness from official staff and colleagues. **Mars in Libra:** Most Miserable life, misunderstanding with co-workers, seniors, loss of wealth, Mortgaging of landed property. **Mars in Scorpio:** Increase in wealth, gain of profit, coordination with office members, and active participation in the work. **Mars in Sagittarius:** Earnings from political and royal sources, destruction of competitors, realisations of all targets. Mars in Capricorn: Success in all pursuits, gains from quadrupeds and increase in wealth, **Mars in Aquarius:** great hardships, obstructions, misunderstandings at work place with superiors, unpleasant relations within family. **Mars in Pisces:** Much wealth, general success, bumper crops in lands, prosperity and acquisition of house and business, great reputation, respect as a learned man.

Rahu (18 Years Dasha):

Rahu in Aries: Auspicious celebration, acquisitions of land, perfect domestic happiness, respect from the rulers.harmony in all prospect. **Rahu** in Taurus: Pilgrimage to holy places. Increase in philosophical knowledge, entertainments, gains from cattle or dealing with quadrupeds. **Rahu** in Gemini: Multiplication of enemies, fear from police and magistrates, sickness to wife and family, misunderstandings and disfavour. **Rahu** in Cancer: serving the suffering, humanitarian propaganda, enthusiasm for nation's service, gain in wealth, increase in reputation, **Rahu** in Leo: Fear of imprisonment, fraud, political offences, destruction of property, mental worries, and distractions. **Rahu** in Virgo: Gain in wealth, general prosperity, success, acquisition of new land, professional gains like promotion or profitability **Rahu** in Libra:Destruction of property, loss of wealth, loss of position, fear of imprisonment, setup in foreign countries- in exile, many miseries, litigation troubles, physical and mental sufferings **Rahu** in Scorpio: If Rahu is debilitated, causes great havoc to the native, chances of abortion, failure in business enterprises and mental discontentment. **Rahu** in Sagittarius: Acquisition of wealth, sudden and unexpected gain, great success in undertakings, mental unrest, **Rahu** in Capricorn: Sudden promotions and success in business, increase of prosperity and wealth, new launching in the company.

Rahu in Aquarius: Orphanages, disappointments, mental derangement, sorrowful news, premature death. **Rahu** in Pisces: fear from thieves, progress in education, mental uneasiness, mixed results depending on the safety, gains in business.

Jupiter: (16 Years Dasha):

Jupiter in Aries: General prosperity, Acquisition of landed property, promotion if in job, increase in business, appointment as a trustee, respect from own office or subordinates.

Jupiter in Taurus: acquisition of wealth, establishment of new business, fresh treaties and pacts. Gains in trade and agriculture, increase in land and cattle, respect from friend and foes. **Jupiter** in Gemini: chances of ill health of wife, loss of children, deportation, fear from rulers, uneasiness. **Jupiter** in Cancer: Jupiter is exalted, all beneficial results, political success, promotions, exalted positions, social work, and acquisition of wealth from unexpected sources. **Jupiter** in Leo: Respect from rulers, gain of lands, realization of one's own desires, ambitions. **Jupiter** in Virgo: Residence in foreign countries, danger to life partner, loss of wealth, misunderstandings among relatives, body ill health, symptoms of nervous troubles and diseases. **Jupiter** in Libra: Residence in desolate place, active part in social reforms, unbearable difficulties, tendency to create fractions in scompany and societies. **Jupiter** in Scorpio: Mental calm amd happiness, acquisition of land, undisturbed progress and religious worship, philosophical discourses. **Jupiter** in Sagittarius: Much of wealth, success in every undertaking, increase in relations and family, fame and perfect happiness. **Jupiter** in Capricorn: ill health of life partner and children, loss of property, reputation adn self respect, irreligious act, predominance of mean motives, liable to prosecution, upspringing of numerous enemies. **Jupiter** in Aquarius: Consumption and fear from animals, bad luck, many hardships, loss of wealth. **Jupiter** in Pisces: Much fame throughout this long period, high education and gain from exhibition of knowledge, acquisition of property, general success.

Ketu (7 years Dasha):

Ketu in Aries: Disgrace, troubles from relatives, loss of property, constant travels, sudden quarrels. **Ketu in Taurus:** Many dangers, acquisition of wealth through evil ways, sorrows, visit to sacred shrines, inclination to religious study. **Ketu in Gemini:** good health, happiness, elevation to responsible and exalted positions, founding of sacred institutions, danger in near relations, loss of self-respect **Ketu in Cancer:** Many dangers, quarrels with colleagues and mates, period of exile, serious sickness to family members. Ketu in Leo: increase of lands, progress in education, debates and discussions on philosophy, travel in mountain tracts and hilly regions. Ketu in Virgo: fear from fires, weapons and poisonous gas, self respect, unrighteous act, venereal complaint. Ketu in Libra: Happiness and prosperity, gains from gold and precious stones, heart disease, nervous breakdown. Ketu in Scorpio: gain from lands, new contract, acquisition of wealth, sycophant acts, suffe from venereal complaints. Ketu in Sagittarius: Perfect domestic harmony, favour from aristocrats, religious discussions, moral elevations. Ketu in Capricorn: Misappropriation of private and governmental funds, anxiety about everything, catching consumption disease.

Ketu in Aquarius: Hating all and hated by all, incurring the displeasure of superiors, happiness and prosperity. Ketu in Pisces: Learned discussions, general success, helping the sufferings, mental depression, promotion, acquisition of fertile lands, travel to different countries through sea routes,

Venus (20 Years Dasha):

Venus in Aries: acquaintances with noble personages, wearing good clothes, spread of fame, much respect, birth of child, praised by friends and enemies. **Venus in Taurus:** A life of great ease, 20 years of enjoyment, company of beautiful friends, travel to foreign countries, and ambitions of life fairly realised**. Venus in Gemini:** Gain of territory, increase of reputation, mental peace, harmony with co workers and clients. Venus in Cancer: Paroxysms of grief and terror, destruction of property, discontent with relations, loss of wealth, and some mental ease at the end. Venus in Leo: mental disease, sorrow, troubles from fires, sickness of life partner or child, aimless wandering, inclination to conceal true identity. Venus in Virgo: quarrels with near and dear ones, mortgage of property, troubles in generative organs, symptoms of arthritis, nervous breakdown. Venus in Libra: great reputation, gain of wealth, comforts, good enjoyment of life, luxurious invitations, period of extreme sensuality. Venus in Scorpio: Hatred for all, many miseries and troubles, serious risks, loss of wealth, mental affliction and imbalance, Venus in Sagittarius, Much name and fame, religious and philosophical studies, regard for holy people, success in all undertakings and general prosperity. Venus in Capricorn: Travel to foreign lands, freedom from domestic worries, progress in general education, perverted ideas, Venus in Aquarius: Acquisition of fresh prosperity, a good library, lands and houses, honours and unexpected gains, domestic happiness and harmony, Venus in Pisces: High political power, promotion to responsible office, acts of charity and generosity, perfect happiness at work place. There are all general principles and shoulod not be applied verbatim. No Dasa or period can prove exclusively harmful or exclusively beneficial as it go through various sub periods of different planets.

9.2 Predictions by Planets in Signs:

Predictions by Planets in Aries:

Sun in Aries: If Sun has strength, he/she will have children, gains through speculation and is fond of war, fierce, good in deeds, roaming, splendour, strong, a quick learner, enthusiastic, curious, and may become a leader. However, if Sun is weak in strength or afflicted by malefic planets, it indicates problem from father, husband, children, obstructions in educational accomplishments, losses in speculation and suffering from heart trouble, frequent stomach disorders, weak eyesight and delay in having children and ordinary person.
Sun in Aries and in aspect to others Planets: If the Sun is aspect by the Moon, he/she will have many servants. If the Sun is aspect by Mars, he/she will display his/her courage in battle, cruel, and strong. If the Sun is aspect by Mercury, he/she will be a servant, will do others' jobs, will not have much wealth, be devoid of power, be subjected too much grief and will possess a dirty body. If Jupiter aspects the Sun, he/she will have plenty of money, will donate, and be a leader or minister, a judge and a supreme person. If Venus aspects the Sun,

he/she will be the husband of a bad woman, will have many enemies, few not well-placed relatives, and will suffer from leprosy. In case Saturn aspects the Sun, he/she will be subjected to grief on account of physical ailments, will have intense passion in his/her undertaking, be dull-witted and a dunce.

Moon in Aries: If Moon is well placed and strong, he/she will have more sons, abundant properties, assets, complete guidance and support from mother, and is excellence in education, domestic peace, devoid of co-born and will be won over by females. He/She makes quick decisions. However if Moon is weak / afflicted by malefic, then there may be delay in marriage, disturbances in education, problem to mother, dispute in assets, respiratory problems, back pain, irregular menstruation, ovarian disorders, excess of phlegm, and disturbed sleep.

Moon in Aries and in aspect to others: If the Moon is in aspect to the Sun, he/she will be honoured by the king, fond of war. If the Moon is aspect by Mars, he/she will suffer from diseases of eyes and teeth, windy diseases and urinary disorders. If Mercury aspect the Moon, he/she will teach various disciplines, will possess good speech, will achieve his desires, be a great poet and be widely famous. If Jupiter aspects the Moon, he/she will be endowed with servants and abundant riches and be a king, or a kings minister. Venus aspecting the Moon denotes that he/she will be lucky, be endowed with sons and wealth, will marry a supreme lady and costly ornaments and will not eat much. The Moon aspect by Saturn indicates that he/she will be jealous, miserable to a great extent, be very poor, dirty and untruthful.

Mars in Aries: If Mars is strong and well placed, it blesses him/her with longevity and authority, interest in sports, good health and recognition in the society; splendour, truthful, valorous, kingdom, and be fond of war, adventurous acts, and be an Army chief, or head of a village, delighted, charitable, endowed with a number of cows, goats and will join many women. However, if it is weak in strength or afflicted by malefic planets then it gives, short temper, indulgence in litigation, bad reputation, obstructions in physical and mental growth, short or medium life span, accidents and health problems like immune deficiency, erratic blood circulation, weak digestion.

Mars in Aries in Aspect to Others: If Mars is aspect by the Sun, he/she will possess wealth, wife and children, be a king's minister, be a justice, be famous and be a charitable king. If Mars is aspect by the Moon, he/she will be bereft of mother, will have a wounded body, will hate his own people, will not have friends, be jealous and will have female children. If Mars is aspect by Mercury, he/she will be an expert in stealing others' money, is a liar, be hostile and will frequently visit prostitutes. If Mars is aspect by Jupiter, he/she will be learned, sweet-spoken, fortunate, and dear to parents, be very affluent and will be a king par excellence. If Mars is aspect by Venus, he/she will be imprisoned due to females and will be deprived of his money on account of females, more than once. If Mars is aspect by Saturn, he/she will be capable of hindering thieves in spite of his being not valorous, will be devoid of his/her men and will maintain another woman.

Mercury in Aries: Mercury, when strong and well placed, blesses him/her with excellent analytical and communicative capabilities, confidence and good health, fond of battles, and be uncompromising, very learned, interested in music and dance, untruthful, attached to sexual pleasures, a writer, will produce fictitious things, eat much, lose hard-earned money, incur debts and imprisonment frequently and be sometimes fickle minded, decisive ways of thinking. If Mercury is weak in strength or afflicted by malefic planets, it causes financial setbacks, lack of analytical and communicative capabilities, intestinal or skin problems, respiratory problems, cervical pain, nervousness, weak resistance power along with chances of loss in case of disputes, litigation or competitive activities.

Mercury in Aries in Aspect to Others: If Mercury is aspect by the Sun, he/she will be truthful, be very happy, be honoured by the king and be patiently disposed. If Mercury is aspect by the Moon, he/she will steal the hearts of the fair sex, will serve others, be dirty and be bereft of virtues. If Mercury is aspect by Mars, he/she will be a liar, a sweet speaker, will promote quarrels, be learned, affluent, dear to king and valorous. If Mercury is aspect by Jupiter, he/she will be happy, will possess a glossy and hairy physique, will have attractive hair, will be very rich, will command others and be sinful. If Mercury is aspect by Venus, he/she will be in royal service, be fortunate, principal among men or in his town will speak skilfully, be trustworthy and will be endowed with a wife. If Mercury is aspect by Saturn, he/she will experience miseries, be fierce, be intent upon doing cruel activities and be devoid of his own men.

Jupiter in Aries: If it is strong by strength it blesses him/her with adequate paternal bliss, favourable fortune, honesty and spirituality, divine grace and possesses innovative nature, be argumentative in disposition, will acquire precious stones and ornaments out of his/her efforts, will be endowed with strength, sons and wealth, will have eminent and famous profession, be splendour, will have many enemies, much expenses and leadership ability. The weak strength of Jupiter or affliction by malefic planets, causes indecisiveness, shrewdness, immorality, and struggles in life, problems to father and lack of fortune. Medically he/she can suffer from liver disorder, jaundice and dark circles under eyes.

Jupiter in Aries in Aspect to Others: If Jupiter is aspect by the Sun, he/she will be charitable, will be truthful, will have famous sons, be very fortunate and will have abundant hair on the body. If Jupiter is aspect by the Moon, he/she will be a historical and poetical writer, will be endowed with many precious stones, be dear to women, be a king and be highly learned. If Jupiter is aspect by Mars, he/she is of royal scion, will be valorous, fierce, endowed with knowledge of politics, be modest, affluent and will have a disobedient wife and disobedient servants. If Jupiter is aspect by Mercury, he/she will be a liar, be crafty, sinful, will be skilful in detecting other's defects, will serve others, be grateful, be modest and be not outspoken. If Jupiter is aspect by Venus, he/she will be very happy in respect of residences, sleeping comforts, robes, scents, garlands, ornaments and spouse and be very timid. If Jupiter is aspect by Saturn, he/she will be dirty, miserly, sharp,

Venus in Aries: If it is strong by strength it blesses him/her with supportive and pleasurable spouse, knowledge of foreign languages, recognition through partners and interest in media, art and sculpture, be night-blind. He/She will have many blemishes, be inimical, will join other housewives, move in forests and hills, will be imprisoned on account of women, an Army chief, or chief of men, eminent, cheerful and positive, full of enthusiasm, fun and exciting,. However its weakness or affliction by malefic planets, causes difference of opinion with partner and spouse, weak health and difficulties in enjoying long-lasting partnerships, lack of awareness. Medically he/she might experience weakness of reproductive organs, diabetes, problem in kidneys, renal diseases, problems in eyes and skin.

Venus in Aries in Aspect to Others: If Venus is in aspect to the Sun, he/she will be miserable on account of women and will lose wealth and happiness on account of them, will be a king and be learned. If Venus is aspect by the Moon, he/she will be imprisoned, very fickle-minded, be libidinous, will marry a base lady and will be bereft of children. If Venus is aspect by Mars, he/she will be devoid of wealth, honour and happiness, will do others' jobs and will perform dirty jobs. If Venus is aspect by Mercury, he/she will be foolish, profligate, unworthy, be not in good terms with his own relatives, and be immodest, thievish, mean and cruel. If Venus is aspect by Jupiter, he/she will be endowed with beautiful eyes, will have a charitable wife, will possess a beautiful and long body and will have many sons. If Venus is aspect by Saturn, he/she will be very dirty, lazy and wandering-natured, will serve others and be a thief.

Saturn in Aries: He/She will be idiotic, wanderer, insincere, peevish, resentful, cruel, fraudulent, immoral, boastful, quarrelsome, gloomy, mischievous, perverse, misunderstanding nature. Saturn, if strong represents earning income of a high order, support and good relationship with friends and elder brother, fulfilment of desires, longevity, be miserable due to his/her vices and hard labour, be deceitful, will hate his relatives, be blameworthy, garrulous, reprobated, poor, bad in appearance, ill-tempered, inimical to his people, jealous and sinful. However its weak strength or affliction by malefic planets might cause dissatisfaction from income, difficulties in fulfilling desire, and losses through employees, differences with elder siblings and friends. In addition to this its weak strength causes arthritis, painful diseases especially in the lower portion of legs, glandular problems.

Saturn in Aries in Aspect to Others: If Saturn is aspect by the Sun, he/she will be interested in agriculture, be very affluent, be endowed with cows, buffaloes and horses, and be fortunate and industrious. If Saturn is aspect by the Moon, he/she will be fickle minded, base, will join mean and ugly women and be devoid of happiness and wealth. If Saturn is aspect by Mars, he/she will kill animals, be base, be a leader of robbers, be famous and be fond of (joining other) women, meat and wine. If Saturn is aspect by Mercury, he/she will be untruthful, not virtuous, will eat much, be a famous thief and be devoid of happiness and riches. If Saturn is aspect by Jupiter, he/she will be endowed with happiness, wealth and fortune, be a king's minister and be chief. If Saturn is aspect by Venus, he/she will be quite unsteady in disposition, very ugly, will join other women and courtesans and be bereft of pleasures.

Predictions by Planets in Taurus:

Sun in Taurus: Sun, if strong, blesses him/her with huge assets, resources, excellent education and complete guidance and support from parents and provides name, fame, recognition in profession and society, will be beautiful, fortunate, nice, handsome, will possess decorum, be wise, will hate barren or confined women, will have knowledge of singing, playing musical instruments and dancing and will face risk from water, be a hard worker. If it weak in strength/afflicted by malefic planets, then it indicates problem to/ from parents, husband, children, obstructions in acquiring education, disputes in properties, assets and domestic peace, delay in marriage, loss of respect, backache, erratic circulation of blood, cardiac problems, weak eyesight and frequent stomach disorders, will have troubles from disease of face and eyes, and will endure difficulties, will have an emaciated body, will not have many sons.

Sun in Taurus and in aspect to others: If Sun in Taurus is aspect by the Moon, he/she will be addicted to prostitutes, will be soft spoken, will have many women, as dependent and will derive livelihood through water. If Mars is aspecting the Sun, he/she will be brave, fond of battle, bright in appearance, will earn wealth and fame out of his valour and will be deformed. Should Mercury aspects the Sun, he/she will be skilful in drawing, writing, poetry, authorship, singing etc. and will possess a good physique. If Jupiter aspect the Sun, he/she will have many foes and friends, be a king's minister, will have beautiful eyes, be splendour and will be a pleased ruler. When Venus aspects the Sun, he/she will be a king or a king's minister, be endowed with wife, wealth and pleasure galore, be wise and timid. If Saturn is aspecting the Sun, he/she will be mean, indolent, will cohabit with aged women, will be wicked and will be troubled by diseases.

Moon in Taurus: When Moon is strong in strength, it gives good physical and mental growth, sound sleep, initiate, learning, writing, communication, new ventures, analytical capabilities, interest in financial matters and cooking, be large-hearted, highly charitable, short-haired, libidinous, famous, brilliant, will have more daughters, will possess eyes resembling that of a bull. However, if it is weak in strength or afflicted by malefic planets, then it causes argumentative behaviour, indulgence in litigation, obstruction in physical and mental growth, lack of confidence in writing or learning, losses in business/profession, problem to mother and younger brother and health problems like respiratory problem, excess of phlegm, chest congestion, gynaecological problems in females and disturbed sleep.

Moon in Taurus and in aspect to others: If the Moon is aspect by Sun; he/she will be a farmer, be very industrious, be very rich with servants and quadrupeds and will lend money on usury. If Mars aspects the Moon, he/she will be highly libidinous, will lose his wife and friends on account of another lady, will steal the heart of the fair sex and will prove adverse for the mother. Should Moon is in aspect by Mercury, he/she will be highly learned, will know the code of speech, be of pleasing disposition, be dear to everyone and will incomparably be of good qualities? If Jupiter aspects the Moon, he/she will have long living wife and children and lasting wealth, be respectfully disposed to his parents, be virtuous and very famous. Should Venus aspect the Moon, he/she will be endowed with ornaments, conveyances and houses and will possess comforts of sleeping and sitting, scents, robes, garlands etc. If Saturn aspects the Moon, he/she will be devoid of wealth, be inauspicious for mother and wife and will be endowed with sons, friends and relatives. Moon in First half of Taurus: Should the Moon be in the first half of Taurus, his/her mother is not long-lived. Similarly his/her father is short-lived, if the Moon occupies the second half of Taurus.

Mars in Taurus: Mars, if strongly placed, blesses him/her with courage, happy marital life, comfortable foreign stays, happiness on account of siblings and excellent executive qualities. He/She will break the vows of chaste women, will eat voraciously, will have little wealth and few sons, be jealous, will maintain many people, will not trust others, play violently, peak very harshly, and be fond of music, inimical to relatives.

However if it is weak in strength/afflicted by malefic planets, he/she may experience difficulties in enjoying happy marital life, troubles in foreign stays, losses, addiction, occasional health problems, hospitalisation, coupled with impatient and temperamental approach.

Mars in Taurus in Aspect to Others: If Mars is aspect by the Sun, he/she will seek to move in forests and hills, will hate women, will have many enemies, will have fierce appearance and be courageous. If Mars is aspect of the Moon, he/she will not honour his/her mother, be wicked, will lord many women and be dear to them and will fear war. If Mars is aspect of Mercury, he/she will promote quarrels, will speak much, will have a soft body, will possess (indifferent or) few sons and little wealth and will be learned in Shashtra. If Mars is aspect of Jupiter, he/she will be skilful in music and in play of musical instruments, be fortunate, be dear to his relatives and will be pure. If Mars is aspect of Venus, he/she will be a king's minister, be liked by the king, be an Army chief, will have famous name (i.e. titles etc.) and will be happy. If Mars is aspect of Saturn, he/she will be happy, famous, and wealthy, be endowed with friends and own men, be learned and will be head of a group of villages/towns, or group of men.

Mercury in Taurus: Mercury in Taurus indicates that he/she will be high position, well built, clever, logical, mental harmony, many children, liberal, persevering, wealthy, practicable, friends among women of eminence, inclination to sensual pleasures, well read, showy. Mercury, when strong, blesses with excellent communication skills, inclination towards higher studies, writing abilities, sharp intellect, adequate knowledge, happiness on account of children, gains through speculation, be skilful, eminently liberal, be famous, will have knowledge of Vedas and Shashtra, be fond of exercises, robes, ornaments and garlands, be firm in disposition, will have sincerely earned wealth, will possess a chaste wife, will be a soft and sweet speaker and be after sexual satisfaction. However, if Mercury is weak in strength/afflicted by malefic planets, it causes delay in childbirth, disturbances in academic and professional studies, losses through speculation, and weak assessment and analytical powers and emotional setbacks in life from children and might cause intestinal problems, respiratory troubles, cervical pain, neurological disorders, and stomach disorders.

Mercury in Taurus in Aspect to Others: If Mercury is aspect by the Sun, he/she will suffer from penury and acute grief, will have a sick physique, be interested in serving others and will be censured. If Mercury is aspect by the Moon, he/she will be trustworthy, affluent, and firmly pious, devoid of sickness, will have lasting family ties, be famous and be a king's minister. If Mercury is aspect by Mars, he/she will be troubled by diseases and enemies, be distressed, will incur royal insult and will be deprived of all worldly objects. If Mercury is aspect by Jupiter, he/she will be highly learned, will fulfil his promise, will be leader of a country/city/group of men and be famous. If Mercury is aspect by Venus, he/she will be fortunate, soft in disposition, be happy, will enjoy good robes make up etc. and will steal the hearts of the fair sex. If Saturn throws his aspect on Mercury, he/she will be devoid of happiness, be dirty, will experience many diseases and evils, will be subjected to grief on account of his/her relatives and be distressed.

Jupiter in Taurus:

Jupiter in Taurus indicates that he/she will be stately, elegant, self-importance, liberal, dutiful sons, just, sympathetic, well read, creative ability, despotic, healthy, happy marriage, liked by all, inclination to self gratification. Jupiter, if strong in strength, blesses with long lifespan, blissful and long lasting marital tie, adequate inheritance, unearned wealth, unexpected gains, inclination towards spirituality and trouble free life, be endowed with a broad body, be corpulent, will honour Brahmins and Gods, be splendour, fortunate, attached to his wife, be endowed with good appearance, profession, cows and abundant wealth. However, it's weak strength/afflicted by malefic planets, causes difficulties in enjoying marital life, lack or loss of inheritance, short or medium life span, humiliation, indecisiveness, shrewdness, immoral activities, selfishness and he/she may suffer from liver disorder, jaundice, flatulence, problem in genitals, piles and dark circles under eyes.

Jupiter in Taurus in Aspect to Others: If Jupiter is aspect by the Sun, he/she will be endowed with attendants and quadrupeds and will wander verily, will have a long body, be learned and be a king's minister. If Jupiter is aspect by the Moon, he/she will be abundantly rich, be very calm, Sweet, be dear to mother and wife and will enjoy much pleasures. If Jupiter is aspect by Mars, he/she will be dear to the fair sex, be learned, courageous, affluent, and happy and be of royal

Scion. If Jupiter is aspect by Mercury, he/she will be learned, Skilful, sweet, fortunate, endowed with riches, be highly virtuous and be splendour. If Jupiter is aspect by Venus, he/she will be very attractive, affluent, will wear excellent ornaments, be merciful and will enjoy excellent sleeping comforts and excellent robes. If Jupiter is aspect by Saturn, he/she will be a scholar, will be endowed with abundant wealth and corns, be excellent among the people of his village/town, be dirty, ugly and be devoid of wife.

Venus in Taurus: Venus in Taurus indicates that he/she will be well built, handsome, pleasing countenance, independent, sensual, love of nature, fond of pleasure, elegant, taste in dancing and music, voluptuous. Venus, when well-placed and strong, blesses him with financial stability, good health, enjoyment of luxuries and material comforts, happy marital life, success in competitions, knowledge about law, global awareness, knowledge of foreign languages and interest in media, art and sculpture, be endowed with many wives and gems, be an agriculturist, will possess scents. However, if it is weak in strength/afflicted by malefic planets, it causes difference of opinion with spouse, financial instability, and lack of awareness, weak health, skin diseases, renal problems, infertility and ovarian problems in women, weak resistance power and losses in case of disputes or litigation.

Venus in Taurus in Aspect to Others: If Venus is aspect by the Sun; he/she will acquire an excellent wife and abundant wealth, will become a great man and will be subdued by women. If Venus is aspect by the Moon, he/she is born of an excellent mother, will be endowed with happiness, wealth, respect and sons, will have excellence and be splendour. If Venus is aspect by Mars, he/she will marry a bad female, will lose home and wealth on account of females and be sensuous. If Venus is aspect by Mercury, he/she will be splendour, sweet, fortunate, happy, bold, wise and virtuous and will possess distinguished strength. If Venus is aspect by Jupiter, one will be endowed with wife, sons, abodes, conveyances, riches etc. and will achieve desired objects. If Venus is aspect by Saturn, he/she will have little happiness, little wealth, will be reprobate, will marry a mean lady and will suffer from diseases.

Saturn in Taurus: Saturn in Taurus indicates that he/she will be dark complexion, deceitful, successful, powerful, unorthodox, and clever, likes solitude, voracious eater,

persuasive cool, contagious diseases, many wives, self-restraint, worried nature. If Saturn is strong, it makes him/her workaholic in analytical technical fields, gives professional opportunities from time to time in contractual type of work and excellent professional skills. He/She will be bereft of wealth, be a servant, will speak undesirable words, be untruthful, will win the hearts of old women, will have bad friends, will be addicted to women, will serve other women, be not outspoken, be a fool, very thrifty, lack in spontaneity and work hard for success. However its weak strength/afflicted by malefic planets, might cause short or medium life span, ups and downs in professional career, losses through servants, labour problems etc. In addition to this he/she might experiences painful diseases especially in joints, glandular problems, flatulence.
Saturn in Taurus in Aspect to Others: If Saturn is aspect by the Sun, he/she will be clear in speech, will lose wealth, be a scholar, will eat in other's houses and be weak in constitution. If Saturn is aspect by the Moon, he/she will gain wealth through women, will be honoured by ministers of kings, be dear to women and be endowed with family. If Saturn is aspect by Mars, he/she will be skilful in war preparations, but will be away from war, will speak much and be endowed with wealth and family. If Saturn is aspect by Mercury, he/she will always be jocularly disposed, be equal to a neuter, will serve females and be base. If Saturn is aspect by Jupiter, he/she will share happiness and misery of others will do others' jobs; be dear to people, charitable and industrious. If Saturn is aspect by Venus, he/she will enjoy and be happy on account of women and wine, be endowed with gems, be very strong and be dear to the king.

Predictions by Planets in Gemini

Gemini people are fortunate one as they have 7 benefic planets -Sun, Moon, Saturn, Venus, Jupiter, Mercury, and Mars.
Sun in Gemini: Sun in Gemini indicates that he/she will be learned, astronomer, scholarly, grammarian, and polite, wealthy, critical, assimilative, good conversationalist, shy, reserved, lacking in originality. Sun, if strong, blesses him with efforts of a high order, good health, vitality, excellent

managerial abilities, recognition through creativeness and support from younger siblings, be a scholar, sweet in speech, affectionate particularly to one's offspring, will have good conduct, be exceedingly affluent, liberal, skilful, will have two mothers, be fortunate and modest, the jack-of-all-trades, be the master of none. However, if Sun is weak in strength/afflicted by malefic planets in their chart, there might be retarded mental and physical growth, obstructions in acquiring higher education, lack or loss of efforts, problem from younger siblings, difficulties in enjoying adequate paternal bliss, loss of respect, difficulties in enjoying favours from superiors, erratic circulation of blood, weak digestive power, weakness of bones, baldness, low immunity, heart problem or weak eyesight.

Sun in Gemini and in aspect to others: If the Sun is aspect by the Moon, he/she will be put to troubles by enemies and relatives, will be distressed by visits to foreign countries, and will in general be wailing. If the Sun is aspect by Mars, he/she will have fear from enemies, will be encountered by quarrels, will be grieved on account of loss in a war, be poor and bashful. If Mercury aspects the Sun, he/she will have a history akin to that of a king, will be famous, be endowed with relatives, free from enemies, but will encounter eye diseases. If Jupiter is aspecting the Sun, he/she will have knowledge of many Shashtra, be a king's messenger or representative, will go to foreign countries, be fierce and be always bewildered. If Venus aspects the Sun, he/she will be endowed with money, wife and sons, will make less friends, be free from sickness, be happy and fickle minded. If Saturn aspects the Sun he/she will have many servants, be anxious, will maintain many relatives, will remain delighted and will be crafty.

Moon in Gemini: Moon in Gemini indicates that he/she will be well read, creative, and fond of women, learned in scriptures, able, persuasive, curly hair, and powerful speaker, clever, witty, dexterous, and fond of music, elevated nose, thought reader, subtle, long life. If Moon is strong, it blesses them with satisfactory financial status, male children, analytical power, and interest in financial matters and recognition in society, have prominent nose and dark eyes, will be skilful in love, poetry, will enjoy sexual pleasures, be very intelligent, splendour, be endowed with happiness, jocular disposition and eloquent speech, be won over by the females, will have a long body, will be friendly neuters and will have two mothers, ability to learn rapidly, quite articulate and witty. If it is weak in strength/afflicted by malefic planets, it causes losses in business or profession, loss of respect, problems in family life, problem to or from mother or wife along with occasional health problems like excess of phlegm, chest congestion, disturbed sleep, stammered speech, gynaecological problems coupled with mental stress.

Moon in Gemini and in aspect to others: If the Moon is aspect by the Sun, he/she will be quite learned, be splendour, very beautiful, be charitable, be very miserable and be not rich. If Mars lends aspect to the Moon, he/she will be very valorous, very learned, be endowed with happiness, conveyances, wealth and beauty. If Moon is aspect by Mercury, he/she will be skilful in producing money, always successful and inviolable king. If Jupiter aspects the Moon, he/she will be a teacher of Shashtra, will be famous, truthful, very beautiful, and honourable and be an eloquent speaker. If Venus aspects the Moon, he/she will be endowed with the company of supreme females, garlands, robes, conveyances, ornaments and jewels and will be sportive. If Saturn aspects the Moon, he/she will be devoid of relatives, wife, happiness and wealth and will be inimical to the public.

Mars in Gemini: Mars in Gemini indicates that he/she will be loving family and children, taste in refinement, scientific, middle stature, well built, earned, ambitious, quick, rash, ingenious, skilled in music fearless, tactless, peevish, unhappy, subservient, diplomat, humiliating, detective. Mars, if strong, blesses with courage, confidence, and gains, help and support from siblings, friends, fulfilment of desire and satisfactory income, be splendour, be capable of enduring miseries, be very learned, be well versed with poetical rules, skilful in various kinds of fine arts, fond of going to foreign countries, highly intelligent, favourably disposed to sons and friends. However a weak Mars/afflicted by malefic planets, causes impatient nature, non- flexible approach, temperamental nature, difference of opinion with siblings, difficulties in maintaining long lasting friendships, obstructions in gains, instability in income. Medically they might suffer from pimples, piles, blood infections, muscle cramps, problem in ankles & feet.

Mars in Gemini in Aspect to Others: If Mars be in aspect to the Sun, he/she will be blessed with learning, wealth and courage, be fond of hills, forests and fortresses and be highly strong. If Mars is aspect by the Moon, he/she will be happy, wealthy, and splendour, will guard women's apartments, will be endowed with women and will manage the king's residence. If Mars is aspect by Mercury, he/she will be skilful in the art of writing, in mathematics and in poetry, be garrulous, be liar, be a sweet speaker, be a messenger (or an ambassador) and will endure lot of misery. If Mars is aspect by Jupiter, he/she will be a king's representative, be bright, will go to foreign countries, as an ambassador, be skilful in all doings and be a leader. If Mars is aspect by Venus, he/she will do the jobs of females, be very fortunate and will enjoy food and robes. If Mars is aspect by Saturn, he/she will be interested in wandering in mines (i.e. places beneath surface), hills and forests, will have husbandry, as his livelihood, be highly miserable, be very valorous, dirty and be devoid of wealth.

Mercury in Gemini:

Mercury in Gemini indicates that he/she will be inclination to physical labour, boastful, sweet speech, tall, active, cultured, tactful, dexterous to mothers, indolent, inventive, taste in literature, arts and sciences, winning manners, liable to throat and bronchial troubles, musician, mirthful, studious. Mercury blesses him with excellent communication skills, confidence, sound educational background, adequate parental bliss, excellent analytical, assessment and decisive powers, enjoying domestic bliss and good properties and assets, auspicious appearance, will speak sweetly, be very affluent, be an able speaker, be honourable, will have two wives, be fond of arguments and will have many sons and friends. However, if Mercury is weak in strength/afflicted by malefic planets, it causes difficulties in enjoying adequate parental bliss, disturbance in acquiring education, lack of confidence, difficulties in forming properties and assets, problem to mother and indecisiveness coupled with nervousness and confusions and causes intestinal problems, respiratory, cervical pain, back pain, neurological disorders, and skin disease.

Mercury in Gemini in Aspect to Others: If Mercury is aspect by the Sun, he/she will speak truth, be fortunate, dear to king, be a Lord himself, be polite in his activities and be liked by all. If Mercury is aspect of the Moon, he/she will be sweet in disposition, garrulous, will promote quarrels, be interested in acquiring great Saint Knowledge, firm and will succeed in all his undertakings. If Mercury is aspect of Mars, he/she will have an injured body, be dirty, be a genius, will serve the king and be dear to him. If Mercury is aspect of Jupiter, he/she will be a king's minister, be excellent, be beautiful, charitable, and rich, be endowed with his own men and be courageous. If Mercury is aspect of Venus, he/she will be highly learned be a royal employee, be a messenger, will honour friendship and will be interested in base women. If Mercury is aspect of Saturn, he/she will be progressive-minded, be modest, will achieve success in undertakings started by him and will be wealthy with money and clothes.

Jupiter in Gemini: Jupiter in Gemini indicates that he/she will be oratorical ability, tall, well-built, benevolent, pure-hearted, scholarly, sagacious, diplomatic, linguist or poet, elegant, incentive. Jupiter, if strong, is favourable for acquiring knowledge, enjoying happy marital life, long-lasting partnerships, and links with foreign land, gains and interest in religious activities, be affluent, scholarly, proficient, will possess attractive eyes, be eloquent, courteous, skilful, virtuous, will honour elders and relatives, be a good poet, have ability to communicate openly, be focus on accumulating wealth. However, it's weak strength/afflicted by malefic planets, causes difficulties in enjoying marital life, loss in partnerships, and delay in marriage, dissatisfaction and fatigue in travelling. Medically they may suffer from liver disorder, flatulence, diabetes, anaemia, jaundice and dark circles under eyes.

Jupiter in Gemini in Aspect to Others: If Jupiter is aspect by the Sun, he/she will be great, important in his village be a householder and will be endowed with wife, sons and money. If Jupiter is aspect by the Moon, he/she will be affluent, dear to mother, fortunate, happy, will have wife and sons and be incomparable. If Jupiter is aspect by Mars, he/she will be easily successful at all times, be ugly, rich and be amiable to all. If Jupiter is aspect by Mercury, he/she will be a skilful astrologer, will have many children and wives and will be exponent of many aphorisms and precepts and speak with great excellence. If Jupiter is aspect by Venus, he/she will undertake acts of temple construction will visit prostitutes and will win women's hearts. If Jupiter is aspect by Saturn, he/she will be the head of a group, state, or

Venus in Gemini: Venus in Gemini indicates that he/she will be rich, gentle, kind, generous, eloquent, proud, respected, gullible, love of fine arts, learned, intelligent, good logician, just, dual marriage, tendencies towards materialism. If Venus is strong, it blesses him/her with intelligence, fine tastes, luxurious and comfortable life, and global awareness, knowledge of foreign languages, higher studies, and interest in media, art and sculpture, intelligent children, gains through speculation, be famous in sciences and Shashtra, be beautiful, libidinous, be skilful in writing and in poetry, be dear to good people, will derive wealth through music and dances, will have many friends, will honour Gods and Brahmins. Whereas, a weak Venus/afflicted by malefic planets, is likely to cause weak decisive power, problem to wife or children, difficulties in enjoying progeny bliss, losses through speculation, lack of awareness, skin disease, renal problems, weak eyesight, kidney failures, infertility.

Venus in Gemini in Aspect to Others: If Venus is aspect by the Sun, he/she will be well disposed towards the king, his own mother and wife is learned and rich. If Venus is aspect by the Moon, he/she will have dark eyes, beautiful hair, be endowed with sleeping comforts, conveyances, be splendour, beautiful in appearance and be fortunate. If Venus is aspect by Mars, he/she will be highly sensuous, be fortunate and will destroy his wealth through women. If Venus is aspect by Mercury, he/she will be learned, sweet in disposition, wealthy and endowed with conveyances and children, be fortunate and be leader of men, or be a king. If Venus is aspect by Jupiter, he/she will be highly happy, be radiant, courageous, learned and be a preceptor. If Venus is aspect by Saturn, he/she will be very miserable, insulted, and fickle-minded.

Saturn in Gemini: Saturn in Gemini indicates that he/she will be wandering nature, miserable, untidy, original, thin, subtle, ingenious, and strategic, few children, and taste for chemical and mechanical sciences, narrow-minded, speculative, logical, desperado. If Saturn is strong, it blesses him/her with affluent father, easy settlement of life, inclination towards religion, career in technical fields, foreign travels, will contract debts and imprisonments, will toil, will have vanity in disposition, will consecrate by hymns and prayers, be bereft of virtues, be always in hide-out, be libidinous, cunning, wicked and fond of wandering and of sports, reasoning and problem solving abilities are exceptional, excellent for the scientific or mathematical fields. However, a weak Saturn afflicted by malefic planets in the nativity, it gives medium or short life span, problem to father, struggles in life, and losses through servants. He/She might suffer from joint pains, glandular problems, breathing trouble, orthopaedic problems.

Saturn in Gemini in Aspect to Others: If Saturn is aspect by the Sun, he/she will be devoid of happiness and wealth, be virtuous, bereft of anger, will endure difficulties and be valorous. If Saturn is aspect by the Moon, he/she will be equal to a king, will possess a bright physique, will earn wealth and honour through women and will do women's jobs. If Saturn is aspect by Mars, he/she will be a famous boxer, be stupefied, will carry heavy loads and will possess an ugly body. If Saturn is aspect by Mercury, he/she will be rich, skilful in war, be a dance master, a skilful singer and an expert in arts. If Saturn is aspect by Jupiter, he/she will be trustworthy in the king's circle, will possess all kinds of (good) qualities, be liked by good men and will earn wealth through his virtues. If Saturn is aspect by Venus, he/she will be skilful in beautifying women, be a teacher of Yoga or a saint and be dear to the fair sex.

Predictions by Planets in Cancer

Sun in Cancer: Sun in Cancer indicates that he/she will be somewhat harsh, indolent, wealthy, unhappy, constipation, sickly, travelling, independent, expert astrologer. Sun, if strong, blesses him/her with good health and vitality, strong will power, paternal bliss, happy family life, fame, interest in politics, male children, wealth, management and administrative qualities, will hate his own men, be unfortunate in respect of wife, have an ugly wife, will be good-looking himself, follow virtuous principles, be honourable, will be eloquent, will be a geographer and a scientist in the matter of atmosphere/space, will hate people from paternal side, attracted to collecting antiques. While its weak strength/affliction by malefic planets, indicates obstructions in acquiring higher education, problem family life, problems to father or children, loss of reputation, difficulties in social esteem, erratic circulation of blood, cardiac problems, weak eyesight, stammered speech and stomach disorders.

Sun in Cancer and in aspect to others: If the Sun is aspect by the Moon, he/she will be a king, or equal to a king, will become rich by business through water and be cruel. If Mars throws his aspect on the Sun, he/she will contract pulmonary consumption and fistula in the anus, or pudendum, be dejected on account of his/her relatives and be a slanderer. If Mercury aspects the Sun, he/she will be famous for his learning and honour, be dear to the king, skilful and will destroy enemies. If Jupiter throws his aspect on the Sun, he/she will be pre-eminent, a king, a minister, or an Army chief, is very famous and learned in arts. If Venus aspects the Sun, he/she will subordinate his wife (or women), will have money through his wife, be helpful to others, fierce in battle and will speak sweetly. Saturn's aspect on the Sun denotes that he/she will suffer from phlegmatic and windy disorders, be wicked and be a tale-bearer.

Moon in Cancer: Moon in Cancer indicates that he/she will be wise, powerful, charming, influenced by women, wealthy, kind, good, a bit stout, sensitive; impetuous, unprofitable voyages, meditative, much immovable property, scientist, middle stature, prudent, frugal, piercing, conventional. Strong Moon, blesses him with good health, sound mind and sleep, good financial status, analytical skills, adequate maternal bliss and recognition in society, be fortunate, valorous, be endowed with residence, friends, journeys and astrological knowledge, be sensuous, grateful, be a minister, be truthful, will live abroad, be passionate, hairy-bodied, interested in construction of houses, are highly sensitive and creative, heart rule head, intuition is quite keen, have suspicious and distrusting nature. However, if it is weak in strength/afflicted by malefic planets, then it causes lack of mental peace, losses in business, loss of self-respect, lack of analytical skills, problems to mother, occasional health problems such as addiction, excess of phlegm, chest congestion, respiratory problems, immune deficiency, menstrual and gynaecological problems, sleeplessness.

Moon in Cancer and in aspect to others: If the Moon is aspect by the Sun, he/she will be in the employ of king, be not rich, is a letter-bearer and will protect forts, i.e. will be a security officer in royal service. If the Moon is aspect by Mars, he/she will be valorous, be deformed, will prove ominous to his mother and be skilful in his jobs. Should Mercury lend his aspect to the Moon, he will be spirited in disposition, be endowed with political wisdom, wealth, wife and sons, will be a king's minister and be happy. Jupiter aspecting the Moon in her own House indicates that he/she will be a king, be endowed with royal qualities, be happy, will have a good wife, will behave well, and be modest and valorous. If Venus aspects the Moon, he/she will be endowed with money, gold, wife, robes and jewels, be head of prostitutes and be splendour. Should Saturn aspect the Moon, he will be of wandering disposition, be miserable, very poor, be a liar, a sinner and be mean.

Mars in Cancer: Mars in Cancer indicates that the he/she will be intelligent, wealthy, rich travels and voyages, wicked perverted, love of agriculture, medical and surgical proficiency, fickle-minded, defective sight, bold, dashing, headlong, speculative unkind, egoistic. If Mars is strong, it blesses with excellent professional skills, executive qualities, gains through profession, interest in armed forces, humble character and support from younger siblings, will like living in others' houses, be deformed, sick, will attain riches through agriculture, will enjoy royal food and robes during childhood, will eat food in others' houses, will become wealthy through the source of water, a quiet, peaceful nature, never too assertive or demanding. However, if Mars is weak or afflicted by malefic planets, in the chart it gives short temperament, impatient nature, problem to younger siblings and dissatisfaction from professional achievements. In addition to this its weak strength might cause blood infections, arthritis, immune deficiency, muscle cramps, problem in ankles and feet.

Mars in Cancer in Aspect to Others: If Mars is aspect by the Sun, he/she will be subjected to bilious diseases, be splendour, be a justice in position and be valorous. If the Moon aspects Mars, he/she will be troubled by various diseases will have mean conduct, will possess an unsightly body and be miserable. If Mercury aspects Mars, he/she will be dirty, sinful, will possess a mean family, will be rejected by his own men and be shameless. If Mars is aspect by Jupiter, he/she will be famous, is king's minister, be learned, charitable, and wealthy and be bereft of carnal pleasures. If Venus aspects Mars, he/she will be grieved on account of women's company, be insulted by women and will lose wealth on account of women. Should Saturn aspect Mars he/she will acquire money through journey in water, be equal to a king, be sportive in his acts and be always bright.

Mercury in Cancer: Mercury in Cancer indicates that he/she will be witty, liking music, disliked by relations, low stature, speculative, diplomatic, discreet, flexible, restless, and sensual though religious, liable to consumption, strong parental love, and dislike for chastity. If Mercury is Strong, it indicates excellent assessment, analytical and decisive powers, communication skills, interest in pursuing higher studies, efforts of a high order, confidence and support from younger siblings, be learned, fond of living in other countries, be interested in cohabiting with women and also in music, be fickle-minded, prattling, inimical to his/her own relatives, be fond of arguments, lose wealth on account of enmity with women, be of bad disposition, interested in many jobs, a good poet. If Mercury is weak/afflicted by malefic planets, due to any reason then the person might experience losses in new ventures, problems in learning, lack of concentration and initiatives, lack of confidence and problem to or from younger brother. In addition to this he/she might suffer from retarded physical and mental growth, intestinal problems, respiratory, cervical pain, depression, neurological disorders, and sleeplessness or skin diseases.

Mercury in Cancer in Aspect to Others: If Mercury is in aspect to the Sun, he/she will be a cloth cleaner, or a gardener, or a house-builder, or gem smith. If the Moon aspects Mercury, he/she will be deprived of wealth (or energy of the physique) on account of women and will be miserable for the same reason. If Mars lends his aspect to Mercury, he will not have much learning, be garrulous, be a great liar, will produce fictitious things, be a thief and will be affectionate in speech. If Jupiter aspects Mercury, he/she will be a great scholar, very dear to all, be fortunate, dear to the king and will cross the boundaries of learning. If Venus aspect Mercury, he/she will be equal to cupid in appearance, will possess attractive physique, be well-versed in the art of singing and in playing musical instruments, be fortunate and softly disposed. If Saturn aspects Mercury, he/she will be fond of vanity, be sinful, will face imprisonment, be devoid of virtues and will hate co-born and elders.

Jupiter in Cancer: Jupiter in Cancer indicates that he/she will be well-read, dignified, wealthy, comfortable intelligent, swarthy complexion, inclined to social gossip mathematician faithful. If Jupiter is Strong, it blesses him/her with interest in studies, knowledge of various subjects, religious nature, financial stability, strong resistance power, divine grace and good health, be a scholar, be beautiful, be highly learned, charitable, good-natured, be very strong, be famous, will possess abundant grains and riches, be endowed with truth and penance, will have long-living sons, be honoured by all, will be a king, will have a distinguished profession, likely to accumulate wealth, very careful about spending money, have an emotional attachment to food and have weight problems. When it is weak in strength/afflicted by malefic planets, it gives instability in finance, health problems, losses in disputes or litigation, weak resistance power, losses through disputes, fire, debts, and problem from children. On health ground they might suffer from liver disorder, jaundice, low sperm count and dark circles under eyes.

Jupiter in Cancer in Aspect to Others: If Jupiter is aspect by the Sun, he/she will be famous, be ahead of others and in the beginning, be devoid of happiness, wealth and spouse, all of which will be acquired by him/her later. If Jupiter is aspect by the Moon, he/she will be very rich, splendour, is a king, will enjoy abundant wealth and conveyances and will have excellent spouse and sons. If Jupiter is aspect by Mars, he/she will marry in the boyhood itself, will be endowed with gold and ornaments, be learned, valorous and will have a wounded physique. If Jupiter is aspect by Mercury, he/she will be endowed with relatives and friends, be rich, will promote quarrels, be bereft of sins, be a minister and be trustworthy. If Jupiter is aspect by Venus, he/she will have many wives, extraordinary riches, various ornaments, are happy and fortunate. If Jupiter is aspect by Saturn, he/she will be important in his/her village, or in the Army, or in his/her town, be talkative, be very affluent, garrulous and will enjoy pleasures in old age.

Venus in Cancer: Venus in Cancer indicates that he/she will be melancholy, emotional, timid, more than one wife, haughty, sorrowful, light character, inconsistent, unhappy, many children, sensitive, learned. If Venus is strong, it blesses with global awareness, knowledge of foreign languages, good education, and interest in media, art and sculpture, good assets, support from mother, luxuries and material comforts, be wise, virtuous, learned, strong, soft, chief among men, will have desired happiness and wealth, be good-looking, very much troubled on account of women and wine and will be miserable with family troubles, deeply sensitive feelings, have a very kind-hearted and sympathetic nature, love relationships with tenderness, be well anchored in a relationship, love is home and hearth. However, if Venus is weak or afflicted by malefic planets, it causes problem to mother or spouse, loss of assets, delay in marriage, lack of domestic peace, and delay in progeny matters, lack of awareness, back aches, skin disease, eye diseases, renal problems, infertility and ovarian problems in women.

Venus in Cancer in Aspect to Others: If Venus is aspect by the Sun, his/her spouse will be attached to her/his duties, will possess a spotless body, will be a king's daughter, i.e. he/she will be of a rich heritage and he/she will be short-tempered and will be endowed with wealth. If Venus is aspect by the Moon, he/she will keep his step-mother happy, will have a daughter first and later on many sons, be happy, fortunate and beautiful. If Venus is aspect by Mars, he/she will be skilful in arts, be very rich, will suffer on account of women, be fortunate and will promote the cause of his/her relatives. If Venus is aspect by Mercury, he/she will marry a learned woman, will suffer miseries on account of his relatives, will wander, be wealthy and learned. If Venus is aspect by Jupiter, he/she will be endowed with many servants, sons, happiness, relatives and friends and be dear to king. If Venus is aspect by Saturn, he/she will be subdued by women, be poor, base, ugly, and fickle-minded.

Saturn in Cancer: Saturn in Cancer indicates that he/she will be poor, weak teeth, pleasure seeking, and few sons, cheeks full, slow, dull, cunning, rich, selfish, deceitful, malicious, and stubborn, devoid of motherly care. Saturn, when strong, provides unearned wealth, unexpected gains, inheritance, and blissful long marital life, and long life span, addition to this they might suffer from joint pains, orthopaedic problems, genital diseases, and piles, will possess a beloved spouse, be devoid of wealth in boyhood, will suffer many diseases, be learned, motherless, soft-spoken, distinguished in acts, will trouble others, be inimical to relatives, crooked, be kingly in his/her mid-life and will enjoy growing pleasures.

When Saturn is weak or afflicted by malefic planets, it gives medium or short life span, delay or difficulties in enjoying marital life, lack or loss of inheritance, humiliation, losses through servants.

Saturn in Cancer in Aspect to Others: If Saturn is aspect by the Sun, he/she will lose his father in his/her boyhood, be bereft of wealth, spouse and happiness, will eat bad food and be sinful. If Saturn is aspect by the Moon, he/she will be a source of evil to his/her mother, be wealthy and will be troubled by his co-born. If Saturn is aspect by Mars, he/she will enjoy king's wealth, be defective bodied, will possess gold and gems, be endowed with a family and will have a bad group of relatives and spouse. If Saturn is aspect by Mercury, he/she will be hard-hearted, garrulous, will conquer his enemies, will show vanity and will do noble acts. If Saturn is aspect by Jupiter, he/she will be endowed with lands, houses, friends, sons, wealth, gems and wife. If Saturn is aspect by Venus, he/she is of a noble descent, but will be bereft of beauty, grace and happiness.

Predictions by Planets in Leo

Sun in Leo: Sun in Leo indicates that he/she will be stubborn, fixed views, strong, cruel, independent, organising capacity and talents for propaganda, humanitarian, frequenting solitary places, generous, famous. If Sun is Strong, it blesses him with good status, recognition, good health, vitality, longevity, interest in administration politics, anger, and will destroy enemies, perform notable acts, eat meat, flesh, wander in forests, hills and fortresses, and will be enthusiastic, valorous, bright in appearance, formidable, restive, strong in a lasting measure, talkative, a king, and plentiful in wealth and famous. The Sun in Leo makes him magnetic, forceful, dominant, possessing good nature a natural leader fine organizer. However, if it is weak in strength/afflicted by malefic planets, due to any reason in that case the person may experience loss of respect or social status, problem to or from father/husband, short or medium life span. In addition to this the weak strength of Sun causes addiction, cardiac problem, and erratic circulation of blood, weak eyesight, weak digestion, and weak resistance power.

Sun in Leo and in aspect to others: If the Sun is aspect by the Moon, he/she will be a scholar, will have a good spouse, will suffer from phlegmatic disorders and will be dear to king. If Mars aspects the Sun, he/she will be interested in others' wives, be courageous, valorous, revolutionary, formidable and chief. If Mercury aspects the Sun, the person will be a scholar, a writer or an exponent, a gambler, be wandering-natured, mean and be endowed with great strength. If the Sun is aspect by Jupiter, he/she will construct temples, gardens and tanks; will have predominant strength, will like loneliness and be highly intelligent. Venus aspecting the Sun will make him/her earn bad name, infamous. He/She will be troubled by leprosy, be unkind and shameless. If Saturn aspects the Sun, he/she will be skilful in creating obstacles, will be a eunuch and will cause grief to others.

Moon in Leo: Moon in Leo indicates that he/she will be bold, irritable, large cheeks, blonde, broad face, brown eyes, repugnant to women, likes meat, frequenting forests and hills, colic troubles, inclined to be unhappy, haughty, mental anxiety, liberal, generous, deformed body, steady, aristocratic, settled views, proud, ambitious. When Moon is strong and well placed and not afflicting any area of his/her house, it provides relations with foreign land, good health, sound sleep, longevity and analysing power, will have sturdy bones, sparse hair, wide face, small and yellowish eyes, will hate women, will suffer from hunger and thirst, will incur stomach disorders and tooth-decay, will eat flesh, be charitable, harsh, will have few sons, will seek sexual union in forests and hills, be respectfully disposed to his/her mother, will have broad chest, be valorous, dutiful and will have majestic looks, have confidence and a desire to lead, self-confident. However, its weak strength/afflicted by malefic planets causes short or medium life span, bad health, hospitalisation, lack of analysing power, problem to mother, losses, loss of bed comforts, addiction, imprisonment. Medically he/she can have excess of phlegm and congestion, respiratory problems, disturbed sleep and gynaecological problems in her.

Moon in Leo and in aspect to others: If the Moon is in Leo in aspect to the Sun; he/she will be equal to a king, will have excellent qualities and majestic voice, will be valorous, be fond of intoxicants and be widely famous. If Mars aspects the Moon in Leo he/she will be an Army chief, will have excellent wife, sons, wealth and conveyances and be superior among people. If Mercury aspects the Moon in Leo; he/she will be endowed with the characteristics of a female and also the grace akin to that of a female, will be in the custody of females, will serve females and will enjoy money, happiness and pleasures. If Jupiter aspects the Moon in Leo, he/she will be excellent among his caste-men, be wide-famed, highly virtuous and will equal a king. If Venus aspects the Moon in Leo, he/she will possess a wife, wealth and high knowledge, be sickly disposed, will be a female's servant and be skilful in sexual union. If the Moon in Leo is aspect by Saturn, he/she will be an agriculturist, be not wealthy, be a liar, will protect forts, be devoid of happiness from spouse and be mean.

Mars in Leo: Mars in Leo indicates that he/she will be tendency to occultism, astrology, astronomy and mathematics, love for parents, regard and respect for elders and preceptors, independent thinking, peevish, liberal, victorious, stomach troubles, worried by mental complaints, generous, noble, author, early in life, successful, combative, restless. Mars good placement and strong strength provides affluent father, vigour and vitality, easy settlement of fortune, interest in religion and joining armed forces, patient nature, be impatient, be valorous, intent upon grabbing others' money and children, will like to live in forests, be fond of eating cow's flesh or beef, will lose his/her first wife, will kill snakes and animals, will be bereft of children, be devoid of charitable acts and be always active in his/her jobs, exceptional will power and creativity, air of confidence, self-sufficiency, and vitality, be well suited for the stage show, producing excellent leadership qualities. However, if it is weak by placement and strength/afflicted by malefic planets, he/she will have short temperament, problem to father and younger brother, unsettled fortune. Medically he/she can suffer from blood infections, problems in pelvic area, immune deficiency, muscular problems, problem in thighs, high BP.

Mars in Leo in Aspect to Others: If Mars is aspect by the Sun, he/she will be humble, helpful to friends, be endowed with own men and be fond of wandering in cow-houses, forests and hills. If the Moon aspects Mars, he/she will be ominous for his mother, be intelligent, will possess a hard body, be widely famous and will obtain money through women. If Mercury aspects Mars, he/she will be clever in many arts, be a miser, be skilful in poetry and fine arts and be wicked. If Jupiter aspects Mars, he/she will be close to the king, be highly learned, be of pure mentality and be an Army chief. If Venus aspects Mars, he/she will have union with many women, be fortunate in respect of his/her wife and will be ever juvenile. Saturn aspecting Mars denotes, that he/she will look, like an old man, be poor, will wander in others' houses/be miserable.

Mercury in Leo: Mercury in Leo indicates that he/she will be few children, wanderer, idiotic, proud, indolent, fond of women, boastful, orator, good memory, two mothers, poor, early marriage, independent in thinking, impulsive, positive will, remunerative profession, likes travelling. Mercury good placement and strong strength is favourable for acquiring satisfactory social status, enjoying happy family life and accumulating wealth, excellent assessment, analytical and decisive powers, be famous in the world, be independent, be favourable to others, a mind with strong will power and stubbornly fixed purposes, comprehension is broad, and expression is authoritative, sometimes even dogmatic, has a very human touch that are accepted and followed by most people. However, its weakness/affliction by malefic planets causes weak analytical and decisive powers, nervousness and confusions, difficulties in acquiring or attaining social status and accumulating wealth. In addition to this the weak strength of Mercury causes difficulties in enjoying family life, problem from children, and loss of reputation. In addition to this they might suffer from intestinal problems, respiratory, cervical pain, neurological disorders, Parkinson disease, skin disease, stammered speech, problem in right eye.

Mercury in Leo in Aspect to Others: If Mercury is aspect by the Sun, he/she will be jealous, rich, virtuous, cruel, mean, fickle-minded and shameless. If the Moon aspects Mercury, he/she will be very beautiful, very skilful, be interested in poetry, fine arts, music and dance, be wealthy and virtuous. If Mars lends his aspect to Mercury, he/she will be base, miserable, physically injured, skilful and impotent. If Jupiter aspects Mercury, he/she will be beautiful, very learned, be a gifted speaker, be very famous and be endowed with attendants and conveyances. If Venus aspects Mercury, he/she will possess unparalleled beauty, will be softly disposed, will have an attractive face, will have many conveyances, be very courageous and be a minister. If Saturn aspects Mercury, he/she will be tall in stature, be splendour less, be ugly, will emanate bad smell from body out of sweat and be miserable.

Jupiter in Leo: Jupiter in Leo indicates that he/she will be commanding appearance, tall, great, easily offended ambitious, active, happy intelligent, wise, prudent, generous broad-minded, literary, harmonious surroundings, likes hills and dales. Jupiter when strong blesses him/her with interest in studies, reading and religion, good conduct and morality, easy settlement of children and gains through speculation, will be lastingly inimical, be strong, courageous, will show abundant friendship, be learned, rich, will have eminent relatives, be a king, will have heroism akin to that of a king, will be recognizable in an assembly, will destroy the entire band of his enemies, will possess a strong physique and will live in hills, fortresses, forests and temples, optimism, self-confidence and generosity, have much physical energy and a strong constitution, excellent leadership qualities, tendency to gamble; cards, horses, business. However, it's weak strength/afflicted by malefic planets causes losses through speculation, emotional setbacks, infertility and problem to father or children. Medically they can suffer from poor liver functioning, jaundice and dark circles under eyes.

Jupiter in Leo in Aspect to Others: If Jupiter is aspect by the Sun, he/she will be dear to good men, be famous, be a king, and be extremely affluent and virtuous. If Jupiter is aspect by the Moon, he/she will be very beautiful, dirty, will be very rich through the fortunes of his/her wife and will conquer his/her five senses. If Jupiter is aspect by Mars, he/she will honour elders at all times, will perform distinguished acts, be very skilful, pure, adventurous and cruel. If Jupiter is aspect by Mercury, he/she will have knowledge of civil works of building construction, will be endowed with profane knowledge, be virtuous, be a sweet speaker, be a minister and be highly learned. If Jupiter is aspect by Venus, he/she will be dear to females, be fortunate, will receive royal honours and be very strong. If Jupiter is aspect by Saturn, he/she will be garrulous, be an eloquent speaker, be devoid of happiness, be sharp and will have mean children and mean wife.

Venus in Leo: Venus in Leo indicates that he/she will be money through women, pretty wife, wayward, conceited, passionate, fair complexion, emotional, zealous, licentious, attracted by the fair sex, premature in conclusions, superior airs, unvanquished by enemies. If Venus is strong, it blesses him with less virility, dear to relatives, miserable in spite of his happiness, help others, respect Brahmins, elders and preceptors and devoid of much discrimination, fickle-minded and unhappy, warm-hearted and fun loving, break in relationship, very hard and bitter one. However, the weak strength/afflicted by malefic planets of Venus might cause lack or loss of efforts, problem from younger brother/wife, delay in marriage and progeny matters, lack of awareness, skin disease, eye diseases, renal problems, infertility and ovarian problems in women .

Venus in Leo in Aspect to Others: If Venus is aspect by the Sun, he/she will be jealous, be dear to the fair sex, libidinous will acquire money through women and will possess elephants. If Venus is aspect by the Moon, he/she will present obsequies water to his/her step mother, will be miserable on account of women, be rich and will have various kinds of mental disposition (i.e. be infirm in disposition). If Venus is aspect by Mars, he/she will be a royal person, be famous, dear to women, be affluent, fortunate and be attached to other's wives. If Venus is aspect by Mercury, he/she will be ever engaged in earning, be miserly in disposition, be addicted to women, will join other's wives, be courageous, crafty, false and wealthy. If Venus is aspect by Jupiter, he/she will be endowed with conveyances, wealth and servants will marry many women and be a king's minister. If Venus is aspect by Saturn, he/she will be a king, or equal to a king, be famous, will have abundant wealth and conveyances, will marry a prostitute (or a widow), be beautiful in appearance and be miserable.

Saturn in Leo: Saturn in Leo indicates that he/she will be middle stature, severe, obstinate, few sons, stubborn, unfortunate, conflicting, hard worker, and good writer evil-minded. Saturn, if strong, provides with pursuing career in technical fields, support from spouse, relation with foreign land, long lifespan, be interested in writing and reading, be disdained, devoid of virtues and wife, will live by servitude, be devoid of his own men and happiness, be interested in doing base acts, ill-tempered, be mad with undue desires, will carry loads, will toil hard and will have a wrinkled body, children may seem a burden, have some difficulties in personal relationships with children as well as his mate, too hard, very reserved and cautious. When it is weak by placement and strength/afflicted by malefic planets in the nativity, it might cause problem to or from spouse, delay in marriage, uncomfortable travel, lack of

technical knowledge, short or medium life span, inharmonious relations with spouse or partner, and losses through servants. Medically they can have joint pains, orthopaedic problems.
Saturn in Leo in Aspect to Others: If Saturn is aspect by the Sun himself, he/she will be devoid of wealth, happiness and nobility, will be a liar and a drunkard, will possess a bad physique, be a servant and be very miserable. If Saturn is aspect by the Moon, he/she will enjoy abundant wealth, precious stones and women, be widely famous and be dear to the king. If Saturn is aspect by Mars, he/she will everyday move from place to place, be unfortunate, will live in fortresses and hills, be base and be bereft of wife and sons. If Saturn is aspect by Mercury, he/she will not be outspoken, be lazy, poor, will do females' jobs, be dirty and miserable. If Saturn is aspect by Jupiter, he/she will be chief and rich in his town, or among his men, will be endowed with progeny and be trustworthy. If Saturn is aspect by Venus, he/she will be averse to women, be splendours, slow (or tardy), happy, rich and will attain a good end.

Predictions by Planets in Virgo

Sun in Virgo: Sun in Virgo indicates that he/she will be linguist, poet, mathematician, taste for literature, well read, scholarly, artistic, good memory, reasoning faculty, effeminate body, and frank, lucid comprehension, learned in religious lore, reserved, wanting adulation. If Sun is strong, it provides establishing relations with foreign land, undertaking foreign journeys, vitality, sound sleep, long lifespan, wisdom and recognition in faraway places, will possess a physique akin to that of a female, be weak, be an expert writer, be learned and will render service to Gods and elders, be expert in repairs of driven vehicles, will be skilful in Vedas, songs and playing. However, if Sun is weak in strength/afflicted by malefic planets it causes short or medium life span, weak health, problem to father/husband/children, unwanted expenses and losses, lack of conjugal bliss, lack of decision, addiction, imprisonment or hospitalisation. In addition to this its weak strength is likely to cause cardiac problem, erratic blood circulation, disturbed sleep, stomach disorders, weak eyesight.

Sun in Virgo and in aspect to others: Should the Sun is aspect by the Moon, he/she will be put to troubles by enemies and relatives, will be distressed by visits to foreign countries, and will in general be wailing. If the Sun is aspect by Mars, it denotes that the subject will have fear from enemies, will be encountered by quarrels, will be grieved on account of loss in a war, be poor and bashful. If Mercury aspects the Sun, he/she will have a history akin to that of a king, will be famous, be endowed with relatives, free from enemies, but will encounter eye diseases. Jupiter aspecting the Sun foretells that he/she will have knowledge of many Shashtra, be a king's messenger (or representative), will go to foreign countries, be fierce and be always bewildered. If Venus aspects the Sun, he/she will be endowed with money, wife and sons, will make less friends, be free from sickness, be happy and fickle minded. If Saturn aspects the Sun, he/she will have many servants, be anxious (as for an absent lover), will maintain many relatives, will remain delighted and will be crafty.

Moon in Virgo: Moon in Virgo indicates that he/she will be lovely complexion, almond eyes, modest, sunken shoulders and arms, charming, attractive, principled, affluent, comfortable, soft body, sweet speech, honest; truthful, modest, virtuous, intelligent, phlegmatic, fond of women, acute insight, conceited in self-estimation, pensive, conversationalist, many daughters, loquacious, astrologer and clairvoyant or attracted towards them, skilled in arts like music and dancing, few sons.

If Moon is auspicious and strong, it indicates earning satisfactory income, fulfilment of desires in life, gains through friends and elder brother, interest in finance and cooking, support from mother and good analytical power, be addicted to women, will have long hands, attractive body and face, beautiful teeth, eyes and ears, be learned, be a religious preceptor teaching Vedas, be an eloquent speaker, be truthful and pure, valorous, be kind to living beings, be interested in others' affairs, be fortunate, will have more daughters, but not many sons. When Moon is weak in strength/afflicted by malefic planets in the nativity it causes fluctuation in income, losses, problem from elder brother and friends, lack of analytical power, dissatisfaction from monetary viewpoint etc. He/She may suffer from excess of phlegm and congestion, respiratory problems, disturbed sleep, mental stress and gynaecological problems in females.

Moon in Virgo and in aspect to others: If the Moon is aspect by the Sun, he/she will be in charge of royal wealth, be famous, will keep up his word and perform distinguished acts. If Mars aspects the Moon, he/she will be skilful in mechanical, or fine arts, be famous, affluent, disciplined, and courageous and will be inauspicious for mother. If Mercury lends aspect to the Moon; he/she will be expert in astrology and literature, be successful in disputes/quarrels and be highly skilful to a surprising extent. Jupiter aspecting the Moon indicates that he/she will be supreme among his relatives, be happy, will carry out royal duties, will keep up his word and will be endowed with wealth. If Venus aspects the Moon, he/she will have many wives, be endowed with many kinds of make-up, pleasures and wealth and will be blessed with fortunes. The Moon is aspect by Saturn, it indicates that he/she will not have firm memory, will suffer from poverty, will not have happiness, will be bereft of mother, will be at the disposal of women or be controlled by them and will derive wealth through females.

Mars in Virgo: Mars in Virgo indicates that he/she will be imitable, explosive, trouble in digestive organs, no marital harmony, general love for the fair sex, revengeful, self-confident, conceited, affable, boastful, materialistic, ceremonial-minded, positive, indiscriminating, pretentious, deceptive, scientific enterprises. Mars when strong blesses him with vigour and vitality, patience, strong marital tie, abundant inheritance, unearned gains, long lifespan, trouble free life, interest in arms and armed forces, be never rich, be very fond of sexual union and music, will have various kinds of expenses, be not much valorous. If it is weak in strength/afflicted by malefic planets it causes short temperament, problem to father and younger brother, delay in marriage, break in marital tie, short or medium life span and loss of inheritance. Medically he/she can suffer from blood infections, obstruction in physical growth, problems in genitals, piles, pimples, muscle cramps, erratic circulation of blood, injuries.

Mars in Virgo in Aspect to Others: If Mars is of the Sun, he/she will be blessed with learning, wealth and courage, be fond of hills, forests and fortresses and be highly strong. If Mars is aspect by the Moon, he/she will be happy, wealthy, and splendours, will guard women's apartments, will be endowed with women and will manage the king's residence. If

Mars is aspect by Mercury, he/she will be skilful in the art of writing, in mathematics and in poetry, be garrulous, be liar, be a sweet speaker, be a messenger or an ambassador and will endure lot of misery. If Mars is aspect by Jupiter, he/she will be a king's representative, be bright, will go to foreign countries, as an ambassador, be skilful in all doings and be a leader. If Mars is aspect by Venus, he/she will do the jobs of females, be very fortunate and will enjoy food and robes. If Mars is aspect by Saturn, he/she will be interested in wandering in mines i.e. places beneath surface, hills and forests will have husbandry, as his livelihood, be highly miserable, be very valorous, dirty and be devoid of wealth.

Mercury in Virgo: Mercury in Virgo indicates that he/she will be learned, virtuous, liberal, fearless, ingenious, handsome, irritable, refined, subtle, intuitive, sociable, no self-control, morbid imaginations, dyspeptic, difficulties, eloquent, author, priest, astronomer. Mercury, if strong, blesses him with charming and attractive personality, confidence, good health and longevity, excellent communication skills and writing, be quite virtuous, eloquent, skilful, will have knowledge of writing and poetry, be learned in fine/mechanical arts, be sweet in disposition, be liked by women, be not much virile, be the eldest son of family, famous, charitable, sharp and intelligent. However, if Mercury is weak in strength/afflicted by malefic planets in the chart it causes loss of reputation, indecisiveness, problem to children, short or medium life span and health problems such as obstruction in mental and physical growth, intestinal problems, respiratory, cervical pain, neurological disorders, constipation, Parkinson disease, skin disease, sleeplessness.

Mercury in Virgo in Aspect to Others: If Mercury is aspect by the Sun, he/she will speak truth, be fortunate, dear to king, be a Lord himself, be polite in his activities and be liked by all. If Mercury receives the aspect of the Moon, he/she will be sweet in disposition, garrulous, will promote quarrels, be interested in acquiring Great Saint knowledge, firm and will succeed in all his undertakings. If Mercury receives the aspect of Mars; he/she will have an injured body, be dirty, be a genius, will serve the king and be dear to him. If Mercury receives the aspect of Jupiter, he/she will be a king's minister, be excellent, be beautiful, charitable, and rich, be endowed with his/her own men and be courageous. Should Mercury receive the aspect of

Venus, he/she will be highly learned be a royal employee, be a messenger, will honour friendship and will be interested in base women. Should Mercury receive the aspect of Saturn, he/she will be progressive-minded, be modest, will achieve success in undertakings started by him and will be wealthy with money and clothes.

Jupiter in Virgo: Jupiter in Virgo indicates that he/she will be middle stature, ambitious, selfish, stoical, resignation, affectionate, fortunate, stingy, and lovable, a beautiful wife, great endurance, learned. Jupiter when strong in strength provides interest in studies, reading and religion, good conduct and morality, domestic peace, gains of assets, support from mother and father, good education, will be a scholar, be virtuous, be skilful in his work, be fond of scents, robes and flowers, will firmly gain in undertakings, be affluent, charitable, pure-hearted, skilful and wonderfully learned. However, it's weak strength/afflicted by malefic planets causes loss of assets, loss of residence and domestic peace, delay in marriage, obstructive studies, problem to mother or father. Medically he/she can suffer from liver disorder, jaundice and dark circles under eyes.

Jupiter in Virgo in Aspect to Others: If Jupiter is aspect by the Sun, he/she will be great, important in his village be a householder and will be endowed with wife, sons and money. If Jupiter is aspect by the Moon, he/she will be affluent, dear to mother, fortunate, happy, will have wife and sons and be incomparable. If Jupiter is aspect by Mars, he/she will be easily successful at all times, be ugly, rich and be amiable to all. If Jupiter is aspect by Mercury, he/she will be a skilful astrologer, will have many children and wives and will be exponent of many aphorisms and precepts and speak with great excellence. If Jupiter is aspect by Venus, he/she will undertake temple construction, will visit prostitutes and win women's hearts. If Jupiter is aspect by Saturn, he/she will be the head of a group, state.

Venus in Virgo: Venus in Virgo indicates that he/she will be petty-minded, licentious, unscrupulous, unhappy, illicit love, agile, loquacious, rich, and learned. Venus, if strong and well placed, gives quite discriminative, be soft in disposition, be skilful, will be helpful to others, will speak sweetly, will earn money through several sources, will cohabit with bad women, be mean, be devoid of happiness and pleasures, will get more daughters and less sons, will visit shrines and be a scholar in an assembly, play safe in affairs of the heart, be shy, but careful and cautious, prefer to be safe than sorry. However, it's weak strength/afflicted by malefic planets causes loss of status and accumulated wealth, delay in marriage, problem in continuation of family life, problem to wife, bad reputation. Medically there may be skin disease, stammered speech, eye diseases or weak right eyesight, kidney failures, renal diseases, infertility and ovarian problems in women.

Venus in Virgo in Aspect to Others: If Venus aspect by the Sun, he/she will be well disposed towards the king, his own mother and wife are learned and rich. If Venus is aspect by the Moon, he/she will have dark eyes, beautiful hair, be endowed with sleeping comforts, conveyances etc., be splendour, beautiful in appearance and are fortunate. If Venus is aspect by Mars, he/she will be highly sensuous, be fortunate and will destroy his wealth through women. If Venus is aspect by Mercury, he/she will be learned, sweet in disposition, wealthy and endowed with conveyances and children, be fortunate and be leader of men, or be a king. If Venus is aspect by Jupiter, he/she will be highly happy, be radiant, courageous, learned and be a preceptor. If Venus is aspect by Saturn, he/she will be very miserable, insulted, and fickle-minded.

Saturn in Virgo: Saturn in Virgo indicates that he/she will be dark complexion, malicious, poor, quarrelsome, erratic, narrow-minded, rude, conservative, taste for public life, weak health. Saturn, if strong, gives resemble a eunuch, be very crafty, will depend on others for food, be addicted to prostitutes, will have a few friends, be unacquainted with arts, be desirous of indulging in ugly acts, be indolent, will intent upon spoiling virgins and be cautious in his actions, be a hard worker and a careful worker, strategies, and record keeping are areas of excellence, may worry too much about trivial problems, cautious investment and conservative planning will probably gain success in life. When it is weak by placement or strength/afflicted by malefic planets in the nativity, it may cause bad health, medium life span, losses in dispute, instability in finance, losses through employees, theft, arson, fire, riots etc. Medically he/she can have joint pains, orthopaedic problems.

Saturn in Virgo in Aspect to Others: If Saturn is aspect by the Sun, he/she will be devoid of happiness and wealth, be virtuous, bereft of anger, will endure difficulties and be valorous. If Saturn is aspect by the Moon, he/she will be equal to a king, will possess a bright physique, will earn wealth and honour through women and will do women's jobs. If Saturn is aspect by Mars, he/she will be a famous boxer, be stupefied, will carry heavy loads and will possess an ugly body. If Saturn is aspect by Mercury, he/she will be rich, skilful in war, be a dance master, a skilful singer and an expert in arts. If Saturn is aspect by Jupiter, he/she will be trustworthy in the king's circle, will possess all kinds of (good) qualities, be liked by good men and will earn wealth through his virtues. If Saturn is aspect by Venus, he/she will be skilful in beautifying women, be a teacher of Yoga or a saint and be dear to the fair sex.

Predictions by Planets in Libra

Sun in Libra: Sun in Libra indicates that he/she will be manufacture of liquors, popular, tactless, base, and drunkard, loose morals, arrogant, wicked, frank, submissive, and pompous. The good placement and strength of Sun blesses him with face frustration, destruction and heavy expenditure, will be intent on living in foreign places out of distress, be wicked, mean, be devoid of affection, will live by selling gold and other metals, be jealous, fond of doing others' jobs, will co-habit with others' wives, be dirty, will incur royal contempt and be shameless, be a courteous. When Sun is weak or afflicted by malefic planets in their birth chart it causes problem to or from father, elder brother and friends, lack of decision, loss of reputation, dissatisfaction from income and losses. In addition to this he/she may suffer from Heart trouble, erratic blood circulation, disturbed sleep, stomach disorders, and weak eyesight.

Sun in Libra and in aspect to others: If Sun is aspect by the Moon, he/she will be addicted to prostitutes, will be soft spoken, will have many women, as dependent and will derive livelihood through water. If Mars is aspecting the Sun in Libra, he/she will be brave, fond of battle, bright in appearance, will earn wealth and fame out of his valour and will be deformed. Should Mercury aspect Sun,, he/she will be skilful in drawing, writing, poetry, authorship, singing etc. and will possess a good physique. If Jupiter lends aspect to the Sun, he/she will have many foes and friends, be a king's minister, will have beautiful eyes, is splendour and will be a pleased ruler. When Venus aspects the Sun, he/she will be a king or a king's minister, be endowed with wife, wealth and pleasure galore, be wise and timid. Saturn aspecting the Sun, he/she will be mean, indolent, will cohabit with aged women, will be wicked and will be troubled by diseases.

Moon in Libra: Moon in Libra indicates that he/she will be reverence and respect for learned and holy people, saints and gods; tall, raised nose, thin, deformed limbs, sickly constitution, rejected by kinsmen, intelligent, principled, wealthy, business-like, obliging, love for arts, far-seeing idealistic, clever, mutable, amicable, losses through women, loves women, just, not ambitious, aspiring.

Moon good placement provides with weak (not prominent) face and emaciated body, will have many wives, many bulls/cattle and abundant landed property, be valorous, will have testicles resembling that of an ox, be skilful in work, will honour Gods and the wise, will be conquered by females, will have an emaciated body, will donate corns, will not be firm in disposition and will be helpful to his relatives, ambitious but very dependent on others. When it is weak in strength/afflicted by malefic planets it gives losses in profession, lack of analysing power, dissatisfaction from professional achievements, problem to mother. Its weak strength causes excess of phlegm and congestion, respiratory problems, disturbed sleep, menstrual and ovarian problems in women.

Moon in Libra and in aspect to others: If the Moon is aspect by the Sun, he/she will be bereft of wealth, be diseased, will wander here and there, be insulted, and be bereft of enjoyment, sons and strength. If Mars aspects the Moon, he/she will be sharp, be a thief, be mean, adulterous, will enjoy scents and garlands, be wise and will suffer from eye diseases. If Mercury aspects the Moon, he/she will be skilful in arts, will have abundant riches and grains, will be an auspicious speaker, be highly learned and be famous in his country. Should Jupiter aspect the Moon, he/she will be always worshipped and be skilful in sales and purchases of gems. If Venus lends his aspect to the Moon, he/she will be beautiful, be free from sickness, fortunate, strong bodied, be learned and will have knowledge of many means. Should Saturn aspect the Moon, he/she will be very affluent, be sweet in speech, be endowed with conveyances, be very much interested in sexual affairs, be devoid of happiness and be favourable to his mother.

Mars in Libra: Mars in Libra indicates that he/she will be tall, body symmetrically built, complexion fair and swarthy, ambitious, self-confident, perceptive faculties, materialistic, live for family, self-earned wealth, affable, warlike, foresight, business-like, deceived by women, sanguine temperament, king, gentle, fond of adulation, easily ruffled, boastful. Mars if strong blesses with vigour and vitality, good support from younger brother, happy marital life, and relation with foreign land, patience, interest in arms and armed forces and interest in pursuing a career in technical studies, will be liable to wandering, be an able speaker, be fortunate, be fond of wars, charming, generous, amiable, and cooperative. When it is weak by placement or strength/afflicted by malefic planets in the nativity, it gives short temperament, delay in marriage, obstruction in settling down overseas and problem to spouse or younger brother. Medically he/she can suffer from blood infections, obstruction in physical growth, muscle cramps, and high/low BP.

Mars in Libra in Aspect to Others: If Mars is aspect of the Sun, he/she will seek to move in forests and hills, will hate women, will have many enemies, will have fierce appearance and be courageous. If Mars is aspect of the Moon, he/she will not honour his mother, be wicked, will lord many women and be dear to them and will fear war. If Mars is aspect of Mercury, he/she will promote quarrels, will speak much, will have a soft body, will possess (indifferent or) few sons and little wealth and will be learned in Shashtra. If Mars is aspect of Jupiter, he/she will be skilful in music and in play of musical instruments, be fortunate, be dear to his relatives and will be pure. If Mars is aspect of Venus, he/she will be a king's minister, be liked by the king, be an Army chief, will have famous name (i.e. titles etc.) and will be happy. If Mars is aspect of Saturn, he/she will be happy, famous, and wealthy, be endowed with friends and own men, be learned and will be head of a group of villages/towns, or group of men.

Mercury in Libra: Mercury in Libra indicates that he/she will be fair complexion, sanguine disposition, and inclination to excesses, perceptive faculties, and material tendencies, frugal, agreeable, courteous, philosophical, faithful, ceremonial-minded, sociable, and discreet. Mercury, is strong, is beneficial for enjoying relations with foreign land, good health, sound sleep, longevity, sharp intellect and communicative abilities, will possess knowledge of arts, be intent upon arguments, be an able speaker, will spend money lavishly, will have business in various directions, will honour wise men, guests, Gods and preceptors (elders), be skilful in pretending to serve others, be amiable, very judicial, hate arguments. When Mercury is weak by placement or strength/afflicted by malefic planets in the nativity it gives short/medium life span, bad health, problem to children, losses, loss of bed comforts, lack of decision, addiction, imprisonment and hospitalisation. Medically he/she can have problems related to nervous system, intestines, respiratory, cervical or Parkinson disease.

Mercury in Libra in Aspect to Others: Should Mercury be aspect by the Sun, he/she will suffer from penury and acute grief, will have a sick physique, be interested in serving others and will be censured. Should Mercury be aspect by the Moon, he/she will be trustworthy, affluent, and firmly pious, devoid of sickness, will have lasting family ties, be famous and be a king's minister. Should Mercury be aspect by Mars, he/she will be troubled by diseases and enemies, be distressed, will incur royal insult and will be deprived of all worldly objects. Should Mercury be aspect by Jupiter, he/she will be highly learned, will fulfil his promise, will be leader of a country/city/group of men and be famous. Should Mercury be he/she aspect by Venus, the person will be fortunate, soft in disposition, be happy, will enjoy good robes, make up and will steal the hearts of the fair sex. If Saturn throws his aspect on Mercury, he/she will be devoid of happiness, be dirty, will experience many diseases and evils, will be subjected to grief on account of his relatives and be distressed.

Jupiter in Libra: Jupiter in Libra indicates that he/she will be handsome, free, open-minded, hasty, attractive, just, courteous, strong, able, exhaustion from over-activity, religious, competent, unassuming, pleasing. Jupiter, when strong and well placed in Libra, provides initiative in learning, interest in studies, reading, writing and marketing, innovative and religious nature, good conduct and morality and support from father/elder brother/younger brother, will be a scholar, will have many sons, be endowed with foreign assignments, will be very affluent, will earn money through dance and drama. However, if it is weak by placement and strength in the nativity then there may be, problems in learning, loss of efforts, retarded physical growth, losses in new initiate, involvement in litigations, problem to younger/elder brother/father, nature of cunningness, shrewdness, greed etc. Medically there may be liver disorder, jaundice and dark circles under eyes.

Jupiter in Libra in Aspect to Others: If Jupiter is aspect by the Sun, he/she will be endowed with attendants and quadrupeds and will wander verily, will have a long body, be learned and be a king's minister. If Jupiter is aspect by the Moon, he/she will be abundantly rich, be very calm, Sweet, be dear to mother and wife and will enjoy much pleasures. If Jupiter is aspect by Mars, he/she will be dear to the fair sex, be learned, courageous, affluent, and happy and be of royal Scion. If Jupiter is aspect by Mercury, he/she will be learned, Skilful, sweet, fortunate, endowed with riches, be highly virtuous and be splendour. If Jupiter is aspect by Venus, he/she will be very attractive, affluent, will wear excellent ornaments, be merciful and will enjoy excellent sleeping comforts and excellent robes. If Jupiter is aspect by Saturn, he/she will be a scholar, will be endowed with abundant wealth and corns, be excellent among the people of his village/town, be dirty, ugly and be devoid of wife.

Venus in Libra: Venus in Libra indicates that he/she will be statesman, poet, intelligent, generous, philosophical, handsome, matrimonial felicity, successful marriage, and passionate, proud, respected, intuitive, sensual, wide travels. Venus, when strong and well placed, blesses him with magnificence, attractive personality, good health and longevity, interest in media, art and architecture, global awareness, knowledge of foreign languages, recognition and happy marital life will acquire hard-earned money, be valorous, endowed with superior robes, interested in living in foreign countries, rich, meritorious, be fortunate. If it is afflicted in the nativity by malefic planets or weak by placement and strength/afflicted by malefic planets then it gives, bad reputation, lack of decision and awareness, medium life span, delay in marriage, problem to wife and health problems like, skin disease, eye diseases, kidney failures, renal diseases, infertility and ovarian problems in women.

Venus in Libra in Aspect to Others: If Venus is aspect by the Sun; he/she will acquire an excellent wife and abundant wealth, will become a great man and will be subdued by women. If Venus is aspect by the Moon, he/she is born of an excellent mother, will be endowed with happiness, wealth, respect and sons, will have excellence and be splendour. If Venus is aspect by Mars, he/she will marry a bad female, will lose home and wealth on account of females and be sensuous. If Venus is aspect by Mercury, he/she will be splendour, sweet, fortunate, happy, bold, wise and virtuous and will possess distinguished strength. If Venus is aspect by Jupiter, he/she will be endowed with wife, sons, abodes, conveyances, riches etc. and will achieve desired objects. If Venus is aspect by Saturn, he/she will have little happiness, little wealth, will be reprobate, will marry a mean lady and will suffer from diseases.

Saturn in Libra: Saturn in Libra indicates that he/she will be famous, founder of institutions and the like, rich, tall, fair, self-conceited, handsome, tactful, powerful, respected, sound judgement, antagonistic, independent, proud, prominent, charitable, sub-servant to females. Saturn when is strong and well placed, it provides intellect, constant gains through speculation and employees, intelligent children, interest in technical studies, will be rich, soft-spoken, will earn money and honours from foreign countries, be a king, or a scholar, will have wealth protected by his/her relatives, be senior in the circle, will attain a high status owing to his gracious speech in an assemblage, be good and have a sense of justice and fairness. However its weak strength/afflicted by malefic planets in the nativity results in problem to children, obstructive higher studies, losses through speculation and employees, emotional setbacks or depression, joint pains, orthopaedic problems.

Saturn in Libra in Aspect to Others: Should Saturn is aspect by the Sun, he/she will be clear in speech, will lose wealth, be a scholar, will eat in other's houses and be weak in constitution. If Saturn is aspect by the Moon, he/she will gain wealth through women, will be honoured by ministers of kings, be dear to women and be endowed with family. If Saturn is aspect by Mars, he/she will be skilful in war preparations, but will be away from war, will speak much and be endowed with wealth and family. If Saturn is aspect by Mercury he/she will always be jocularly disposed, be equal to a neuter, will serve females and be base. If Saturn is aspect by Jupiter, he/she will share happiness and misery of others will do others' jobs; be dear to people, charitable and industrious. If Saturn is aspect by Venus, he/she will be happy on account of women and wine (i.e. will enjoy these), be endowed with gems, be very strong and be dear to the king.

Predictions by Planets in Scorpio

Sun in Scorpio: Sun in Scorpio indicates that he/she will be adventurous, bold, fearing thieves and robbers, reckless, cruel, stubborn, unprincipled, impulsive, idiotic, indolent, surgical skill, dexterous, military ability. Strong Sun provides command in profession, wisdom, recognition through good deeds, interest in politics, be cruel and be attached to mean women, will follow bad course, be a miser, be fond of promoting quarrels, be unfortunate in respect of parents, possessed with uncontrollable jealousies. However, if it is weak or afflicted by malefic planets in the nativity, it causes losses in profession, dissatisfaction, bad reputation or humiliation, problem to father or children. Medically, he/she can suffer from heart problem, low or high BP, problem in knee joints, stomach disorders, weak eyesight.

Sun in Scorpio and in aspect to others: Should the Sun be aspect by the Moon, he/she will be interested in giving away gifts, will have many servants, be charming, be dear to fair sex and will have a soft physique. Should the Sun be aspect by Mars, he/she will display his courage in battle, be cruel, will possess eyes, hands and legs of blood-red colour, and be splendour and strong. If the Sun is aspect by Mercury, he/she will be a servant, will do others' jobs, will not have much wealth, be devoid of strength, be subjected too much grief and will possess a dirty body. Should Jupiter aspects the Sun, he/she will have plenty of money, will donate, and be a king's minister, a judge and a supreme person. If Venus aspects the Sun, he/she will be the husband of a bad woman, will have many enemies, but few relatives, be poor and will suffer from leprosy. In case Saturn aspects the Sun, he/she will be subjected to grief on account of physical ailments, will have intense passion in his undertaking, and is dull-witted.

Moon in Scorpio: Moon in Scorpio indicates that he/she will be broad eyes, wide chest, round shanks and thighs, isolation from parents or preceptors, brown complexion, straight-forward, frank, open-minded, cruel, simulator, malicious, sterility, agitated, unhappy, wealthy, impetuous, obstinate.

Moon, if strong strength, blesses with affluent father and mother, a miser, coarse physique and nose, will be cruel, be a thief, be sick in childhood, will have spoiled chin and nails, but beautiful eyes, will be plentiful, industrious, skilful, fond of others' housewives, devoid of relatives, insane or infatuated with passion, valorous, will lose wealth due to royal wrath and will have a big abdomen and a big head. But if it is weak by placement and strength/afflicted by malefic planets, it causes lack of analysing power, obstructions in early settlement of life and problem to father or mother. Medically he/she can suffer from immune deficiency, excess of phlegm and congestion, respiratory problems, disturbed sleep, menstrual and ovarian problems in women.

Moon in Scorpio and in aspect to others: If the Moon is aspect by the Sun, he/she will hate people, be learned, wandering nature and rich, but be not happy. Should Mars aspect the Moon, he/she will have incomparable courage, be equal to a king, be endowed with wealth, be valorous, unconquerable in battle and be a voracious eater. If Mercury aspects the Moon, he/she will not be skilful, be hard in speech, will obtain twins, be tricky, will produce fictitious things and be an expert singer. If Jupiter lends aspect to the Moon, he/she will be interested in performing his duties, will be biased towards people, be wealthy and be beautiful. If Venus aspects the Moon, he/she will be highly intelligent, be fortunate, be endowed with riches, conveyances and beauty and will lose strength on account of women. If Saturn aspects the Moon, he/she will have base sons, be a talebearer; be sick, poor and untruthful.

Mars in Scorpio: He/She will be middle stature, clever, diplomatic, positive tendency, indulgent, tenacious memory, and malicious, aggressive, proud, haughty, great strides in life. Strong Mars provides vigour, vitality, agility, decisiveness, financial stability, excellent decisive power, good health, longevity, interest in judiciary, armed forces, sports, will be attached to trade, be leader of thieves, skilful in his duties, be interested in wars, be highly sinful, will do big crimes, will be perfidious towards his enemies, will betray, be disposed towards killing, be unhelpful, will be endowed with lands, sons and wife and be troubled by poison, fire, weapons and wounds. When it is afflicted by malefic planets or weak by placement and strength in the nativity, it causes health problems, instability in financial position, short or medium life span, losses through disputes, fire, arson, and riots. Finally, the weak strength of Mars causes blood infections, high or low BP, obstruction in physical growth, muscle cramps, diabetic problems, and stomach disorders.

Mars in Scorpio in Aspect to Others: If Mars is aspect by the Sun, he/she will possess wealth, wife and children, be a king's minister, be a justice, be famous and be a charitable king. If Mars is aspect by the Moon, he/she will be bereft of mother, will have a wounded body, will hate his own people, will not have friends, be jealous and will have female children. If Mars is aspect by Mercury, he/she will be an expert in stealing others' money, be a liar, highly libidinous, be hostile and will frequently visit prostitutes. If Mars is aspect by Jupiter, he/she will be learned, sweet-spoken, fortunate, and dear to parents, be very affluent and will be a king par excellence. If Mars is aspect by Venus, he/she will be imprisoned due to females and will be deprived of his money on account of females, more than once. If Mars is aspect by Saturn, he/she will be capable of hindering thieves in spite of his being not valorous, will be devoid of his men and will maintain another woman.

Mercury in Scorpio: He/She will be short, curly hair, incentive to indulgence, liable to disease of the generative organ, general debility, crafty, malicious, selfish, subtle, indiscreet, bold, and reckless. Strong Mercury blesses with sharp intellect, fulfilment of desires in life, gains through friends and elder brother and lot of communicative and counselling abilities, will experience troubles, grief and evils, will hate the virtuous, will be devoid of truth, religion and shame, be a dunce, be not virtuous, be a miser, will cohabit with wicked women, be fond of giving cruel punishments, be interested in blameworthy jobs, will incur debts, will join base men and will steel other's properties, a secretive. If Mercury is weak in strength/afflicted by malefic planets in the nativity it gives problem from elder brother and friends, indecisiveness, obstructions in gains, dissatisfaction and losses. Medically he/she can have problems related to nervous system, intestines, respiratory, cervical, constipation or Parkinson disease.

Mercury in Scorpio in Aspect to Others: If Mercury is aspect by the Sun, he/she will be truthful, be very happy, be honoured by the king and be patiently disposed. If Mercury is aspect by the Moon, he/she will steal the hearts of the fair sex, will serve others, be dirty and be bereft of virtues. If Mercury is aspect by Mars, he/she will be a liar, a sweet speaker, will promote quarrels, be learned, affluent, dear to king and valorous. If Mercury is aspect by Jupiter, he/she will be happy, will possess a glossy and hairy physique, will have attractive hair, will be very rich, will command others and be sinful. If Mercury is aspect by Venus, he/she will be in royal service, be fortunate, principal among men or in his town will speak skilfully, be trustworthy and will be endowed with a wife. If Mercury is aspect by Saturn, he/she will experience miseries, be fierce, be intent upon doing cruel activities and be devoid of his own men.

Jupiter in Scorpio: He/She will be tall, somewhat stooping, elegant manners, serious, exacting, well built, superior air, selfish, imprudent, weak constitution, sub-servant to women, passionate, conventional, proud, zealous, ceremonious, unhappy life. Jupiter, if strong and well placed, it provides innovative nature, social status through knowledge and teaching, interest in marketing, male offspring, religious nature, good conduct and morality, be a king, will be a commentator of many Languages, be skilful, will construct temples and towns, will have many wives, but few sons. If it is afflicted by malefic planets or weak by placement and strength then it causes loss of reputation, nature of cunningness, shrewdness, greed etc, delay in marriage, problem in continuation of family life, problem to father/husband/children, health problems like, liver disorder, jaundice, diabetes, weak right eyesight and dark circles under eyes.

Jupiter in Scorpio in Aspect to Others: If Jupiter is aspect by the Sun, he/she will be charitable, will be truthful, will have famous sons, be very fortunate and will have abundant hair on the body. If Jupiter is aspect by the Moon, he/she will be a historical and poetical writer, will be endowed with many precious stones, be dear to women, be a king and be highly learned. If Jupiter is aspect by Mars, he/she is of royal scion, will be valorous, fierce, endowed with knowledge of politics, be modest, affluent and will have a disobedient wife and disobedient servants. If Jupiter is aspect by Mercury, he/she will be a liar, be crafty, sinful, will be skilful in detecting other's defects, will serve others, be grateful, be modest and be not outspoken. If Jupiter is aspect by Venus, he/she will be very happy in respect of residences, sleeping comforts, robes, scents, garlands, ornaments and wife and be very timid. If Jupiter is aspect by Saturn, he/she will be dirty, miserly, sharp,

Venus in Scorpio: He/She will be broad features, quarrelsome, medium stature, and independent, artistic, unjust, proud, disappointed in love, haughty, not rich. Strong Venus blesses him with beautiful and supportive wife, relations with foreign land, longevity, global awareness, jealous, be very malicious, be not religious, be argumentative, be crafty, be not attached to brothers, be not fortunate, will be troubled by enemies, be distressed, will be inimical to unchaste women, be skilful in killing, will incur heavy debts, will suffer penury, be proud and will contract venereal diseases. When Venus is weak by placement and strength/afflicted by malefic planets in the nativity it gives short or medium life span, bad health, problem from wife, losses, loss of bed comforts, addiction, imprisonment and hospitalisation. Medically he/she may suffer from diabetes, problems in eyes, reproductive organs, menstrual and ovarian problems in women.

Venus in Scorpio in Aspect to Others: If Venus is in aspect to the Sun, he/she will be miserable on account of women and will lose wealth and happiness on account of them, will be a king and be learned. If Venus is aspect by the Moon, he/she will be imprisoned, very fickle-minded, be libidinous, will marry a base lady and will be bereft of children. If Venus is aspect by Mars, he/she will be devoid of wealth, honour and happiness, will do others' jobs and will perform dirty jobs. If Venus is aspect by Mercury, he/she will be foolish, profligate, unworthy, be not in good terms with his/her own relatives, and be immodest, thievish, mean and cruel. If Venus is aspect by Jupiter, he/she will be endowed with beautiful eyes, will have a charitable wife, will possess a beautiful and long body and will have many sons. If Venus is aspect by Saturn, he/she is very dirty, lazy, thief, wandering-natured and serves others.

Saturn in Scorpio: He/She will be rash, indifferent, hard-hearted, adventurous, petty, self-conceited, reserved, unscrupulous, violent, unhappy, danger from poisons, fire and weapons, wasteful, unhealthy. Strong Saturn blesses him with good education, gains of assets, interest in technical fields, support from mother, will be hostile, be crooked, affected by poison and weapons, very ill-tempered, miserly, egoistic, rich, capable of stealing others' money, malicious, very miserable and will face destruction, misery and diseases, impatience, have terrific willpower and much energy, determination, secretive and unforgiving. If it is weak by placement and strength/afflicted by malefic planets in the nativity then there may be, obstructions in studies, difficulties in enjoying domestic peace, loss of assets, delay in marriage, problem from mother, losses through employees or servants and may be cardiac problem High or low BP, joint pains, orthopaedic problems.

Saturn in Scorpio in Aspect to Others: If Saturn is aspect by the Sun, he/she will be interested in agriculture, be very affluent, be endowed with cows, buffaloes and horses, and be fortunate and industrious. If Saturn is aspect by the Moon, he/she will be fickle minded, base, will join mean and ugly women and be devoid of happiness and wealth. If Saturn is aspect by Mars, he/she will kill animals, be base, be a leader of robbers, be famous and be fond of joining other women, meat and wine. If Saturn is aspect by Mercury, he/she will be untruthful, not virtuous, will eat much, be a famous thief and be devoid of happiness and riches. If Saturn is aspect by Jupiter, he/she will be endowed with happiness, wealth and fortune, be a king's minister and be chief. If Saturn is aspect by Venus, he/she will be quite unsteady in disposition, very ugly, will join other women and courtesans and be bereft of pleasures.

Predictions by Planets in Sagittarius

Sun in Sagittarius: He/She will be short-tempered, spoil, reliable, rich, obstinate, and respected by all, happy, popular, religious, wealthy, musician. Sun, when strong and well placed, blesses him with excellent administrative qualities, wisdom, recognition, interest in politics, affluent father, and timely settlement in life and interest in religion, will be endowed with wealth, be dear to king, learned, will respect Gods and Brahmins, be skilful in rendering training in use of weapons and arrows and breeding of elephants, be always peaceful, be rich, will possess a broad and beautiful physique. When it is weak in strength/afflicted by malefic planets, it causes obstructions in settlement in life, dissatisfaction, loss of respect, problem from father/children, can suffer from cardiac problem, problem in thighs, stomach disorders, and weak eyesight.

Sun in Sagittarius and in aspect to others: If the Sun is aspect by the Moon, he/she will be endowed with eloquent speech, wisdom, wealth and sons, be equal to a king and be devoid of misery, possess a pleasing body. If Mars lends aspect to the Sun, he/she will earn fame through battle, be endowed with clarity of speech, money and happiness and be short-tempered. Mercury aspecting the said Sun denotes that he/she will possess sweet speech and will have knowledge of writing, literature, arts, assembly, Journey and minerals. If Jupiter aspects the Sun, he/she will move in royal palaces, or be a king himself, will possess elephants, horses and wealth and be ever after learning. If Venus aspects the Sun, he/she will enjoy women of superior class, be endowed with scents and garlands and be peaceful. If Saturn aspects the Sun, he/she will be unclean, will eat other's food, will join bad men and will breed animals.

Moon in Sagittarius: He/She will be face broad, teeth large, skilled in fine arts, indistinct shoulders, disfigured nails and arms, deep and inventive intellect, yielding to praise, good speech, upright, help from wife and women, happy marriage, many children, good inheritance, benefactor, patron of arts and literature, ceremonial-minded, showy, unexpected gifts, author, reflective mentality, inflexible to threats. Strong Moon is favourable for enjoying strong marital tie, longevity, trouble free life, good analysing power, abundant inheritance, unearned wealth, interest in occult science, sound sleep, interest in finance and cooking, will be dwarfish, will have round eyes, big heart, waist and hands, be a good speaker, will have prominent shoulders and neck, will live near watery zone, be very strong, be grateful and distinguished. However its weak strength/afflicted by malefic planets, may cause lack of analysing power, humiliation, short/medium life span, problems in enjoying happy marital life, lack or loss of inheritance coupled with ups and downs in life. On health ground he/she can suffer from mental stress, immune deficiency, excess of phlegm and congestion, respiratory problems, disturbed sleep, problem in pelvic area, genitals and gynaecological problems in females.

Moon in Sagittarius and in aspect to others: If the Moon is aspect by the Sun, he/she will be a king, be affluent, valorous, and famous and will have incomparable happiness and conveyances. Should Mars aspect the Moon, he/she will be an

Army chief, will be very rich, fortunate, and famous for his valour and will have a large working force. If the Moon is aspect by Mercury, he/she will have many servants, will be an expert astrologer and artist and be a skilful dancer. If Jupiter aspects the Moon, he/she will possess very attractive physique, will be a king's minister and will be endowed with wealth, virtues and happiness. Should Venus aspect the Moon, he/she will be happy, beautiful and fortunate, be endowed with sons, wealth and sexual desires and will have good friends and wife. If Saturn aspects the Moon, he/she will be sweet in speech, will speak good words, will have wide knowledge of Shashtra, be truthful, soft and be a king's man

Mars in Sagittarius: He/She will be gentlemanly, many foes, famous minister, statesman, open, frank, pleasure loving, few children, liable to extremes, conservative, indifferent, exacting, impatient, severe, quarrelsome, litigation troubles, good citizen. Strong Mars is favourable for enjoying vigour and vitality, gains through speculation, easy settlement of life of children, interest in higher studies and sports, have highest levels of physical energy and enthusiasm, sense of justice is very strong, be outspoken debater, having a good fun, be a born gambler, approach sex as a sport, and hard to tie you down to one person. When Mars is weak by placement and strength/afflicted by malefic planets, in the nativity, it causes losses through speculation, temperamental nature, emotional setbacks, and problem from children, lack of intellect and obstructions in pursuing higher studies. In addition to this the weak strength of Mars causes blood infections, erratic blood circulation, obstructions in physical growth, muscle cramps, backache, diabetes, stomach disorders.

Mars in Sagittarius in Aspect to Others: If Mars is aspect by the Sun, he/she will be world honoured, be fortunate, will live in forests, hills and fortresses and be cruel. If Mars is aspect by the Moon, he/she will be deformed, belligerent, learned and be always inimical to the king. If Mars is aspect by Mercury, he/she will be a scholar, will be skilful, learned in fine arts and be highly learned in general. If Mars is aspect by Jupiter, he/she will be devoid of wife and happiness, will be beyond the reach of enemies, be wealthy and be fond of exercises. If Mars is aspect by Venus, he/she will be very much dear to women, be interested in cosmetics, make-up, be charitable, be libidinous and be fortunate. If Mars is aspect by Saturn, he/she will have defective body, is sinful, wandering-nature, devoid of happiness and interested in others' religion.

Mercury in Sagittarius: He/She will be taste in sciences, respected by polished society, tall, well built, learned, rash, superstitious, vigorous, executive, diplomatic, cunning, just, and capable. Mercury, if good placement and strength, blesses with sharp intellect, recognition, success of a high order in professional career, communicative and counselling abilities, will be famous, liberal, will have knowledge of Vedas and Shashtra, be valorous, will practice abstract meditation, be a minister, or family priest, be chief among his race-men, will be very rich, be interested in performing Yagnas and teaching Vedas, be a skilful speaker, be charitable and be an expert in writing and fine arts. When Mercury is weak by placement and strength/afflicted by malefic planets, in the nativity it causes indecisiveness, lack of communication skills, loss of reputation, dissatisfaction and losses in professional career etc. Medically he/she can suffer from arthritis, problem related to nerves, intestines, respiratory tract troubles, cervical, constipation or Parkinson disease.

Mercury in Sagittarius in Aspect to Others: If Mercury is aspect by the Sun, he/she will suffer from urinary diseases and epilepsy and be peaceful in disposition. If the Moon aspects Mercury, will be a writer, be beautiful par excellence, be very affluent, trustworthy, amiable and happy. If Mars aspects Mercury, he/she will be a leader of townsmen, or thieves, will reside in forests and will be a famous writer. Should Jupiter aspect Mercury, he/she will be rich by memory, intelligence and decadency, be beautiful, noble and knowledgeable, be a king's minister or his treasurer and be a writer. If Venus aspect Mercury, he/she will educate boys and girls, be wealthy, soft in disposition and brave. If Saturn aspect Mercury, he/she will be intent upon living in forests, will eat much, be wicked, dirty and will be unsuccessful in all his undertakings.

Jupiter in Sagittarius: He/She will be pretty inheritance, wealthy, influential, handsome, noble, trustworthy, charitable, good executive ability, weak constitution, artistic qualities, poetic, open-minded, good conversationalist. Strong Jupiter blesses him with innovative nature, status through knowledge, teaching and consultancy, interest in marketing, good health and longevity, blesses with male offspring, religious nature, good conduct and morality, will be a preceptor, will conduct religious vows, initiations, sacrifices, will have lasting wealth, be charitable, be friendly to his own men, be the head of a zone, or a minister, will live in many countries, fond of sports and travel. If it is weak in strength/afflicted by malefic planets, it causes short or medium lifespan, bad health, bad reputation, and nature of cunningness, shrewdness, greed, problem to father/husband/children and health problems like, stomach disorder, jaundice, anaemia, diabetes, headaches and dark circles under eyes.

Jupiter in Sagittarius in Aspect to Others: If Jupiter is aspect by the Sun, he/she will be inimical to the king and will be miserable being bereft of wealth and relatives. If Jupiter is aspect by the Moon, he/she will enjoy many kinds of happiness, be very fortunate in respect of wife and will have the pride of honour, wealth and possessions. If Jupiter is aspect by Mars, he/she will be injured in war, be cruel, will cause torture, will harm others and will lose children. If Jupiter is aspect by Mercury, he/she will be a minister, or a king, will be happy with wealth, fortunes and progeny and will enjoy all kinds of delights. If Jupiter is aspect by Venus, he/she will be happy, learned, be devoid of blemishes, be long-lived, fortunate and will be blessed by Goddess Lakshmi. If Jupiter is aspect by Saturn, he/she will be dirty, will be fear-stricken, be neglected by the people of his village/town and be devoid of happiness, pleasures and virtues.

Venus in Sagittarius: He/She will be medium height, powerful, wealthy, respected, impertinent, generous, frank, happy domestic life, and high position, philosophical. Venus, when strong and well placed, blesses him with satisfactory income, beautiful and supportive wife, global awareness, gains, support from elder brother or friends, fulfilment of desire in life, interest in media, art and sculpture, will have dutifulness and wealth, be dear to all people, be splendour, be an excellent personage, will shine like the Sun before his family members, be a scholar, will enjoy wealth, wife and fortunes, be a king's minister, be skilful, will have a stout and long physique and be respected by all. When Venus is weak by placement and strength/afflicted by malefic planets, in the nativity it causes problem to/ from wife, losses, dissatisfaction from income, problem to from elder brother or friends. As far as health is concerned, the weak strength of Venus causes diabetes, kidney stone or Gall bladder, renal problems, eyes, weakness of reproductive organs, low sperm count, menstrual and ovarian problems in women and problems in lower portion of legs.

Venus in Sagittarius in Aspect to Others: If Venus is aspect by the Sun, he/she will be very short tempered, learned, wealthy, be liked by all and will go to foreign countries. If Venus is aspect by the Moon, he/she will be famous, be kingly, will eat rich food, be distinguished and will possess incomparable strength. If Venus is aspect by Mars, he/she will be very much ill-disposed towards women, will be both happy and miserable, be rich and will possess cows. If Venus is aspect by Mercury, he/she will be endowed with all kinds of robes, ornaments, foods and drinks and will possess plenty of money and horses and cows, will possess many wives and many sons, be happy and be very wealthy. If Venus is aspect by Saturn, he/she will be always amassing financial gains, be happy, will enjoy pleasures and abundant wealth and be fortunate.

Saturn in Sagittarius: He/She will be Pitiful, artful, cunning, famous, peaceful, faithful, pretentious, apparently generous, troubles with wife, courteous, dutiful children, generally happy. Strong Saturn blesses him with efforts of a high order, learning, new ventures, communication and expression powers, interest in technical fields, support from younger siblings, will be skilful in behaviour, teaching, be famous due to virtuous children, family profession and his own virtues, will enjoy excellent affluence in his old age. If it is weak by placement and strength/afflicted by malefic planets, in the nativity then there is lack of interest in learning, lack or loss of efforts, lack of confidence, losses through new ventures/employees or problem to / from younger siblings, may suffer from respiratory problems, retarded physical and mental growth, joint pains, and orthopaedic problems.

Saturn in Sagittarius in Aspect to Others: If Saturn is aspect of the Sun he/she will be the father of others' children and through these children he will attain wealth, name, fame and honour. If Saturn is aspect by the Moon, he/she will be bereft of mother, will have two names and be endowed with wife, children and wealth. If Saturn is aspect by Mars, he/she will be troubled by windy diseases, will dislike people, be sinful, mean, blameworthy etc. If Saturn is aspect by Mercury, he/she will be equal to a king, be happy, be a preceptor, and be honourable, rich and fortunate. If Saturn is aspect by Jupiter, he/she will be a king, or equal to a king, or a minister, or an Army chief and be free from all kinds of danger. If Saturn is aspect by Venus, he/she will have two mothers and two fathers, will live in forests and hills, will be unsteady and be endowed with many kinds of assignments.

Predictions by Planets in Capricorn

Sun in Capricorn: He/She will be mean-minded, stubborn, ignorant, miserly, pitiful, unhappy, boring, active, meddlesome, obliging, humorous, witty, affable, prudent, and firm. Sun, if good placement and strength, is favourable for enjoying strong marital tie, long lifespan, trouble free life, recognition, abundant inheritance, undue favours, unearned wealth, and interest in mystical sciences, will be interested in bad women, be greedy, will advance with mean jobs, be timid, devoid of relatives, fickle-minded, fond of wandering, will lose everything due to conflicts with his relatives. When it is weak or afflicted by malefic planets, it causes loss of respect, humiliation, short or medium life span, difficulties in enjoying marital life, lack or loss of inheritance, problem to father or children and upheavals in life. On health ground its weak strength causes heart problem, low or high Blood pressure, problem in pelvic region or genitals, piles, stomach disorders, weak eyesight, weakness of bones, low immunity.

Sun in Capricorn and in aspect to others: If the Moon aspects the Sun, he/she will be highly cunning in disposition and will lose his wealth and happiness on account of his befriending females. Mars aspecting the Sun, denotes that he/she will be troubled by diseases and enemies will be wounded by weapons on account of quarrelling with others and be deformed. If Mercury aspects the Sun, he/she will be brave, will have a eunuch's nature, will steal other's wealth and will have all limbs devoid of strength. Should Jupiter aspect the Sun, he/she will undertake to do auspicious deeds, be wise, will patronize all, widely famous and intelligent. If Venus aspects the Sun, he/she will deal with conch, coral and ruby will derive abundant wealth through prostitutes and females and be happy. If Saturn aspects the Sun, he/she will destroy his enemies and will have royal honours.

Moon in Capricorn: He/She will be ever attached to wife and children, virtuous, good eyes, slender waist, quick in perception, clever, active, crafty, somewhat selfish, sagacious, strategic, liberal, merciless, unscrupulous, inconsistent, low morals, niggardly and mean. Strong Moon governs the house of marriage, partnerships, tie-ups, foreign travel. It blesses him with suitable life partner, happy marital life, support from spouse, gains through partners or partnerships, foreign travel, relations with overseas, sound mind and sleep, interest in finance and cooking, will be a singer, will have stout body, , be distinguished, famous, less irascible, be libidinous, unkind and shameless. However, if it is weak in strength/afflicted by malefic planets then it causes losses through partnerships, delay in marriage, and problem from spouse, marital discord. Medically he/she may suffer from weak immunity, excess of phlegm, fevers, upper respiratory tract problems, sinusitis, disturbed sleep, mental stress and gynaecological problems in women.

Moon in Capricorn and in aspect to others: If the Moon is aspect by the Sun, he/she will be penniless, miserable, wandering-nature, interested in others' work, dirty and clever. If Mars aspects the Moon, he/she will enjoy abundant riches, be highly liberal, be fortunate, wealthy, will have conveyances and be brave. If Mercury aspects the Moon, he/she will be dunce, be interested in living in other places, be bereft of wife, be fickle minded and be devoid of happiness and money. Should Jupiter aspect the Moon, he/she will be a king, be incomparably brave, will have royal qualities and will possess many wives, children and friends. If Venus aspects the Moon, he/she will join others' wives, be endowed with wealth, ornaments, conveyances, garlands, be blameworthy and be issue less. If Saturn aspects the Moon, he/she will be indolent, dirty, be endowed with money, be troubled by sexual feelings, will join others' housewives and be untruthful.

Mars in Capricorn: He/She will be rich, high political position, many sons, brave, generous, love for children, middle stature, industrious, indefatigable, successful, penetrating, bold, tactful, respected, generous, gallant, influential. A well placed and strong Mars blesses him/her with enterprising efforts, good educational background, adequate support from mother and younger siblings, gains of assets, interest in sports, be wealthy, be endowed with happiness and pleasures, be famous, be an

Army chief, or a king, will possess a good wife. If it is weak in the chart by placement or strength/afflicted by malefic planets, there may be obstructions in acquiring education, lack of efforts and confidence, difficulties in forming properties, loss of assets, and lack of domestic peace along with problem to mother or younger brother. Medically he/she may suffer from blood infections, erratic blood circulation, obstructions in physical growth, muscle cramps, back pain, chest congestion, and fever.

Mars in Capricorn in Aspect to Others: If Mars is aspect by the Sun, he/she will have very dark body, be courageous, will have many wives, sons and abundant wealth and be very sharp. If Mars is aspect by the Moon, he/she will be fickle minded, be not well disposed towards his mother, be fond of beautification, be charitable, be not firm in friendship and be rich. If Mars is aspect by Mercury, he/she will walk very slowly, be not rich, will not have any profession, be not strong, be not outspoken and be not virtuous. If Mars is aspect by Jupiter, he/she will be very beautiful, will possess kingly qualities, will fulfil his undertakings, be long-lived and be endowed with relatives. If Mars is aspect by Venus, he/she will be endowed with various kinds of (carnal) pleasures, be interested in fostering women and be belligerent. If Mars is aspect by Saturn, he/she will be a king, be very affluent, will hate women, will possess many children, be learned, be devoid, of happiness and be timid in war.

Mercury in Capricorn: He/She will be selfless, business tendencies, economical, debtor, inconsistent, low stature, cunning, inventive, active, restless, auspicious, and drudging. If it is strong in strength, it blesses him with affluent father, good settlement of life, sharp intellect, excellent communication and expression power, an excellent memory and attention to the business at hand; a down-to-earth and practical thinker, very good powers of concentration, dignity, and an earnestness, planning and organizing is special strengths, task will surely be completed and done well. When it is weak in strength/afflicted by malefic planets it causes obstructions in settlement of life, dissatisfaction, weak analytical and assessment power and problem to father, suffer from neurological, intestinal, skin or respiratory problems, constipation, forgetfulness, Parkinson disease.

Mercury in Capricorn in Aspect to Others: If Mercury be aspect by the Sun, he/she will be a boxer, be very strong, one will eat abundantly, be censured, will speak sweetly and be famous. If Mercury be aspect by the Moon, he/she will derive his livelihood through water, be plentiful will sell flowers, liquor and bulbs, vegetables, will have fierce appearance and be not moving much. If Mercury be aspect by Mars, he/she will be infirm in speech, be calm in disposition, bashful and happy. If Mercury is aspect by Jupiter, he/she will be endowed with abundant money and grains, be honoured in his village/town and by his men, be happy and be famous. If Mercury is aspect by Venus, he/she will be the husband of a base woman, be ugly, be unintelligible, be troubled by sexual passion and will have many sons. If Mercury is aspect by Saturn, he/she will be sinful, very poor, miserable and mean.

Jupiter in Capricorn: He/She will be a weak body; not very intelligent, not much stamina or virility; will have to put in much hard labour for low financial returns; a little sense of cleanliness and lives at places other than his native place and is timid and melancholy. Jupiter, when strong, blesses him with happy marital life, sound health, foreign settlement, longevity, interest in religion and religious activities, reading and teaching, will be less virile, will experience much grief and difficulties, will be mean in conduct, be a dunce, will meet a bad end, will suffer from penury, will be bereft of auspiciousness, mercy, purity, affection to his relatives, will have an emaciated body, be timid. However if Jupiter is weak in strength/afflicted by malefic planets in the birth chart it causes losses, bad health, disturbed sleep, problem to father or elder brother, difficulty in settling down in overseas, marital discord. Medically he/she can might suffer from have problem related to diabetes, jaundice, liver disorder, dark circle under eyes and swollen feet.

Jupiter in Capricorn in Aspect to Others: If Jupiter is aspect by the Sun, he/she will be learned, be a king, be rich by birth, will enjoy various kinds of pleasures and be very courageous. If Jupiter is aspect by the Moon, he/she will be devoted to his/her parents, be superior by birth, and be learned rich, virtuous and very charitable. If Jupiter is aspect by Mars, he/she will be valorous, be an Army chief with the king, splendour, well-dressed, famous and honoured. If Jupiter is aspect by Mercury, he/she will be highly libidinous, important among his folk, rich by conveyances and wealth, famous and will have many friends. If Jupiter is aspect by Venus, he/she will be endowed with food, drinks, excellent residence, sleeping comforts, wealth, conveyances, excellent wife, ornaments and robes. If Jupiter is aspect by Saturn, he/she will be endowed with incomparable learning, be supreme, be a king of country, has attendants and quadrupeds and will enjoy pleasures.

Venus in Capricorn: He/She will be fond of low class women, imprudent, ambitious, unprincipled, licentious, boastful, subtle, learned, weak body. If Venus is strong, it is favourable for enjoying luxuries, comforts, romantic pleasures, interest in media, film, music, art and sculpture, beautiful and supportive wife, global awareness, and fulfilment of desire in life, intellect, recognition and success in professional career, have an emaciated body, be fond of aged women, suffer from heart diseases. When Venus is weak by placement or strength/afflicted by malefic planets in the nativity it causes problem to wife, losses, dissatisfaction from professional achievements, lack of awareness. As far as health is concerned, its weak strength causes diabetes, renal disease, eyes, knees, and weakness of reproductive organs, low sperm count and urinary tract troubles.

Venus in Capricorn in Aspect to Others: If Venus is in aspect to the Sun, he/she will be firm, very dear to females, very wealthy, very happy and truthful and be courageous. If Venus is aspect by the Moon, he/she will be very splendour, very valorous, very affluent and fortunate. If Venus is aspect by Mars, he/she will lose his spouse, will suffer many evils and diseases, will undergo difficulties and later on be happy. If Venus is aspect by Mercury, he/she will be learned, rich, skilful in sacred precepts, highly scholastic, truthful and happy. If Venus is aspect by Jupiter, he/she will enjoy robes and garlands, be beautiful, will be skilful in music and musical instruments and will posses a good wife. If Venus is aspect by Saturn, he/she will be endowed with servants, conveyances and wealth, be dirty and will possess a black, beautiful and broad physique.

Saturn in Capricorn: He/She will be intelligent, harmony and felicity in domestic life, selfish, covetous, peevish, intellectual, learned, suspicious, reflective, revengeful, prudent, melancholy, inheritance from wife's parties. Saturn good placement and strength is favourable for enjoying satisfactory social status, gains, good health and longevity, male children, good conduct and nature of possessiveness, will lord over the lands of others' females, be honourable, will respect others and is famous. However, if it is weak in strength/afflicted by malefic planets then it causes short or medium life span, loss of respect, difficulties in enjoying family life, problem to or from father or children, losses through servants/employees. As far as health is concerned, its weak strength may cause health problems related to teeth, throat infections, problem in right eye, arthritis, weak constitution, orthopaedic problems, and anaemia.

Saturn in Capricorn in Aspect to Others: If Saturn is aspect by the Sun, he will be sick, will have an ugly wife, will eat other's food, be miserable, wandering in nature and will carry loads. If Saturn is aspect by the Moon, he/she will be fickle-minded, untruthful, and sinful, will not have good terms with his mother, be rich and be sorrowful due to wandering. If Saturn is aspect by Mars, he/she will be very valorous, famous, be superior among great men and sharp. If Saturn is aspect by Mercury, he will carry loads, be clouded in mentality or aggressive in disposition, be good-looking, wandering-natured, learned, be not quite wealthy and be fortunate. If Saturn is aspect by Jupiter, he/she will be famous for his good qualities, be a king, or be of royal scion, long-lived and be free from diseases. If Saturn is aspect by Venus, he/she will be rich, addicted to other women, fortunate, happy and will enjoy food and drinks.

Predictions by Planets in Aquarius

Sun in Aquarius: He/She will be poor, unhappy, unlucky, unsuccessful, medium height, rare faculties, and self-esteem. Sun, when strong, is favourable for enjoying pleasures or recognition through supportive spouse, foreign travel, gain and recognition through partner, harmonious marital life, strong will power and social status of a high order, have enormous strength (courage), be very short-tempered, and be fortunate through other housewives. However its weak strength/affliction by malefic planets causes bad health of spouse, losses through partnerships, troubles in foreign travel, problem to from father/husband, delay in marriage, and difference of opinion with life partner. Its weak strength causes cardiac problems, low or high Blood Pressure, stomach disorders, weak eye sight, weakness of bones.

Sun in Aquarius and in aspect to others: If the Moon aspects the Sun, he/she will be highly cunning in disposition and will lose his/her wealth and happiness on account of his/her befriending females. Mars aspecting the Sun denotes that he/she will be troubled by diseases and enemies will be wounded by weapons on account of quarrelling with others and be deformed. If Mercury aspects the Sun, he/she will be brave, will have a eunuch's nature, will steal other's wealth and will have all limbs devoid of strength. If Jupiter aspects the Sun, he/she will undertake to do auspicious deeds, be wise, will patronize all, widely famous and intelligent. If Venus aspects the Sun, he/she will deal with conch, will be derived of wealth through prostitutes and females. If Saturn aspects the Sun, he/she will destroy his enemies and will prosper have royal honours.

Moon in Aquarius: He/She will be fair-looking, well-formed body, tall, large teeth, belly low, youngish, sensual, pure-minded, artistic, intuitive, diplomatic, lonely, energetic, emotional, esoteric, mystical, grateful, healing power. Moon, if strong, is favourable for enjoying good health and financial position, gains through litigation and disputes, have elevated nose, will be addicted to intoxicants, will be averse to the virtuous, will obtain illegal sons, ugly/diseased eyes, will have bad mentality, be miserable and will be very poor, unusual and unpredictable, sometimes seem eccentric and unconventional. However its weak strength/affliction by malefic planets might cause losses through fire, theft, arson, riots, disputes, debts. Medically he/she may suffer from weak resistance power, excess of phlegm and congestion, respiratory problems, diabetes, infertility, mental and gynaecological problems.

Moon in Aquarius and in aspect to others: If the Moon is aspect by the Sun, he/she will be very dirty in disposition, be valorous, will be like a king in guise, be virtuous and be an agriculturist. If Mars aspects the Moon, he/she will be quite truthful, will not inherit money from mother and elders; be indolent and mysterious and interested in others' work. If Mercury aspects the Moon, he/she will be skilful in treating others well, while hosting them, will be proficient in music, will be liked by the fair sex and will possess less money and less happiness. Should Jupiter aspect the Moon, he/she will own villages, agriculture lands and trees, be endowed with superior castles and superior ladies and will be devoted to sensual enjoyment, rich person, or a king. If Venus aspects the Moon, he/she will be base, issueless, friendless, timid, be censured by preceptors, be sinful, will have a bad wife and will be least happy. If the Moon has Saturn's aspect, he/she will have prominent nails and hair, be dirty, will seek union with other women, be a dunce, be irreligious and be rich possessing many immovable.

Mars in Aquarius: He/She will be unhappy, miserable, poor, not truthful, independent, unwise, wandering, impulsive, controversial, combative, well-versed in dialects, free, quick in forgiving and forgetting, conventional, danger on water, morose, meditative. If Mars is strong and well placed, it blesses him with agility, quick mobility, excellent communication, expression & writing abilities, support from younger siblings and interest in sports and armed forces, will be devoid of both affection and purity, will look like an old person, will die a bad death, will have spite, jealousy, untruthful disposition, afflicted speech and lost wealth, will be ugly, will have abundant hair on the body, will lose money in gambling, will have miserable profession, be fond of liquor and be unfortunate. However, a weak Mars/afflicted by malefic planets is likely to cause obstructions in learning, lack or loss of initiative and confidence along with problem from younger siblings. Its weak strength may cause blood infections, obstructions in physical growth, muscle cramps, respiratory problems, chest congestion, and fever.

Mars in Aquarius in Aspect to Others: If Mars is aspect by the Sun, he/she will have very dark body, be courageous, will have many wives, sons and abundant wealth and be very sharp. If Mars is aspect by the Moon, he/she will be fickle minded, be not well disposed towards his mother, be fond of beautification, be charitable, be not firm in friendship and be rich. If Mars is aspect by Mercury, he/she will walk very slowly, be not rich, will not have any profession, be not strong, be not outspoken and be not virtuous. If Mars is aspect by Jupiter, he/she will be very beautiful, will possess kingly qualities, will fulfil his undertakings, be long-lived and be endowed with relatives. If Mars is aspect by Venus, he/she will be endowed with various kinds of carnal pleasures, be interested in fostering women and be belligerent. If Mars is aspect by Saturn, he/she will be a king, be very affluent, will hate women, will possess many children, be learned, be devoid, of happiness and be timid in war.

Mercury in Aquarius: He/She will be middle stature, licentious, proud, quarrelsome frank, sociable, rapid strides in life, famous, scholar, cowardly, weak constitution. If Mercury is strong and well placed, it blesses him with strong marital tie, abundant inheritance, and long life span, easy gains in life, timely marriage, undue favours, will be bereft of good disposition and good deeds, will give up doing worthy things, be insulted by others, be impure, be not virtuous, be very wicked, be inimical to wife, be devoid of carnal pleasures, be very unfortunate, very timid, impotent, dirty and modest, do things that are unusual or savant grade just to shake up the establishment and create controversy, have no respect for tradition, love to rebel against the system. When it is weak in strength/afflicted by malefic planets it causes difficulties in enjoying happy and prosperous marital life, short/medium life span, humiliation, extra marital affairs, and weak analytical and decisive powers. Its weak strength causes problems related to nervous system, intestinal or skin problems, respiratory tract troubles, piles, problem in genitals, pelvic area, cervical, weak memory, Parkinson disease.

Mercury in Aquarius in Aspect to Others: If Mercury be aspect by the Sun, he/she will be a boxer, be very strong, one will eat abundantly, be censured, will speak sweetly and be famous. If Mercury is aspect by the Moon, he/she will derive his livelihood through water, be plentiful will sell flowers, liquor and bulbs, vegetables, will have fierce appearance and be not moving much. If Mercury is aspect by Mars, he/she will be infirm in speech, be calm in disposition, bashful and happy. If Mercury is aspect by Jupiter, he/she will be endowed with abundant money and grains, be honoured in his village/town and by his men, be happy and be famous. If Mercury is aspect by Venus, he/she will be the husband of a base woman, be ugly, be unintelligible, be troubled by sexual passion and will have many sons. If Mercury is aspect by Saturn, he/she will be sinful, very poor, miserable and mean.

Jupiter in Aquarius: He/She will be learned, not rich, controversial figure, philosophical, popular, compassionate, sympathetic, amiable, prudent, humanitarian, melancholic, meditative, dreamy, dental troubles. Jupiter good placement and strength is favourable for earning satisfactory income, gains, fulfilment of desires, support and co-operation from elder siblings / father / friends, interest in religion reading, teaching and marketing, will be a tale-bearer, interested in evil jobs, be always attached to bad men, be malicious, miserly, will suffer from diseases, and will lose wealth. However, if it is weak in strength/afflicted by malefic planets it causes dissatisfaction from income, losses, obstruction in gains, strained relations with father or elder siblings or friends. Medically he/she can have problem related to diabetes, jaundice, flatulence, liver disorder, dark circle under eyes and swollen foot.

Jupiter in Aquarius in Aspect to Others: If Jupiter is aspect by the Sun, he/she will be learned, be a king, be rich by birth, will enjoy various kinds of pleasures and be very courageous. If Jupiter is aspect by the Moon, he/she will be devoted to his parents, be superior by birth, and be learned rich, virtuous and very charitable. If Jupiter is aspect by Mars, he/she will be valorous, be an Army chief with the king, splendour, well-dressed, famous and honoured. If Jupiter is aspect by Mercury, he/she will be highly libidinous, important among his folk, rich by conveyances and wealth, famous and will have many friends. If Jupiter is aspect by Venus, he/she will be endowed with food, drinks, excellent residence, sleeping comforts, wealth, conveyances, excellent wife, ornaments and robes. If Jupiter is aspect by Saturn, he/she will be endowed with incomparable learning, be supreme, be a king of country, has attendants and quadrupeds and will enjoy pleasures.

Venus in Aquarius: He/She will be liked by all, middle stature, handsome, affable, persuasive, witty, timid, chaste, calm, helpful and humanitarian. Venus strong strength blesses with romantic pleasures, beautiful and supportive wife, affluent father, and early settlement in life, interest in media, film, music, art and sculpture, will suffer from fear and excitement, be not successful in undertakings, will go to other women, be not virtuous, be inimical to elders and to his children, be bereft of ablutions, like bathing and will not be endowed with good robes, ornaments and be dirty, romantic side is cool, calm and detached. However, if it is weak in strength/afflicted by malefic planets, the native may experience problem from wife or father, lack of awareness, struggles in life. Medically he/she can suffer from diabetes, renal problems, weak eyesight, and weakness of, reproductive organs, low sperm count.

Venus in Aquarius in Aspect to Others: If Venus is in aspect to the Sun, he/she will be firm, very dear to females, very wealthy, very happy and truthful and be courageous. If Venus is aspect by the Moon, he/she will be very splendours, very valorous, very affluent and fortunate. If Venus is aspect by Mars, he/she will lose his spouse, will suffer many evils and diseases, will undergo difficulties and later on be happy. If Venus is aspect by Mercury, he/she will be learned, rich, skilful in sacred precepts, highly scholastic, truthful and happy. If Venus is aspect by Jupiter, he/she will enjoy robes and garlands, be beautiful, will be skilful in music and musical instruments and will posses a good wife. If Venus is aspect by Saturn, he/she will be endowed with servants, conveyances and wealth, be dirty and will possess a black, beautiful and broad physique.

Saturn in Aquarius: He/She will be practical, able, diplomatic, ingenious, a bit conceited, prudent, happy, reflective, intellectual, and philosophical, vanquished by enemies. If Saturn is well placed as well as strong in strength, it blesses him with wise nature, sharp intellect, technical knowledge, good health and longevity. It also blesses the native with good conduct and the nature of possessiveness, will be a great liar, be eminent, be addicted to women and wine, be wicked, crafty, will fall prey to evil friendship, be very ill-tempered, be averse to knowledge, conversation and traditional law, be addicted to other women, be harsh in speech. If it is weak in strength/afflicted by malefic planets then it causes short/medium life span, weak health, loss of respect and health problems like, cancerous disease, early ageing process, teeth infections, weak constitution, and orthopaedic problems.

Saturn in Aquarius in Aspect to Others: If Saturn is aspect by the Sun, he will be sick, will have an ugly wife, will eat other's food, be miserable, wandering in nature and will carry loads. If Saturn is aspect by the Moon, he/she will be fickle-minded, untruthful, and sinful, will not have good terms with his mother, be rich and be sorrowful due to wandering. If Saturn is aspect by Mars, he/she will be very valorous, famous, be superior among great men and sharp. If Saturn is aspect by Mercury, he will carry loads, be clouded in mentality or aggressive in disposition, be good-looking, wandering-natured, learned, be not quite wealthy and be fortunate. If Saturn is aspect by Jupiter, he/she will be famous for his good qualities, be a king, or be of royal scion, long-lived and be free from diseases. If Saturn is aspect by Venus, he/she will be rich, addicted to other women, fortunate, happy and will enjoy food and drinks.

Predictions by Planets in Pisces

Sun in Pisces: He/She will be pearl merchant, peaceful, wealthy, uneventful, religious, prodigal, and loved by women. A strong Sun blesses him with good health, vitality and strong financial position, gains through disputes, litigation, satisfactory social status and interest in finance and administration, be friendly, will have tendency to amass, be fond of women and happy, and be learned, will destroy many enemies and be wealthy and rich, be endowed with wife, good sons and servants, will have wealth by means of smuggler, be an eloquent speaker, but a liar. However, if Sun is weak in strength/afflicted by malefic planets, it causes weak resistance power and losses through fire, theft, disputes, debts, litigation etc and problem to or from father. On the health ground the weak strength of Sun causes heart problem, erratic blood circulation, stomach disorders, weak eyesight, infertility, low immunity and weakness of bones.

Sun in Pisces and in aspect to others: If the Sun is aspect by the Moon, he/she will be endowed with eloquent speech, wisdom, wealth and sons, be equal to a king and a pleasing body. If Mars lends aspect to the Sun, he/she will earn fame through battle, be endowed with clarity of speech, money and happiness and be short-tempered. Mercury aspecting the Sun denotes that he/she will possess sweet speech and will have knowledge of writing, literature, arts, assembly, Journey and minerals. If Jupiter aspects the Sun, he/she will move in royal palaces, or be a king himself, will possess elephants, horses and wealth and be ever after learning. If Venus aspects the Sun, he/she will enjoy women of superior class, be endowed with scents and garlands and be peaceful. If Saturn aspects the Sun, he/she will be unclean, will eat other's food, will join bad men.

Moon in Pisces: He/She will be dominant, dealer in pearls and fond of wife and children, perfect build, long nose, bright body, annihilating enemies, subservient to opposite sex, handsome, learned, steady, simple, good reputation, loose morals, adventurous, many children, spiritually inclined later in life. When Moon is strong and well placed, it blesses him with sound mind, exceptional decisive and analytical power, happiness on account of children, gains through speculation, fulfilment of desire relationship, interest in cooking and finances, beautiful bodied, proficient in music. If it is weak in the chart/afflicted by malefic planets, then the native may experience weak analytical power, emotional setbacks, problem to or from children, losses through speculation, weak decisive power, mental stress and problem to or from Mother. The weakness of moon causes weak immunity, excess of phlegm and congestion, respiratory problems, stomach disorders, lack of mental peace, backache, diabetes, slip disc, infertility and gynaecological problems in women.

Moon in Pisces and in aspect to others: If the Moon is aspect by the Sun, he/she will be highly libidinous, be happy, is an Army chief, be very affluent and will have delighted wife. If Mars aspects the Moon, he/she will be insulted, be devoid of happiness, is an unchaste woman's son, will be interested in sins and will be valorous and a king. If the Moon is aspect by Mercury; he/she will be highly intelligent, be happy and will be surrounded by supreme females and be controlled by them. If Jupiter aspects the Moon, he/she will be beautiful, fierce, head of a district, be very affluent, beautiful and will be surrounded by many women. If Venus aspects the Moon, he/she will be interested in dance, instrumental music and songs and will steal the hearts of the fair sex. Should Saturn aspect the Moon, he/she will be deformed, be unfavourable to mother, be sexually distressed, be devoid of sons, wife and intelligence and will be attached to mean and ugly females.

Mars in Pisces: A strong and well-placed Mars blesses him/her with satisfactory social status through intellect, knowledge of technical fields, gains through recognition, male children and affluent father, decisive power, interest in forces, will have indifferent children, will live in foreign countries, be insulted by his/her own relatives, will lose all his/her wealth by his/her cunning and cheating disposition, will be depressed, be very miserable, will disrespect elders and Brahmins, be unkind, will conceive others). If it is weak in strength/afflicted by malefic planets, then it causes difficulties in acquiring or attaining satisfactory social status, difficulties in accumulating wealth, delay in marriage, loss of reputation, impatient nature and short temperament. Its weak strength causes muscular problems, blood infection, circulatory problems, and obstructions in physical growth, cramps, stammered speech and fever.

Mars in Pisces in Aspect to Others: If Mars is aspect by the Sun, he/she will be world honoured, be fortunate, will live in forests, hills and fortresses and be cruel. If Mars is aspect by the Moon, he/she will be deformed, belligerent, learned and be always inimical to the king. If Mars is aspect by Mercury, he/she will be a scholar, will be skilful, learned in fine arts and be highly learned in general. If Mars is aspect by Jupiter, he/she will be devoid of wife and happiness, will be beyond the reach of enemies, be wealthy and be fond of exercises. If Mars is aspect by Venus, he/she will be very much dear to women, be interested in cosmetics, make-up, be charitable, be libidinous and be fortunate. If Mars is aspect by Saturn, he/she will have defective body, is sinful, wandering-nature, devoid of happiness and interested in others' religion.

Mercury in Pisces: He/She will be a dependent, serves others, dexterous, petty-minded, respect for goods, and Brahmins. Strong Mercury blesses him with suitable life partner, happy marital life, foreign travels, gains & recognition through partner and long lasting partnerships, will be fond of good conduct and purity, will live in foreign countries, be issue less, will have a chaste wife, be virtuous, fortunate, be devoid of religion, be devoid of profane knowledge, will be proficient in bagging other's wealth and will be devoid of wealth. When it is weak in strength in the birth chart then the native may have weak analytical, assessment and decisive powers. In addition to this its weakness/afflicted by malefic planets, causes lack of confidence, problem to/ from spouse, losses in partnerships, obstructions in foreign travel, delay in solemnising marriage, difficulties in enjoying happy marital life. Medically he/she can suffer from problems related to nervous system, intestinal or skin problems, respiratory troubles, cervical spondolysis, insomnia, weak memory or Parkinson disease.

Mercury in Pisces in Aspect to Others: If Mercury is aspect by the Sun, he/she will suffer from urinary diseases and epilepsy and be peaceful in disposition. If the Moon aspects Mercury, he/she will be a writer, be beautiful par excellence, be very affluent, trustworthy, amiable and happy. If Mars aspects Mercury, he/she will be a leader of townsmen, or thieves, will reside in forests and will be a famous writer. If Jupiter aspects Mercury, he/she will be rich by memory, intelligence and decadency, be beautiful, noble and knowledgeable, be a king's minister or his treasurer and be a writer. If Venus aspects Mercury, he/she will educate boys and girls, be wealthy, soft in disposition and brave. If Saturn aspects Mercury, he/she will be intent upon living in forests, will eat much, be wicked, dirty and will be unsuccessful in all his undertakings.

Jupiter in Pisces: He/She will be good inheritance, medium height, two marriages and if with malefic, enterprising, political diplomacy, high position. Strong Jupiter is favourable for enjoying good health, acquiring knowledge, recognition and settlement of life through profession in finance, teaching, marketing, innovation and consultancies, will be honoured by friends and virtuous people, will be a headsman in the king's employ, be praiseworthy, unconquerable, rich, devoid of fear, be proud, firm in undertakings, be a king, be skilful in policies, training, behaviour and war tactics. When Jupiter is weak in strength/afflicted by malefic planets, in the birth chart, it causes dissatisfaction from professional achievements, problems to father, obstructions in satisfactory settlement of life and difficulties in enjoying adequate progeny bliss. The weak strength of Jupiter causes diabetes, jaundice, and liver disorder, dark circle under eyes, arthritis and swollen feet.

Jupiter in Pisces in Aspect to Others: If Jupiter is aspect by the Sun, he/she will be inimical to the king and will be miserable being bereft of wealth and relatives. If Jupiter is aspect by the Moon, he/she will enjoy many kinds of happiness, be very fortunate in respect of wife and will have the pride of honour, wealth and possessions. If Jupiter aspect by Mars, he/she will be injured in war, be cruel, will cause torture, will harm others and will lose children. If Jupiter is aspect by Mercury, he/she will be a minister, or a king, will be happy with wealth, fortunes and progeny and will enjoy all kinds of delights. If Jupiter is aspect by Venus, he/she will be happy, learned, be devoid of blemishes, be long-lived, fortunate and will be blessed by Goddess Lakshmi. If Jupiter is aspect by Saturn, he/she will be dirty, will be fear-stricken, be neglected by the people of his village/town and be devoid of happiness, pleasures and virtues.

Venus in Pisces: He/She will be tactful, learned, popular, modest, refined, powerful, exalted, and respected and pleasure seeking. Venus, when strong and well-placed blesses him with strong marital tie, abundant inheritance, and long lifespan, easy gains in life, happy marital life and luxurious life style, will be courteous, liberal, virtuous, very wealthy, will destroy enemies, be famous in the world, excellent, distinguished, dear to king, be endowed with good speech and wisdom, be liberal, will derive wealth and respect from the virtuous, will keep up his/her promise, will maintain his family members and will be endowed with knowledge mind. When it is weak in strength/afflicted by malefic planets, it causes difficulties in enjoying marital bliss of a high order, short life span, humiliation, extra marital affairs, lack or loss of inheritance and problem to/from spouse. Medically its weak strength causes diabetes, kidney failure, renal disease, and weakness of lens of eye, weakness of reproductive organs, piles, and genitals, low sperm count.

Venus in Pisces in Aspect to Others: If Venus is aspect by the Sun, he/she will be very short tempered, learned, wealthy, be liked by all and will go to foreign countries. If Venus is aspect by the Moon, he/she will be famous, be kingly, will eat rich food, be distinguished and will possess incomparable strength. If Venus is aspect by Mars, he/she will be very much ill-disposed towards women, will be both happy and miserable, be rich and will possess cows. If Venus is aspect by Mercury, he/she will be endowed with all kinds of robes, ornaments, foods and drinks and will possess plenty of money and horses and cows, will possess many wives and many sons, be happy and be very wealthy. If Venus is aspect by Saturn, he/she will be always amassing financial gains, be happy, will enjoy pleasures and abundant wealth and be fortunate.

Saturn in Pisces: He/She will be clever, pitiful, gifted, polite, happy, good, wife trustworthy, scheming, wealthy, and helpful. If Saturn is strong and un-afflicted, it is favourable for forming foreign relations, good span of life, good health, sound sleep, bed comforts, and protection from losses or expenses, gains through labour class, employees, servants and contract work, be chief among his relatives and friends, will have increasing wealth, be skilful in policy-making, be capable of diamond testing, be virtuous, modest and will later on acquire an authoritative position. When Saturn is weak in strength/afflicted by malefic planets, in the nativity it causes losses, obstruction in settling overseas, short life span, bad health, imprisonment, marital discord. Medically he/she can have problems such as arthritis, cancerous diseases, early ageing process, teeth infections, and painful diseases.

Saturn in Pisces in Aspect to Others: If Saturn is in aspect of the Sun, he/she will be the father of others' children and through these children he will attain wealth, name, fame and honour. If Saturn is aspect by the Moon, he/she will be bereft of mother, will have two names and be endowed with wife, children and wealth. If Saturn is aspect by Mars, he/she will be troubled by windy diseases, will dislike people, be sinful, mean, blameworthy etc. If Saturn is aspect by Mercury, he/she will be equal to a king, be happy, be a preceptor, and be honourable, rich and fortunate. If Saturn is aspect by Jupiter, he/she will be a king, or equal to a king, or a minister, or an Army chief and be free from all kinds of danger. If Saturn is aspect by Venus, he/she will have two mothers and two fathers, will live in forests and hills, will be unsteady and be endowed with many kinds of assignments.

10

Predictions by Sun and Moon 144-Combinations in Signs

General

The Sun and Moon are the most important Planets and provides horoscope predictions for 144 possible combinations of Sun in different Signs combined with Moon in various Rashi.

14.1 Predictions by Sun in Aries with Moon placed in Twelve Signs

1-1 Sun in Aries with Moon in Aries: He/She has a truly explosive, dynamic, hard hitting, powerful, and magnetic personality. Mind is always active in reading, talking and discussing. He/She is getting things done rapidly. This position provides exceptional leadership traits. 1-2 Sun in Aries with Moon in Taurus: He/She has forcefulness, enough tact, the charm and dash of manner, drive, determination, aggressiveness and rough edges and softness. He/She can be an agreeable and polite person, with tactful behaviour. He/She has a powerful magnetic quality, coupled with a very strong will and determination for leadership. He/She is natural leaders, and thinks highly of self, and never suffers from an inferiority complex. 1-3 Sun in Aries with Moon in Gemini: He/She has the qualities of most verbally active, dominating any conversation, fast mind, fast making decisions. He/She is likely to get far in the world because of quick wit and good intentions. In business, he/she commands facts and figures. He/She may aspire to become an executive. 1-4 Sun in Aries with Moon in Cancer: He/She has sensitivity, sympathy, and intuition. He/She is apt to put more thought and understanding into actions and have a knack for handling people. He/She is a humanitarian, self-centred, talented enough and sensitive, damaging to others and hurting them. 1-5 Sun in Aries with Moon in Leo: He/She has drive for success. He/She is an interesting, active, adventurous, and exciting person and has great energy, excitement, a sense of adventure, and boundless enthusiasm and magnetic personality. He/She is a natural born leader and is willing to listen and follow self. He/She has a warm heart and an affectionate nature. 1-6 Sun in Aries with Moon in Virgo: He/She is very cooling emotionally, practical and has a sensible attitude, driving determination to get things done, and to get them done right. He/She is definitely an achiever, combining cool-headed judgment. He/She is potentially a very fine executive and can keep making too many enemies with his aggressive and critical nature. He/She is not a romantic or a socially outgoing person, and may even be something of a 'lone wolf' type. 1-7 Sun in Aries with Moon in Libra: He/She has contradictions, emotional nature and is ambitious and eager to please people and is highly dependent and daydreamer. 1-8 Sun in Aries with Moon in Scorpio: quick to judge people and situations. He/She is romantic and highly idealistic e/She has a strong drive to control, a pleasant, smiling face and is a very competent and never inclined to give up. He/She is always ready to improve, and revolutionize, and seeking a new mountain to climb or a new challenge to meet. He/She is self-contained and capable. 1-9 Sun in Aries with Moon in Sagittarius: He/She has independent thought, action, and speech, popular appeal, strong marked executive powers, controlling power with ideas and principles. He/She is tough and does not have human frailties of pettiness, emotionalism, and jealousy but open and frank personality. He/She is definitely a leader. 1-10 Sun in Aries with Moon in Capricorn: He/She has talents for managing and directing, inner drive for success and knows how to judge people, spot problem, and problematic situations, and works always hard, and fast, compared to normal standards. He/She is full of energy and determination, and does always productive activities. 1-11 Sun in Aries with Moon in Aquarius: He/She is very sociable and magnetic, but not a tolerant person, quick to judge people and romantic and has a good sense, innate social skills, a good deal of self-esteem and material success. 1-12 Sun in Aries with Moon in Pisces: He/She is introverted, self-analysing, soul-searcher, very determined, timid, a little afraid of the world, and has some difficulty in concentrating, not executive type and isn't well suited for business.

14.2 Predictions by Sun in Taurus with Moon placed in Twelve Signs

2-1 Sun in Taurus with Moon in Aries: He/She is strong, determined to achieve the goal, very dogmatic in efforts and has congenial, peaceable, and placid attributes, sparkling enthusiasm of the bull and natural leadership traits. 2-2 Sun in Taurus with Moon in Taurus: He/She has concentration, staying power, authority and is second to none, honest and loyal. He/She enjoys good friends, good food, good music, and good life. 2-3 Sun in Taurus with Moon in Gemini: He/She has good verbal speaking and writing skills, stubbornness and jumps to conclusions and is mentally alert, restless and fickle and lacks memory usually. 2-4 Sun in Taurus with Moon in Cancer: He/She is sensitive, easily hurt, and not one to be pushed around easily. He/She relies on diplomacy, changes tactics to suit his purposes, and to avoid conflict. In business he/she is a shrewd planner and he never gives up on a job. 2-5 Sun in Taurus with Moon in Leo: He/She has a strong determined will, domineering and inflexible in attitude, well defined opinions, and a very forthright, honest, and fearless personality and is a good manager or boss. 2-6 Sun in Taurus with Moon in Virgo: He/She has charm, poise, balance and a good deal of common sense. He/She has charisma and magnetism combined with the ability to produce a comfortable lifestyle. 2-7 Sun in Taurus with Moon in Libra: He/She has more wit, ambition, intellect, a nice home and a close family, social contact and involvement, sense of humour and an easygoing outlook with optimism. 2-8 Sun in Taurus with Moon in Scorpio: He/She has more emotional intensity, emotionally temperament, and insists for black is black and white is white and is very headstrong, self-willed, and independent. 2-9 Sun in Taurus with Moon in Sagittarius: He/She is extroverted and sociable, and can get along with all types of people with the ability to say the right thing at the right time. He/She is a pretty good, law abiding, conventional human being. 2-10 Sun in Taurus with Moon in Capricorn: He/She is gentle, easygoing, stubborn, and can handle much stress and strain in life and is practical. He/She has much executive talent and is keen understanding of human nature. 2-11 Sun in Taurus with Moon in Aquarius: He/She is an easy; friendly, charming manner, pleasant demeanour and has diplomacy and friendly dealings and knows how to control wisely the people. He/She has an excellent, very studious mind, curious and inventive. 2-12 Sun in Taurus with Moon in Pisces: He/She has creativity and imagination and strong powers of restraint and has discrimination in selecting friends.

14.3 Predictions by Sun in Gemini with Moon placed in Twelve Signs

3-1 Sun in Gemini with Moon in Aries: He/She is mentally 'quick as a cat', good leader, witty, versatile, and mobile. He/She is naturally alerted, active, and aggressive and has a vast fund of information, stories, and quips. 3-2 Sun in Gemini with Moon in Taurus: He/She is easygoing, cool-headed, intellectual, impressive, and has driving personality, able to move mountain in record time and his judgment and business sense are inevitably sound and accurate. 3-3 Sun in Gemini with Moon in Gemini: He/She has very high static charges of electricity bouncing from one place to another, constantly active. He/She has tremendous mental capabilities, and a great business or law head. 3-4 Sun in Gemini with Moon in Cancer: He/She is full of stimuli and fascinations, and has a great need for peace, quiet, and stability in the life. 3-5 Sun in Gemini with Moon in Leo: He/She has a happy-go-lucky nature and turns from one subject to the next, from one relationship to another. 3-6 Sun in Gemini with Moon in Virgo: He/She is dominated, moody and temperamental. 3-7 Sun in Gemini with Moon in Libra: He/She is ethereal, light-hearted, and bright and totally enjoyable and doesn't have many tastes for hard work, but leads an active business and pleasure life. He/She is very busy in travelling and loves to travelling, socializing, working and playing. 3-8 Sun in Gemini with Moon in Scorpio: He/She is an exceptionally forceful, magnetic, and versatile person, especially quick and active. 3-9 Sun in Gemini with Moon in Sagittarius: He/She is frank, honest, and very restless and has highly individualistic nature. His honesty and frankness can lead him into difficulties at times. 3-10 Sun in Gemini with Moon in Capricorn: He/She is worldly-wise and shrewd and respects for tradition, law, and authority, and can meet any challenge. His business judgment is amazingly sharp. 3-11 Sun in Gemini with Moon in Aquarius: He/She has quick mind and intelligence, foresight and insight, and is a true progressive. He/She is genuinely humanitarian, generous, and possesses a desire to help others, an excellent teacher and has a romantic outlook. 3-12 Sun in Gemini with Moon in Pisces: He/She has a high degree of sixth sense and is intellectual, literary, aesthetic or artistic.

14.4 Predictions by Sun in Cancer with Moon placed in twelve signs

4-1 Sun in Cancer with Moon in Aries: He/She has a quick mind, remarkable memory and is very sensitive to everyone, and becomes easily aroused, quickly cooled and back to normal. He/She succeeds in artistic endeavour, rather than in a business. People and ideas appeal him more than money. 4-2 Sun in Cancer with Moon in Taurus: He/She is honest, very accommodating and gracious, determined, very charming, sincere and loyal, sensitive and adaptable individual and knows how to accumulate wealth and make wise investments. 4-3 Sun in Cancer with Moon in Gemini: He/She is highly sensitive, intuitive, flexible, adaptable, tuned in to the world and always eager to please. 4-4 Sun in Cancer with Moon in Cancer: He/She is extremely emotional, timid, and aloof. He/She is extremely domestic and a very protective parent. 4-5 Sun in Cancer with Moon in Leo: He/She is positive, confident, proudly, dignified and self-respected, open-hearted, amiable and outgoing. 4-6 Sun in Cancer with Moon in Virgo: He/She manages his affairs with a cool, calm and practical head. He/She respects principles, codes, laws and traditions. 4-7 Sun in Cancer with Moon in Libra: He/She needs companionship. He/She is always a winner in a large gathering. He/She likes to have people around him. 4-8 Sun in Cancer with Moon in Scorpio: He/She has magnetic appeal and commanding demeanour, forceful and temperamental personality, intense and dramatic actions. 4-9 Sun in Cancer with Moon in Sagittarius: He/She has a more sociable and progressive demeanour and can inspire confidence in people. His communication is frank, open. 4-10 Sun in Cancer with Moon in Capricorn: He/She has ambition and determination, affectionate, a high degree of self-discipline, shrewdness. 4-11 Sun in Cancer with Moon in Aquarius: He/She has reputation and appeal. He/She has a good deal of pride and has a way of achieving power over people. 4-12 Sun in Cancer with Moon in Pisces: He/She has fluid nature, peaceable and amenable personality, and is agreeable, friendly and diplomatic, affable and pleasant. He/She can be popular in a large broad way with groups of people.

14.5 Predictions by Sun in Leo with Moon placed in Twelve Signs

5-1 Sun in Leo with Moon in Aries: He/She has intense emotions, strong drive, and quick temper and is rash and impulsive and fears only his superior. 5-2 Sun in Leo with Moon in Taurus: He/She has very strong personality, fixed attitude, and is successful in business and lacking in innate diplomacy. 5-3 Sun in Leo with Moon in Gemini: He/She possesses action and ideas, vitality, personal warmth, generosity, authoritativeness, intelligence, cleverness, flexibility and adaptability. 5-4 Sun in Leo with Moon in Cancer: He/She has well-integrated personality, and enjoys good company and a good time. Romance and love are very important to him. 5-5 Sun in Leo with Moon in Leo: He/She has a forceful, ambitious, and concentrative personality; and gets success in the world. He/She has tremendous control, good deal of pride and vanity. 5-6 Sun in Leo with Moon in Virgo: He/She is warm, impetuous, and dynamic and the cool, cautious, and introverted. He/She can be extremely stubborn and opinionated. 5-7 Sun in Leo with Moon in Libra: He/She is a very independent, idealistic, highly honourable and has integrity, romance warm-heart and sentimental approach to life. 5-8 Sun in Leo with Moon in Scorpio: He/She has a vital and highly charged temperament, passion and is high in moral attitudes, and judges others on his stated standards. 5-9 Sun in Leo with Moon in Sagittarius: He/She has a constant passion for travel, adventure, change, and excitement. He/She is continuously looking for another mountain to climb or stone to turn, and vitality, authoritativeness, magnanimity and warmth, honesty, directness, versatility, and sincerity, yielding a very optimistic and enthusiastic temperament. 5-10 Sun in Leo with Moon in Capricorn: He/She has good deal of mental ability and quickness. His reputation and the impression are vitally important to him. 5-11 Sun in Leo with Moon in Aquarius: He/She has very attractive personality, the vitality, warmth, generosity, magnanimity, and pride. He/She is highly social, likable, and extroverted. He/She is very philosophical and inclined to mental expansion and creativity. 5-12 Sun in Leo with Moon in Pisces: He/She has very kind-hearted and likable personality. It's really hard for him to remain cool and aloof and is a very honest person.

14.6 Predictions by Sun in Virgo with Moon placed in Twelve Signs

6-1 Sun in Virgo with Moon in Aries: He/She has conflicting nature and high-spirited, combative, and critical personality. He/She is demanding and exacting and a perfectionist. 6-2 Sun in Virgo with Moon in Taurus: He/She has a very solid, dependable personality, and will lead to success in business and in personal relationships alike. 6-3 Sun in Virgo with Moon in Gemini: He/She has greatest mind power; quick, brilliant reasoning and is rational, capable of figuring things out analytically. 6-4 Sun in Virgo with Moon in Cancer: He/She has a sensitive, but harmonious personality; and is reserved and a little timid and attached to traditions, convention, and habit. 6-5 Sun in Virgo with Moon in Leo: He/She has a very earnest, analytical insight and discrimination, the vitality, confidence and authoritativeness and a very confident and positive personality, often proving assertive, sometimes aggressive. 6-6 Sun in Virgo with Moon in Virgo: He/She is very loyal and sympathetic, but never very romantic and has an inferior mind. He/She is very finicky in the tastes in matters of people, dress, food, and even household furnishings. 6-7 Sun in Virgo with Moon in Libra: He/She has a discriminating, intellectual, and peace loving personality, the intelligence, analytical insight, the emotional balance, courtesy and friendliness. 6-8 Sun in Virgo with Moon in Scorpio: He/She is intellectual, emotional, genius, and successful in business. 6-9 Sun in Virgo with Moon in Sagittarius: He/She thinks in large broad terms; and is philosophic and introspective, and has a good talent for acting and uses of body language and gestures. 6-10 Sun in Virgo with Moon in Capricorn: He/She is well-integrated and has creative mind, moral support, and ambition, splendid abilities, business shrewdness, a calculating and realistic mind, and the ability to organize and sound enterprise. 6-11 Sun in Virgo with Moon in Aquarius: He/She is very earth bound and practical understanding with knowledge. He/She strives to be aloof from the mundane elements of everyday life. 6-12 Sun in Virgo with Moon in Pisces: He/She has insight, an understanding mind, and a profound personality and is both intellectual and intuitive, cold logic, learn facts, and absorb figures. He/She has intelligence, analytical insight and discrimination with the emotional sensitivity and intuitive understanding. If born during a full Moon, he expresses a profundity of mind, studious inclination, and deep understanding that sets him apart from most people. Success in business or a profession is aided.

14.7 Predictions by Sun in Libra with Moon placed in Twelve Signs

7-1 Sun in Libra with Moon in Aries: He/She has a personality not easily tied down, fenced in. He/She doesn't like obligation, duty, or possessiveness. He/She is a true soldier for justice, self-sufficient and mentally independent. 7-2 Sun in Libra with Moon in Taurus: He/She has a charming personality with plenty of independence, personal charm and attractiveness and purpose, which is his greatest asset and has the emotional balance, courtesy and friendliness with the stability, firmness, determination, and set purpose and gets success in business. The tactful handling of business contacts, associates and partners attracts opportunities for him. 7-3 Sun in Libra with Moon in Gemini: He/She has a very airy and magnetic, voluble, expressive, and talkative, with an ability to hold an audience, dramatics personality. 7-4 Sun in Libra with Moon in Cancer: He/She has malleable manner and is very skilled at avoiding conflicts and clashes of temperament, a true lover of freedom, justice, equality, and independence. He/She minds his own business and doesn't impose on the rights of others. 7-5 Sun in Libra with Moon in Leo: He/She has a very romantic personality and makes a career of love affairs; and is refined, devoted, and discreet. 7-6 Sun in Libra with Moon in Virgo; He/She is very self-sufficient, analytical, and critical, a good listener always ready for a social gathering. 7-7 Sun in Libra with Moon in Libra: He/She has a tendency to be too trusting and thrives on love, attention, and dreams, and often seems to live in a dream world. He/She is naturally very refined, highly artistic, intuitive and peaceful. 7-8 Sun in Libra with Moon in Scorpio: He/She is independent and has action, thought and expression, a detective approach to things, a strong sense of justice, and is a born idealist and rebel, a radical, an enthusiast in all matters. 7-9 Sun in Libra with Moon in Sagittarius: He/She is aspiring, expansive, highly adventurous, open and accepting of others, and apt to move about and seek a variety of experiences, outgoing and eager to meet people. 7-10 Sun in Libra with Moon in Capricorn: He/She possesses a good deal of self-esteem, recognition and admiration and is willing to work hard to achieve the rewards and a bit like a machine. 7-11 Sun in Libra with Moon in Aquarius: He/She has social circle, and is extroverted and outgoing, sought out and popular for his sympathy and warm, yet detached, understanding, unselfish. He/She has splendid foresight, natural refinement, and an inspirational outlook; a happy and optimistic personality. 7-12 Sun in Libra with Moon in Pisces: He/She is broad, quiet, pensive, and is always watching, staring, observing, and mulling over, extremely sensitive, quite timid, aloof or distant, reserved and shy.

14.8 Predictions by Sun in Scorpio & Moon placed in Twelve Signs

8-1 Sun in Scorpio with Moon in Aries: He/She has definite, highly personal and strong ambition with a little feeling for right, wrong, justice, philosophy. He/She is extremely independent and highly impulsive. He/She is a strong person, and has little compassion or concern for others. 8-2 Sun in Scorpio with Moon in Taurus: He/She is solid and down-to-earth, with a great deal of strength, intensity and concentrative power, with the realities and facts. His thoughts, emotions, plans and aims are wide and he thinks and acts in the grand manner. 8-3 Sun in Scorpio with Moon in Gemini: He/She has a highly versatile and clever personality, and is flair for people, very popular, clever and quick with words and he may achieve success in analytical and investigative skills. 8-4 Sun in Scorpio with Moon in Cancer: He/She is proud and aloof, tending to rely on himself more than others. He/She is a magnetic, self-confident, very powerful, suspicious, shrewd, defensive, and in affairs of heart, very sensitive and jealous. 8-5 Sun in Scorpio with Moon in Leo: He/She has a strong, positive, fixed opinions and emotions, very strong character and is full of personal confidence, persistence, determination and courage. 8-6 Sun in Scorpio with Moon in Virgo: He/She has emotional forces, powers of will, determination, the intellectuality, discrimination, and critical analysis and does not provide a shoulder to cry on. He/She gets success in many businesses and professions that require employment of the scientific methods, persistence, and attention to detail. 8-7 Sun in Scorpio with Moon in Libra: He/She is socially warm, likable and charming, enthusiastic and has a good deal of charisma and poise, an abundance of energy and vitality. In romantic relationships he is keenly tuned-in to the other. 8-8 Sun in Scorpio with Moon in Scorpio: He/She is executive and commanding, very emotional, but he conceals or hides his feelings. He/She has a powerful sex nature. 8-9 Sun in Scorpio with Moon in Sagittarius: He/She is assertive and expressive, emotional, idealistic, aloof, social-minded and eager to help others, affectionate, loyal and sincere, almost to a fault. He/She can do well in job such as medicine, religion, or public service, for business. 8-10 Sun in Scorpio with Moon in Capricorn: He/She has honour; integrity and authority, business strategy and management organization. There is very strong ambition, urge for recognition and high office, steadfast in opinions, control, and to be a C.E.O. 8-11 Sun in Scorpio with Moon in Aquarius: He/She has a great deal of mental pressure and a constant flow of undercurrents. But he can be a tough boss but doesn't believe in discriminating against anyone. 8-12 Sun in Scorpio with Moon in Pisces: He/She has a terrific intuition and sense that ordinary people simply doesn't have and is overly dramatic and frequently exaggerates problems. He/She is very creative, anxious and worried.

14.9 Predictions by Sun in Sagittarius with Moon in Twelve Signs

9-1 Sun in Sagittarius with Moon in Aries: He/She has ultimate energy and pioneering spirit and is defiant, extremely self-confident and apt to work hard. Excitement, adventure, advancement are what he wants. He/She is a leader and things move rapidly when he is in charge and is a natural executive. 9-2 Sun in Sagittarius with Moon in Taurus: He/She is warm, genial, romantic, very idealistic and aspiring rather than ambitious. 9-3 Sun in Sagittarius with Moon in Gemini: He/She is leader, very adventurous, both physically and mentally, naturally lucky in a financial way, and is never really very concerned about security. He/She will have variety; travel, changes, mental stimulation, and a wide range of personalities around him. He/She wants to get to the top and he very well may. 9-4 Sun in Sagittarius with Moon in Cancer: He/She is idealistic, very much romantic, a dreamer, a visionary and may often attempt to reach too high or strive for too much recognition. 9-5 Sun in Sagittarius with Moon in Leo: He/She is a very warm-hearted, honest, open and has good judgment and doesn't take shortcuts; does not lie, no cheating, and doesn't do anything underhanded in order to gain an advantage. He/She will have happiness and success and the prestige of the position is more important than the salary. 9-6 Sun in Sagittarius with Moon in Virgo: He/She has power, excellent mind, broad concepts and ability and is achiever, outstanding, diplomatic, and charming with the authority of a director or tutor. 9-7 Sun in Sagittarius with Moon in Libra: He/She has peace, confidence, very good understanding, good will, a cheerful nature, is most tolerant and excels as mediator. 9-8 Sun in Sagittarius with Moon in Scorpio: He/She has self-reliance, optimism, determination, tact and diplomacy, developed life, and is a true go-getter. 9-9 Sun in Sagittarius with Moon in Sagittarius: He/She has an explosive personality and is a hair-trigger and rapid-fire type. 9-10 Sun in Sagittarius with Moon in Capricorn: He/She is a vigorous thinker and rises to a high position; has self-esteem, personal dash, optimism, enthusiasm, foresight, and directness of manner common with the ambition, shrewdness, realism and practicability and deals in business affairs. He/She is highly executive, and definitely an executive with outstanding balance. 9-11 Sun in Sagittarius with Moon in Aquarius: He/She is independent, idealistic, romantic and highly expressive and has the dash, optimism, enthusiasm and foresight combined with the originality, great sense of timing and persuasive abilities, leadership role and is suited for politics, public, civic or community service, or in organizations or movements. 9-12 Sun in Sagittarius with Moon in Pisces: He/She is forgiving, incapable of holding a grudge, and utterly free of meanness, spite, jealousy, or vengefulness and not practical minded.

14.10 Predictions by Sun in Capricorn with Moon in Twelve Signs

10-1 Sun in Capricorn with Moon in Aries: He/She has a very forceful and strong drive to succeed, prestige, diplomacy, vigorous invulnerability and is willing to work very hard, shrewd quality and deals with people very effectively. His brand of tact is to speak softly and carry a big stick. 10-2 Sun in Capricorn with Moon in Taurus: He/She has determined mind, stability and enjoys facing challenges and is a strong executive or professional, manager, an organizer, with solid and practical ideas. 10-3 Sun in Capricorn with Moon in Gemini: He/She has talent for self-expression in public speaking and is flexible, versatile, and adaptable, fast learner, swiftly assimilating impressions and ideas and making quick decisions, good planner and organizer. 10-4 Sun in Capricorn with Moon in Cancer: He/She has a highly refined intuitive insight and has a knack for making money. 10-5 Sun in Capricorn with Moon in Leo: He/She is in-charge, very self-sufficient, aristocratic; a sort of aloof and detached and capable of great popularity and gets much success in the world and likes to be the leader. 10-6 Sun in Capricorn with Moon in Virgo: He/She has difficulty making decisions and is well endowed mentally, possessing a logical, systematic, analytical and discriminating mind, aloof, reserved, and cautious. 10-7 Sun in Capricorn with Moon in Libra: He/She is drawn in two opposite directions such as disciplined and cautious, but rash and impulsive, humanitarian, not very romantic, extremely social, and likes to have people around him, proudly, a bit suspicious, and he remains aloof. 10-8 Sun in Capricorn with Moon in Scorpio: He/She has strong inner drive, a keen sense of an individual, inner pride, and is very difficult; very honourable, loyal, and devoted, and a very reserved individual. 10-9 Sun in Capricorn with Moon in Sagittarius: He/She has a high-powered physic and mind, boundless energies, and loves the camaraderie of getting together with the gang and is political; diplomatic or shrewd, as well as frank and sincere. 10-10 Sun in Capricorn with Moon in Capricorn: He/She is a taskmaster, an overseer and has a heart of gold, and demands realism, truth and facts, rejecting anything that is superficial or frivolous. 10-11 Sun in Capricorn with Moon in Aquarius: He/She is thoughtful, sincere, responsible, friendly, very popular, and enjoys involvement in a good bit of public activity. 10-12 Sun in Capricorn with Moon in Pisces: He/She is honest, trustworthy, harmonious, interesting and has temperament, intuitive insight and would never resort to trickery or deceit.

14.11 Predictions by Sun in Aquarius with Moon in Twelve Signs

11-1 Sun in Aquarius with Moon in Aries: He/She has aggressive mind, intelligence, leadership potential, judicial attitude, superiority complex, and stubborn disposition and is a very solid thinker, accurate, well thought out, and well stated. 11-2 Sun in Aquarius with Moon in Taurus: He/She is never petty or small, interested in people, and likes to be around them, executive, devoid of missionary spirit, willing to live and let live. 11-3 Sun in Aquarius with Moon in Gemini: He/She has a romantic personality, idealism and is quick, apt, intuitive, and finds himself drifting in life and is a natural for advertising, writing and public speaking. 11-4 Sun in Aquarius with Moon in Cancer: He/She possesses a restless and changeable nature, susceptibility to surroundings, considerable personal charm, proving gracious and can be very popular in business. 11-5 Sun in Aquarius with Moon in Leo: He/She helps people in need; a true humanitarian, proud, independent, and self-reliant and projects his personality into the world or affairs. 11-6 Sun in Aquarius with Moon in Virgo: He/She is very outstanding, believable, and sincere and has a great intelligence and true mental prowess and plans precisely, and proceeds with confidence and assurance. 11-7 Sun in Aquarius with Moon in Libra: He/She is gust of wind, but lacks constancy and is apt to leave the project half finished and tires of concentrating. He/She may abandon a relationship without as much as a 'fair thee well.' Marriage is very important to him, but settling down to a stable relationship will be difficult. 11-8 Sun in Aquarius with Moon in Scorpio: He/She has appeals to others, and commands a lot of respect and is very strong, and possesses tremendous powers. 11-9 Sun in Aquarius with Moon in Sagittarius: He/She is very active, and has independence of mind and action, thought and expression. He/She admires honesty and straightforward people. 11-10 Sun in Aquarius with Moon in Capricorn: He/She has authoritative personality; a very solid and practical individual qualities; deep determination, talent, sound judgment and realistic evaluation of conditions or situations. 11-11 Sun in Aquarius with Moon in Aquarius: He/She is philosophical, open-minded, a trailblazer, original and progressive in his thinking, new ideas and advanced techniques in business. 11-12 Sun in Aquarius with Moon in Pisces: He/She has a very original mind, independence, emotional sensitivity and imagination and is interested in detail, method and order and spends days just dreaming and speculating.

14.12 Predictions by Sun in Pisces with Moon placed in Twelve Signs

12-1 Sun in Pisces with Moon in Aries: He/She has a strong urges for personal advancement, fame or recognition, and is very self-reliant; and seems to be very assertive, competitive and determined. 12-2 Sun in Pisces with Moon in Taurus: He/She is highly sociable, a sensitive person, and possesses a good bit of social tact and avoids hurting anyone and has talent in an artistic or musical sense. 12-3 Sun in Pisces with Moon in Gemini: He/She has a variable nature, the tendency to go off in different directions, or face real mental binds and problems. He/She may feel one way, go another way and is variable, adaptable, changeable, and always a bit indecisive. 12-4 Sun in Pisces with Moon in Cancer: He/She is very complex, sensitive, alert with sort of a sixth sense and tackles any job with vigour, discipline, and drive. 12-5 Sun in Pisces with Moon in Leo: He/She has a personality that seems to be on a perpetual seesaw. He/She goes up when he would prefer to go down and vice versa. 12-6 Sun in Pisces with Moon in Virgo: He/She is constructive, versatile, flexible and highly adaptable, and has a deep sense of the truth and an ability to face facts. 12-7 Sun in Pisces with Moon in Libra: He/She is very pleasant, but not an intellectual scholar; and has a very reliable sixth sense and extremely accurate perceptions. 12-8 Sun in Pisces with Moon in Scorpio: He/She is a very emotional and sensitive and possesses a strong, fixed, determined, persistence, determination and courage in personality. 12-9 Sun in Pisces with Moon in Sagittarius: He/She is never content with earthy, materialistic, or mundane goals, not a materialist and perfectly satisfied. 12-10 Sun in Pisces with Moon in Capricorn: He/She has a pleasant, sweet-tempered outlook and is practical, realistic, objective and persevering. He/She has a tendency to underrate himself and worry much too much. 12-11 Sun in Pisces with Moon in Aquarius: He/She is crusaders of all sorts, easygoing, charming and versatile. He/She can find success in business and has a very original way of thinking. 12-12 Sun in Pisces with Moon in Pisces: He/She is highly introspective, and has a very warm and likable quality and is very much in need of relationships and company. Mentally he is quick, perceptive, and intuitive.

11

Gulika and Upagraha

Gulika: There are 5 Kaal Vela, such as Gulika/Mandi, Kaal, Mrityu, Yamaghantak, and Ardhaprahar. Gulika/Mandi is most commonly known as Upagraha. Divide the duration of the day by 8. The lord of the first part of the day duration is the lord of the weekday in question. The lords of the rest six parts of the day are the rulers of the balance week days in order. **Example I:** The Sun is the ruler of the 1st part on Sunday - day duration. Moon is the ruler of the 2nd part of Sunday - day duration. The Mars is ruler of the 3rd part, Mercury of the 4th part, Jupiter of the 5th part, Venus of the 6th part, and Saturn of the 7th part of the day duration. The 8th part of the day duration has no ruler. Now, the 7th part of the day duration ruled by the Saturn is called "Gulika".

Similarly, divide the duration of the night times by 8. The lord of the first part of the night is the lord of the 5th day from the weekday in question. The lords of the rest six parts of the night are the rulers of the balance weekdays in following order after the 5th day.

Example II: The Jupiter is the ruler of the 1st part of the night time's birth on Sunday and not the Sun, counting in order the 5th day from Sunday. Venus will be the ruler of the 2nd part of night time's birth on Sunday. Accordingly, the Saturn will be ruler of the 3rd part, Sun of the 4th part, Moon of the 5th part, Mars of the 6th part, and Mercury of the 7th part of the night on Sunday. The 8th part of the night time's birth has no ruler. Here, the 3rd part of the nighttimes birth is ruled by the Saturn and is called "Gulika". The degree, ascending at the time of start of Gulika portion, will be the longitude of Gulika at a given place.

The cusp of the sign rising at the beginning of the Gulika segment is considered as Gulika. Gulika is known as the son of Saturn.

Gulika: The 7th part of the day time or the 3rd part of night time duration of the weekday is ruled by Saturn and is known as Gulika/Mandi.

Kaal: The one-eighth part of the day time or night time duration ruled by Sun is called Kaal.

Yamaghanta: The one-eighth part of the day time or night time duration ruled by the Jupiter is called Yamaghantakala.

Mrityu: The one-eighth part of the day time or night time duration ruled by the Mars is called Mrityu.

Ardhaprahara: The one-eighth part of the day time or night time duration ruled by the Mercury is called Ardhaprahara.

17.1 Predictions by Gulika/ Mandi

Calculation Method 1: The standard length of day and night each is 30 Ghati (60 Ghati make 24 hours or 2.5 Ghati = 1 hour). During a standard Day or Night of 30 Ghati each, the rising time for Gulika during the day time birth, for the seven days from Sunday to Saturday birth, is at 26, 22, 18, 14, 10, 6 and 2 Ghati respectively from the time of sunrise. For night time birth, the order from Sunday to Saturday birth is 10, 6, 2, 26, 22, 18 and 14 Ghati respectively from the time of sunset. But, actually, the Day and Night duration varies in a year. The Day duration on June 22 is longest, i.e. more than 30 Ghati. Similarly, the Night duration on December 22 is longest, i.e. more than 30 Ghati. Accordingly, the rising time for Gulika for its position is to be corrected.

Longitude of Gulika/Mandi: The Gulika Rising Time = "A" = Sun Set Time + M hours. The Ascendant or Rising Sign Time (Time of Birth) is given, say, = "B". The difference in Rising Sign Time (Time of Birth) and Gulika Rising Time = ("B" - "A") hours. The difference in Longitude of Rising Sign and Gulika Rising = ("B" - "A") Hrs x 360/24 degree. Rising Sign longitude from Aries (zero degree of zodiac) is given in the Kundali, say = "X" degree. The Longitude of Gulika = {Rising Sign Longitude ± ("B" - "A") x 360/24} ={X ± ("B" - "A") x 360/24} degree. (In case of "±", it is subtracted, i.e. (-) symbol is used, if the resultant is positive. It is added, i.e. (+) symbol is used, if the resultant is negative.

Correction of Gulika Timing: The correct (actual) position of Gulika is mathematically ascertained for predictions purposes. The following formula helps in working out the correct (actual) position of Gulika/Mandi: D or N (as the case may be) X I / 30 = M. Where, M is the correct rising time (birth Ghati) of Gulika after sunrise or sunset (as the case may be). D is actual duration of day expressed in Ghati on that day. N is actual duration of night expressed in Ghati on that night. "I" is the standard rising time of Gulika as per the table. Indian ephemeris (Panchang) gives ahas (daytime) for the day as well as (night time) for the night. If it is not available there, the actual Day or Night duration on that particular day can be ascertained from the time of Sun Rise and Sun Set given in the Kundali or Panchang or other sources.

Calculation Method 2: Divide Ishtaghati by 8 = 4 Ghati 2 Pala 30 Vighati Pala. Divide the length of the day given in Ghati Pala for day time birth or the length of the night given in Ghati Pala for night time birth (Ishtaghati) by 8 and multiply the result by a constant given in the table below. The time thus calculated is known as Gulikeshta Kala (Rising time of Gulika) and the Ascendant, which corresponds to this time, is known as Gulika.

Day	Sunday	Monday	Tuesday	Wednesday
Night	3	2	1	7
Day	7	6	5	4

Day	Thursday	Friday	Saturday
Night	6	5	4

Day	3	2	1

The Sun's longitude and the corresponding Lagna longitude for this moment are given in the Kundali. Gulika/Mandi and all these Upagraha are broadly classified as malefic and inauspicious or a dreaded malefic, but it gives good results in the 6th and 11th houses.

A Special Rule of Gulika: A point 180° from Gulika is also considered as acutely malefic in the natal chart. This point must be taken into consideration when the lethal potential of a planet is to be determined. This is better known as the Pramaana Gulika in Vedic astrology. The dispositor of Gulika or Pramaana Gulika may prove fatal during its Dasha, displacing the other Maraka or killer planets. Gulika is said to represent that entire vulgar, base, sordid and disgusting and so his association with any planet may bring out these qualities in the situation concerned with the Karkatva of the planet. Also, the Gulika aspect (Drishti) alone is harmful in the similar way. When a house affected by Vishanadi, the native even if born as a ruler of the earth, he will become a beggar. This disposition of Gulika is treated as so damning as capable of destroying everything in the native's life. Major Planets, in association with Gulika, said to produce malefic effects and stop his Signification or the Karkatva much in the same manner as any other malefic planet does.

Predictions by Gulika/Mandi:

Gulika/Mandi in a cardinal or a Trine house with reference to ascendant, bestows plenty of riches and fame on the individual. Its association with a planet makes the planet inauspicious. These Upagraha causes destruction of the family, longevity and wisdom, if they are in combination with the Sun, the Moon or the Ascendant. Gulika in the first house reduces the good results of the chart. The impact is strong when the cusp of the Lagna and the Gulika are close in degrees. In such a situation, the Raja-yoga or other benefic yoga loses his potency. When the Nakshatra of the Lagna and Gulika are mutually trine, the basic benevolence of the Lagna - Nakshatra Lord (LNL) is lost. Nakshatra 1, 10 and 19 are mutually trine. Gulika spoils the benefic significations of most of the houses by occupying them. Life will be full of troubles when Gulika occupies the Lagna. In the second house, it curtails family comforts and financial savings. Poor social status results when Gulika occupies the fourth house. In the fifth, it leads to troubles from progeny. And so on. Only in houses 3, 6, 10 and 11 does Gulika generate beneficial results. In the tenth house, however, it indicates delay in getting established in a career. The good results of Gulika in houses 3, 6, 10 and 11 are lost when the Rashi and Navamsa lords of Gulika are Neecha (debilitated) or combust. The association of Gulika with a natural Karaka (significator) always destroys the good effects of that Karaka. Gulika conjunct with the Sun creates lack of comforts for father, with the Moon it is bad for mother, with Mars it is adverse for brother, with Mercury, he/she becomes mentally disturbed, with Jupiter, he/she becomes a hypocrite, with Venus brings troubles from women and ruins marital life, with Saturn leads to disease and skin disorder, with Rahu it leads to proneness to infection, and with Ketu fear from fire. Gulika can alter the results of other planetary yoga in the chart. All good yoga is lost when the birth time and the Gulika-Kaala coincide together with another inauspicious factor like Mahapaata, Gandanta, and Visha-Ghati.

Example: In one Chart of a native born on May 6, 1982; at 11:30 hours; in Nepal, the Lagna is Karka (Cancer), at 23° 06'. Gulika is 2S 11° 54', associated with Rahu. He is mentally retarded. A Mahapaata birth is confirmed here.

Hazardous Timing Predictions: This is as per Dasha and Transits of Sign-Lord of Gulika. The Dasha periods of the dispositor of Gulika or its Navamsa lord may prove hazardous. Trouble starts during the Dasha of a planet that happens to be an associate of Gulika in the Natal chart. Severe troubles are also likely during the Dasha of any planet associating with Gulika in the Navamsa or the Dvadasamsa charts. Especially adverse results ensue during the Dasha period of Jupiter or Saturn when they associate with Gulika in the Navamsa, or the Dasha of the Sun when it associates with Gulika in the Dvadasamsa, or that of the Moon associating with Gulika in the Trimsamsa chart.

The transit of Saturn and Jupiter over the Navamsa Sign lord of Gulika, the Sun over the Dvadasamsa Sign lord of Gulika, and that of the Moon over the Trimsamsa sign lord of Gulika, put him/her in trouble.

Example: In one Chart of Smita Patil, a famous actress is born on October 17, 1955; at 17:30 hours; at Pune. Her Lagna is Meena (Pisces) while Gulika occupies Makara (Capricorn) in the eleventh house. She died of meningitis in the Saturn-Mars period. Saturn is the Sign lord of Gulika while Mars is its Navamsa Sign lord of Gulika.

Accidents Predictions by Gulika: Gulika is extremely important as an active killer in Vedic astrology. It is used in the judgement of longevity. The Dasha and Transit of Sign-Lord of Gulika in houses 1, 5 or 9 authorises the Lagna lord to disburse the Maraka effect during its Dasha. All Trikona (1, 5, and 9) signs should be considered in transits for application of the Maraka effect. The Lagna, the Moon and Gulika falling in Dwiswabha (dual) or Sthira (fixed) signs prompt multiple ailments and fatality. In Chara (movable) signs, it leads to good health and long life. The Lagna, the Moon and Gulika falling in mutual Trikona in the Navamsa chart, particularly in Rashi 4, 8 and 12 are fatal. In such case, he/she is disease prone and accident prone.

Longevity Predictions by Gulika: Although Gulika is favoured in the third house, it turns death-inflicting when joined by or aspect by Saturn. Gulika in the third with the Moon or receiving the full aspect of the Moon causes tuberculosis or other internal fatal disease. For the calculation of longevity, the Lagna, the Moon, the Sun and Gulika are all important. A special rule for assessment of longevity involves using the cusp of the Lagna and the longitudes of the Sun, the Moon and Gulika. This is as follows: Lagna ´ 5 + Gulika = Prana Sphant. Moon ´ 8 + Gulika = Deha Sphant. Gulika ´ 7 + Sun = Mrityu Sphant. Note: Lagna ´ 5 means Longitude of Lagna divided by 5. When the sum of Prana and Deha Sphant is more then the Mrityu Sphant, he/she is likely to live long. However, if Mirtyu is bigger than the Sun of Prana and Deha, a rudder end of life is indicated.

Example 1: In Chart of Late Indian Prime Minister Indira Gandhi born on November 19, 1917; at 23:11 hours IST; at

Allahabad, the Lagna is 3S27°22', the Sun is 7S04°7', the Moon 9S05°35' and Gulika 1S29°23'. The sum of Prana and Deha Sphant is 1S2°16' while Mrityu Sphant is 2S15°04'. She met a tragic end.

Major Ventures Predictions by Gulika: Add the cusp of the Lagna to the 'longitude' of Gulika and find out the Resultant Sign. The month when the Sun transits the resultant sign will lead to troubles and miseries to him/her. The most hazardous day is when the Moon transits the Resultant Sign obtained by adding the longitude of the Moon to the 'longitude' of Gulika. Add the 'longitude' of Lagna, the Moon and Gulika and get the Resultant Sign, which is considered to be fatal for him/her. The month, the day and the Lagna indicated by the sum obtained as above must be avoided in all his/her major ventures.

Example: In case of Late Indian Prime Minister Indira Gandhi, Gulika is at 1S 29° 23', which is in Kanya (Virgo) Navamsa, in Mesha (Aries) Dvadasamsa and in Vrischika (Scorpio) Trimsamsa. Indira Gandhi was shot dead on October 31, 1984 at around 10:00 hours IST, at Delhi. In her Lagna Chart, the ascendant is Vrischika; Saturn is in Tula (Libra) along with the Sun, Jupiter and Gulika are in Dhanu (Sagittarius) in Lagna. The Moon is in Makara (Capricorn) while the Navamsa Lagna is Meena (Pisces). Adding the Moon and Gulika of Indira Gandhi of Lagna Chart, we get Meena (Pisces) Sign. This is in time with the rising Lagna as well as with its Navamsa Lagna. The Gulika was in Aries Dvadasamsa. At the time of the fatal mishap, Jupiter and Gulika were in Sign of Mesha.

Progeny Predictions by Gulika: Gulika in the fifth house, particularly in Signs 3, 6, 10 and 11, leads to childless because of some defect in his/her generative organs. Gulika may ensure successful conception when: (i) Gulika and the Moon are in the same sign; (ii) Gulika is with the fifth lord; (iii) Gulika is aspect by the fifth lord; (iv) Gulika in the other sign of the fifth lord; (v) Navamsa lords of Gulika and the Moon are related mutually. A successful conception results when Jupiter transits the times of Gulika Rashi or Gulika Navamsa Rashi. When Gulika is in any of the first six Rashi (Mesha to Kanya), consider the transit of Jupiter from the Gulika Rashi. When it is in the last six signs (Tula to Meena), consider the transit from Gulika Navamsa Rashi.

Special Lagna Predictions by Gulika: Gulika in houses 1 or 7 in the Lagna Pad causes grey hair, ill health and serious disease of the stomach. When Atma Karaka and Gulika occupy the Karkamsha Lagna and both are aspect by the waxing Moon, they ensure losses from repeated theft. He/She dies of poisoning, if Gulika, in Karkamsha Lagna, is not aspect by any planet. These effects are applicable to the natal chart as well to all the special Lagna charts like Karakamsha, Pad, Upa-pad, and Navamsa.

Raja Yoga Predictions by Gulika: The Phaladeepika states that the dispositor of Gulika and its Navamsa lord placed in a Kendra or Trikona or in its own sign or in Exaltation, mollifies the adverse effects of Gulika and gives Raja yoga effects, though it's lethal propensity (markatwa) would remain intact. If the Gulika be in a Kendra, a Trikona, or be strong, in its own house, or in Exaltation or in a friendly house or in the eleventh house, a potent Raja-yoga results from this disposition. If Gulika-Sign Lord, Venus, is strong, in own Navamsa, the Gulika Navamsa sign lord in time shows a good and strong Raja-yoga.

Example: In the horoscope of Mr. Chandra Babu Naidu, born on April 27, 1951; at 6:30 hours IST; at Hyderabad, the Lagna is 0S22°16'. Gulika is 1S08°37'. Gulika-Sign lord, Venus, and its Navamsa lord, Jupiter, are in there in own signs. Its Dvadasamsa lord Sun is exalted. A strong Raja-yoga is formed and he became powerful during Jupiter Dasha. During Jupiter-Venus, the Raja-yoga results were enhanced.

Example: Ms Jai Lalitha was born on February 24, 1948; 14:34 hours IST; at Chennai. Gulika is at 1S20°14' in the twelfth house. However, Gulika-Sign lord, Venus, is exalted in the tenth house. This exhibits a potential Raja-yoga to Ms Jai Lalitha.

Additional Effects by Gulika: If Gulika is with the Sun, he/she will hate his father and his/her dynasty will decline; with the Moon, he/she will cause distress to mother; with Lagna, it will destroy his/her longevity and wisdom; with the Mars, he/she will have no younger brother or deprived of siblings and gives lack of happiness from younger siblings or younger brothers or sisters or their absence, even though the Mars is Karaka for brothers; with the Mercury, he/she will be insane; with the

Jupiter, he/she will heretic; with the Venus, it will deprive him/her of happiness from spouse and will suffer from venereal disease and a profligate; with the Saturn, he/she will be fond of pleasure and enjoyment; with the Rahu, he/she will be prisoner; with Ketu, he/she will be militant or terrorist.

Gulika in 12 houses:

As per Parasara or even Jataka Parijata, the results of the Gulika/Mandi in the 12 houses are as below:

First house: He/She is afflicted by diseases and will be lustful, sinful, crafty, vicious and very miserable. When Gulika is in association with the malefic, it makes him/her deceitful, lustful and depraved.

Second house: He/She will be ugly or Vikrutah, miserable, mean, given to vices, shameless and penniless. When Gulika is in association with the malefic, it makes him/her without wealth or even a semblance to education and learning, i.e. uneducated.

Third house: He/She will be charming in appearance, head of a village, virtuous, liked by good people, honoured by the ruler. Gulika makes him/her stand out due to his/her aloofness, pride, conceit and such qualities. He/She will be highly ill tempered and very busy with the acquisition of wealth. He/She will be free from distress and will be without brothers and sisters.

Fourth house: He/She is afflicted by bilious and gastric disorders and will be sickly, unhappy, always engaged in evil (sinful) deeds. Gulika makes him/her bereft of education, wealth, House, happiness, lands and vehicles and will become a wanderer.

Fifth house: He/She will be insignificant, poor, short-lived, mean and spiteful, eunuch, hen-pecked and atheist. Gulika makes him/her immoral, irresolute, and evil minded, with few sons and short-lived.

Sixth house: He/She will be Invincible, robust physique and sturdy limbs, brilliant, beloved of one's spouse, optimistic, endowed with fortitude and helping nature. Gulika will destroy hosts of foes, and he/she will be interested in black magic and similar occult pursuits and will be brave.

Seventh house: He/She will be Unhappy in marriage, given to evil deeds, debauchee, having no friends and living off his wife.

Gulika makes him/her quarrelsome, have a bad wife, inimical to many people, and could be anti-social, stupid and ungrateful.
Eighth house: He/She will be Starving, miserable, cruel, wrathful, wicked, poor and without any good quality. He/She will be deformed in his face, have weak eyes and weak body.
Ninth house: He/She will be many hardships, emaciated, doing evil deeds, dull and a talebearer. He/She will be engaged in vile deeds and may even kill him/her parents and preceptors.
Tenth house: He/She will be many sons, happy, enjoyments, pious and God-fearing, righteous life. He/She will give up or disregard all duties and obligations of him/her caste, status and relationship and will have no self respect.
Eleventh house: He/She will be fond of aristocratic ladies, leader, working for welfare of relatives, short and ruler of men. He/She will have much happiness, wealth, power and beauty, but will lose elder brother.
Twelfth house: He/She will be engaged in vile deeds or occupation, sinful, physical deformity in limb, lethargic and drawn to base female company. He/She will move around giving the impression of an ascetic and will be able to extract money from others through him/her powers of speech.

Mitigation of the Evil Effects of Gulika: The evil effects of Gulika must be neutralised by prescribed remedies. He/She should worship lord Shiva regularly in the evening, bow down to the Sun-god and lord Vishnu in the morning, and light a holy lamp of 'ghee' before a Shiva temple. This would defy the evil arising out of an adverse disposition of Gulika.

17.2 Predictions by Upagraha:

Calculation: There are few Non-luminous Upagraha (Sub-Graha), such as Dhooma, Vyatipat, Parivesha, Chap (Indradhanu) and Ketu (Upketu), which are similar to the Nodes and are also based on the position of the luminary Sun at a particular point of time. Upagraha's position is calculated as per following method: Dhooma Longitude = Sun's Longitude + 133^0 20'. Vyatipat Longitude = 360^0 - Dhooma Longitude. Parivesha Longitude = 180^0 + Vyatipat Longitude. Indradhanu Longitude = 360^0 - Parivesha Longitude. Upketu Longitude = 16^0 40' + Indradhanu Longitude. Finally, the Sun's Longitude = 30^0 + Upketu Longitude. We must get back the Sun longitude, which shows that the Ascendant is correct. These are the Upagraha, devoid of splendour, and are malefic by nature and cause affliction to the main planets.

-- Pisces	Mrityu ; Aries	Yamaghantak Taurus	Parivesha; Kaal Gemini
Upketu Aquarius	**Exaltation Sign**		Gulika Cancer
-- Capricorn			Dhoom Leo
Indrachap Sagittarius	Vyatipat; Ardhaprahar Scorpio	-- Libra	-- Virgo

Pisces	--	Vyatipat; Ardhaprahar	Indrachap
Dhooma Aquarius	**Debilitation Sign**		-- Cancer
Gulika Capricorn			Upketu Leo
Parivesha; Kaal Sagittarius	Yamaghantak Scorpio	Mrityu Libra	-- Virgo

-- Pisces	-- Aries	-- Taurus	Vyatipat; Ardhaprahar Gemini
Gulika Aquarius	**Own Sign**		Indrachap; Upketu; Kaal Cancer
Dhooma; Kaal Parivesha; Capricorn			-- Leo
Yamaghantak Sagittarius	Mrityu Scorpio	-- Libra	-- Virgo

-- Pisces	-- Aries	-- Taurus	-- Gemini
Gulika Aquarius	**Moolatrikona Sign**		-- Cancer
Kaal Capricorn			--) Leo
-- Sagittarius	-- Scorpio	-- Libra	-- Virgo

Strength of Upagraha: They are Exaltation, Debilitation, Own Rashi respectively in Rashi, such as Dhooma in Simha, Kumbh, Makara; Vyatipat in Vrischika, Vrishabh, Mithuna; Parivesha in Mithuna, Dhanu, Makara; Indrachap in Dhanu, Mithuna, Karka; Upketu in Kumbh, Simha, Karka; Gulika in Karka, Makara, Kumbh; Yamaghantak in Vrishabh, Vrischika, Dhanu; Ardhaprahar in Vrischika, Vrishabh, Mithuna; Kaal in Mithuna, Dhanu, Makara; Mrityu in Aries, Tula, Vrischika respectively.

Out of the 5 Kaal Velas, four have their own Rashi system in the respective Rashi, ruled by their fathers except Kaal. Gulika, son of Sani, has Kumbh, as his own Bhava. Guru's son, Yamaghantak, has it in Dhanu. Ardhaprahar, Buddha's son, is in own Rashi, if in Mithuna. Mrityu, son of Mangal, has Vrischika, as own Bhava. It is not known, why Kaal, a son of Surya shifted to Makara, a Rashi of his brother (Sani), leaving his father's Simha. Obviously, Sani has given his Moolatrikona to his son Gulika, while he gave Makara (a secondary Rashi) to his 'brother' Kaal. These factors can be kept in mind while interpreting the Chart and it's Sign for the best results of the native.

Predictions by Upagraha:

Dhum in Various Bhava: 1st House: He/She will be valiant, unkind, wicked and highly short-tempered and endowed with beautiful eyes, stupefied in disposition. **2nd House:** He/She will be sickly, wealthy, devoid of a limb, will incur humiliation at royal level, be dull witted and be a eunuch. If Dhum is in Sahaj Bhava; he/she will be intelligent, very bold, delighted, and eloquent and be endowed with men and wealth. If Dhum is in Bandhu Bhava, he/she will be grieved on account of being given up by his female, but will be learned in all Shashtra. If Dhum is in Putra Bhava, he/she will have limited progeny, be devoid of wealth, be great, will eat anything and be bereft of friends and Mantras. If Dhum is in Ari Bhava, he/she will be strong, will conquer his enemies, be very brilliant, famous and free from diseases. If Dhum is in Yuvati Bhava; he/she will be penniless, be ever sensuous, skilful in going to others' females and be always devoid of brilliance. If Dhum is in Randhra Bhava, he/she will be bereft of courage, but be enthusiastic; be truthful, disagreeable, hardhearted and selfish. If Dhum is in Dharma Bhava, he/she will be endowed sons and fortunes, be rich, honourable, kind, religious and well disposed to his relatives. If Dhum is in Karma Bhava, he/she will be endowed with sons and fortunes, be delighted, intelligent, happy and truthful. If Dhum is in Labh Bhava, he/she will be endowed with wealth, grains and gold, be beautiful, will have knowledge of arts, be modest and be skilful in singing. If Dhum is in Vyaya Bhava, he/she will be morally fallen, will indulge in sinful acts, be interested in others' wives, and addicted to vices, unkind and crafty.

Vyatipat in Various Bhava: If Vyatipat is in Thanu Bhava, he/she will be troubled by miseries, be cruel, will indulge in destructive acts, be foolish and will be disposed to his relatives. If Vyatipat is in Dhan Bhava, he/she will be morally crooked, be bilious, will enjoy pleasures, be unkind, but grateful, and be wicked and sinful. If Vyatipat is in Sahaj Bhava, he/she will be firm in disposition, be a warrior, be liberal, very rich, and dear to the king and be head of an Army. If Vyatipat is in Bandhu Bhava, he/she will be endowed with relatives etc., but not sons and fortunes. If Vyatipat is in Putra Bhava, he/she will be poor, be charming in appearance, will have imbalances of phlegm, bile and wind, and be hard-hearted and shameless. If Vyatipat is in Ari Bhava; he/she will destroy his enemies, be physically mighty, skilful in use of all kinds of weapons and in arts and be peaceful in disposition. If Vyatipat is in Yuvati Bhava, he/she will be bereft of wealth, wife and sons, will subdue to females, be miserable, sensuous, shameless and friendly to others. If Vyatipat is in Randhra Bhava; he/she will have deformity of eyes, be ugly, unfortunate, and spiteful to Brahmins and be troubled by disorders of blood. If Vyatipat is in Dharma Bhava, he/she will have many kinds of business and many friends; he will be very learned, well disposed to his wife and he will be eloquent. Vyatipat is in Karma Bhava, he/she will be religious, peaceful, skilful in religious acts, very learned and far-sighted. Vyatipat is in Labh Bhava, he/she will be extremely opulent, be honourable, truthful, firm in policy, endowed with many horses and be interested in singing. Vyatipat is in Vyaya Bhava, he/she will be given to anger, associated with many activities, disabled, irreligious and hate his relatives.

Paridhi (or Parivesha) in Various Bhava: If Paridhi is in Thanu Bhava, he/she will be learned, truthful, peaceful, rich, endowed with sons, pure, charitable and dear to elders. If Paridhi is in Dhan Bhava, he/she will be wealthy, charming, will enjoy pleasures, be happy, very religious and be a Lord. If Paridhi is in Sahaj Bhava, he/she will be fond of his wife, be very charming, pious, and well disposed to his men, be a servant and be respectful of his elders. If Paridhi is in Bandhu Bhava, he/she will be wonder-struck, helpful to enemies as well, kind, endowed with everything and be skilful in singing. If Paridhi is in Putra Bhava, he/she will be affluent, virtuous, splendorous, affectionate, religious and dear to his wife. If Paridhi is in Ari Bhava; he/she will be famous and wealthy, be endowed with sons and pleasures, be helpful to all and will conquer his enemies. If Paridhi is in Yuvati Bhava, he/she will have limited number of children, be devoid of happiness, be of mediocre intelligence, very hard-headed and will have a sickly wife. If Paridhi is in Randhra Bhava, he/she will be spiritually disposed, peaceful, strong-bodied, firm in decision, religious and gentle. If Paridhi is in Dharma Bhava; he/she will be endowed with sons, be happy, brilliant, and very affluent, be devoid of excessive passion, be honourable and be happy with even an iota. If Paridhi is in Karma Bhava; he/she will be versed in arts, will enjoy pleasures, be strong-bodied and be learned in all Shashtra. If Paridhi is in Labh Bhava; he/she will enjoy pleasures through women, is virtuous, intelligent, and dear to his people and will suffer disorders of digestive fire. If Paridhi is in Vyaya Bhava, he/she will be a spendthrift, miserable, firm and will dishonour elders.

Indrachap (Indradhanu, or Kodanda): If Chap is in Thanu Bhava, he/she will be endowed with wealth, grains and gold and be grateful, agreeable and devoid of all actions. If Chap is in Dhan Bhava, he/she will speak affably, be very rich, modest, learned, charming and religious. If Chap is in Sahaj Bhava; he/she will be a miser, be versed in many arts, will indulge in thieving, be devoid of some limb and be unfriendly. If Chap is in Bandhu Bhava, he/she will be happy, endowed with quadrupeds, wealth, grains etc., be honoured by the king and be devoid of sickness. If Chap is in Putra Bhava; he/she will be splendorous, far-sighted, pious, and affable and will acquire prosperity in all his undertakings. If Chap is in Ari Bhava, he/she will destroy his enemies, be happy, affectionate, and pure and will achieve plentiful ness in all his undertakings. If Chap is in Yuvati Bhava, he/she will be wealthy, learned, religious, and agreeable and endowed with all virtues. If Chap is in Randhra Bhava; he/she will be interested in others' jobs, be cruel, interested in others' wives and have a defective limb. If Chap is in Dharma Bhava; he/she will perform penance, will take to religious observations, be highly learned and be famous among men. If Chap is in Karma Bhava; he/she will be endowed with many sons, abundant wealth, cows, buffaloes etc. and will be famous among men. If Chap is in Labh Bhava; he/she will gain many treasures, will be free from diseases, very fiery in disposition, affectionate to his wife and will have knowledge of Mantras and weapons. If Chap is in Vyaya Bhava; he/she will be wicked, very honourable, evil in disposition, shameless, will go to other's females and be ever poor.

Dhwaj (Sikhi, or Upketu) in Various Bhava: If Dhwaj is in Thanu Bhava; he/she will be skilful in all branches of learning, be happy, efficient in speech, agreeable and be very affectionate. If Dhwaj is in Dhan Bhava; he/she will be a good and affable speaker, be splendorous, will write poetry, be scholarly, honourable, and modest and endowed with conveyances. If Dhwaj is in Sahaj Bhava, he/she will be miserly, cruel acts, thin-bodied, poor and will incur severe diseases. If Dhwaj is in Bandhu Bhava; he/she will be charming, very virtuous, gentle, interested in Vedic Knowledge and be always happy. If Dhwaj is in Putra Bhava; he/she will be happy, will enjoy pleasures, be versed in arts, skilled in expedients, intelligent, eloquent and will respect elders. If Dhwaj is in Ari Bhava, he/she will be ominous for material relatives, will win over his enemies, and be endowed with many relatives, valiant, splendour and skilful. If Dhwaj is in Yuvati Bhava; he/she will be interested in gambling, be sensuous, will enjoy pleasures and will befriend prostitutes. If Dhwaj is in Randhra Bhava; he/she will be interested in base acts, be sinful, shameless, will blame others, will lack in marital happiness and will take other's side. If Dhwaj is in Dharma Bhava, he/she will wear badges, be delighted, helpfully disposed to all and he will be skilled in religious deeds. If Dhwaj is in Karma Bhava, he/she will be endowed with happiness and fortunes, be fond of females, be charitable and will befriend Brahmins. If Dhwaj is in Labh Bhava, he/she will ever acquire gains, be very religious, honourable, affluent, fortunate, valiant and skilled in sacrificial rites. If Dhwaj is in Vyaya Bhava; he/she will be interested in sinful acts, be valiant, untrustworthy, and unkind, interested in others' females and be short-tempered.

12

Bhava Pad, Upa Pad, Dara Pad and Pranapad

12.1 Predictions by Bhava Pad / Lagna Pad

The Bhava Pad (Arudha Bhava) will correspond to the house arrived at by counting so many houses from that Brava's Lord, as he is away from that Bhava. Take the number of houses that a lord of a given House/Bhava has moved from that house and then take as many houses from its Lord itself. The house arrived at is called the Arudha Pad or Bhava Pad of the Natal Chart. There are exceptions to this rule that the Arudha Pad of any Bhava cannot be in same Bhava or in the 4th or in the 7th from there. In such case, if the lord of a Bhava is placed in same Bhava; take the 10th from there as the Arudha Pad. If the lord of a Bhava is placed in the 4th from that Bhava, take the 4th itself from there as Arudha Pad. In case the lord of a Bhava is placed in the 7th from that Bhava, take the 10th from there as Arudha Pad. Names of the 12 Pads are Lagna Pad of Thanu Bhava, Dhan Pad of Dhan, Vikram Pad of Sahaj, Matru Pad of Bandhu, Mantra Pad of Putra, Rog pad of Ari, Dara Pad of Yuvati, Maran pad of Randhra, Pitru pad of Dharma, Karma Pad of Karma, Labh Pad of Labh, and Vyaya Pad of Vyaya Bhava.

(a) Predictions by Lagna Pad:

Financial gain (by 2nd / 7th from Lagna Pad): If one, two, or all three of Guru, Sukra and Chandra are in the 7th or 2nd from Lagna Pad, he/she will be very wealthy. Whether a benefic, or a malefic, if any planet is exalted in the 7th or 2nd from Lagna Pad, he/she will be affluent and be famous. If anyone of Mercury, Jupiter, and Venus is exalted in the 2nd from Lagna Pad and is with strength, it will make him/her very rich.

Financial gain (by 11th / 12th from Lagna Pad): If the 11th or 12th from Lagna Pad is occupied, or receives a Drishti from any Graha he/she will be happy and rich; wealth will come through various means. If a benefic is related, as above, it will confer wealth through good means. A malefic, as above, will confer wealth through questionable means. If there be both a benefic and a malefic, it will be through both means. If the Graha in question be in exaltation, or in own Rashi, there will be plenty of gains and plenty of happiness. The quantum of gains will correspond to the number of Graha in, or giving a Drishti to the 11th or 12th from Lagna Pad. If there is Argala (defined in Chapter 8.7) for the said 11th or 12th, there will be more gains, while a benefic Argala will bring still more gains. If the said benefic, causing Argala is in his exaltation Rashi, the gains will be still higher. If the said 11th or 12th receives a Drishti from a benefic from Lagna, the 9th, gains will increase in the ascending order. In all these cases, the 12th from Pad should simultaneously be free from malefic association. A benefic, placed in Lagna, giving a Drishti to the 11th or 12th from Arudha Lagna will be still beneficial. If the Drishti is from the 9th from Lagna, it will confer much more gains. If the 12th from Lagna Pad does not receive a Drishti, as the 11th from Lagna Pad receives a Drishti from a Graha, then the gains will be uninterrupted.

Financial Loss (by 12th from Lagna Pad): If the 12th from Lagna Pad is conjunct by both benefic and malefic, there will be abundant earnings but plenty of expenses. The benefic will cause through fair means, malefic through unfair means and mixed planets through both fair and unfair means. If the 12th from Lagna Pad is conjunct the Sun, Venus and Rahu, there will be loss of wealth through the king. The Moon aspecting (the said trio in the said house) will specifically cause more such losses. If Mercury is in the 12th from the Lagna Pad and be with or aspect by benefic similarly there will be expenses through paternal relatives. A malefic so related to the said Mercury will cause loss of wealth through disputes. If Jupiter is in the 12th from Lagna Pad, aspect by others, the expenses will be through taxes and on one's own. If Saturn is in the 12th from the Lagna Pad along with Mars and is aspect by others, the expenses will be through one's co born.

Financial Loss (by 7th from Lagna Pad): If Rahu and Ketu are placed in the 7th from Lagna Pad, he/she will be troubled by stomach disorders, or by fire. If there is Ketu in the 7th from the Lagna Pad and is aspect to or be conjunct another malefic, he/she will be adventurous; will have (prematurely) grey hair and big male organ.

(b) Predictions by Bhava Pad:

The lord of the A10 is exalted and well-placed; it will show fortunate, intelligence and energy. The 12th Lord from Lagna, if exalted, it will instil delay, separation and difficulty. The 12th Lord may send him/her off to foreign lands for official work. If lord of A10 happens to be in the 7th from AL, it will give Raja Yoga for undying fame. Exalted planets in the 7th from the AL show that personal interaction is elevated and the fame of the native shall remain even after he has passed away. It show gain from partnerships or gains through the marriage. If Mercury is in the 2nd from Arudha Lagna, he/she will lord over the whole country. Venus in the 2nd from Lagna Pad will make one a poet or a speaker. If the Dara Pad (i.e. the Pad of the 7th house) falls in an angle or in a trine counted from the Lagna Pad or if the Lagna Pad and Dara Pad both have strong planets, he/she will be rich and be famous in his country. If the Dara Pad falls in the 6th, 8th, or 12th from the Lagna Pad, then he/she will be poor. If the Lagna Pad and the 7th there from or an angle/a trine/ an Upachaya there from be occupied by a strong planet, there will be happiness between the husband and wife. If the Lagna Pad and Dara Pad are mutually angular or trine, there will be amity between the couple; if these be in mutually 6th, 8th or 12th, doubtlessly mutual enmity will crop up. Similarly mutual relationship or gain or loss through son etc, be known based on Lagna Pad and the relative Bhava Pad. If the Lagna Pad and Dara Pad are mutually angular or 3rd and 11th or trine, he/she will be a king ruling the earth. Similar deductions are made with reference to mutual positions of Lagna Pad and Dhan Pad.

Calculation of Pad Strength: If the Bhava happens to be in Kanya, Mithuna, Tula, Kumbh, or the first half of Dhanu, deduct Yuvati Bhava from the Bhava. If Mesh, Vrishabh, Simha, or first half of Makara or the second half of Dhanu happens to be the Bhava, deduct Bandhu Bhava from it. If the Bhava is in Karka, or in Vrischika, deduct from it Lagna. Deduct Karma Bhava from the Bhava, happening to fall in Makara second half, or Meena. Convert the product so obtained into degrees and divide by 3 to get Bhava Bala. If the balance in the process of deducting Nadira, Meridian, Lagna, or Yuvati exceeds 6 Rashi, deduct it again from 12 Rashi, before converting into degrees and dividing by 3. The product after division should be increased by one fourth, if the Bhava in question receives a benefic Drishti. If the Bhava receives a malefic Drishti, one fourth should be reduced. If Guru or Buddha gives a Drishti to a Bhava, add that Graha's Drik Bala also. And add the strength acquired by the Lord of that Bhava. This will be the Net Bhava Bala.

12.2 Predictions by Upa Pad (A12)

Calculation of Upa Pad (A12): The Bhava following the Lagna at birth, i.e. the Vyaya Bhava (12th house) in the case of an odd Rashi, or, it is the Dhan Bhava (2nd Hose) in the case of an even Rashi, is taken as Rising Sign and is called Upa Pad or Arudha Pad (A12). This Upa Pad is also called Gaun Pad.

Predictions:

If Upa Pad is yuti with, or receives a Drishti from a benefic Graha, he/she will obtain full happiness from progeny and spouse. If the Upa Pad is in a malefic Rashi, or receives a Drishti from, or is yuti with a malefic, he/she will become an ascetic and go without a wife. If there be a benefic Drishti on Upa Pad, or the related malefic, or a yuti, deprive of spouse will not come to pass. In this case Surya, being exalted, or in a friendly Rashi, is not a malefic. He is a malefic, if in debilitation, or in an enemy's Rashi.

Effect from the 2nd from Upa Pad: If the 2nd from Upa Pad is a benefic Rashi, or receives a Drishti from, or is yuti with a benefic, the same good results as for wife and sons will come to pass. If there is a Graha in the 2nd from Upa Pad in its debilitation Rashi, or debilitation sign, or is yuti with a debilitated or malefic Graha, there will be destruction of wife. If the said occupant be in its exaltation Rashi, or Navamsa, or receives a Drishti from another Graha, there will be many charming and virtuous wives. If Mithuna happens to be the 2nd from Upa Pad, then also there will be many wives. If the Upa Pad, or the 2nd there from be occupied by its own Lord, or, if the said Lord is in his other own Bhava, the death of wife will be at advanced age.

Effects on Wife from the 2nd of Upa Pad: If the 7th Lord or Sukra is in its own Bhava, there will be loss of wife only at a later stage. If the Lord of Upa Pad or the constant significator of wife is in exaltation, the wife will be from a noble family. Reverse will be the case, if he is debilitated. If the 2nd from Upa Pad is related to a benefic, the wife will be beautiful, fortunate and virtuous. Should Sani and Rahu be in the 2nd from Upa Pad, he/she will lose his wife on account of calumny, or through death? His/Her wife will be troubled by disorder of blood, leucorrhoea (Pradar) etc., if Sukra and Ketu are in the 2nd from Upa Pad. Buddha with Ketu in the 2nd from Upa Pad will cause breakage of bones, while Rahu, Sani and Surya will cause distress of bones.

Buddha and Rahu in the 2nd from Upa Pad will give a stout-bodied wife. If the 2nd from Upa Pad happens to be one of Buddha Rashi and is tenanted by Mangal and Sani, his/her wife will suffer from nasal disorders. Similarly a Rashi of Mangal, becoming the 2nd from Upa Pad and occupied by Mangal and Sani, will cause nasal disorders to one's wife. Guru and Sani will, if be in the 2nd from Upa Pad, cause disorders of ears and/or eyes to the wife. If Buddha and Mangal are placed in the 2nd from Upa Pad other than their own Rashi, or, if Rahu is with Guru in the 2nd from Upa Pad, his/her wife will suffer from dental disorders. Sani and Rahu together in one of Sani Rashi, which is the 2nd from Upa Pad, will cause lameness or windy disorders to his/her spouse. These evils will not come to pass, if there happens to be a Yuti with or a Drishti from a benefic (or from another benefic in the case of affliction being caused by a benefic him). All these effects are deduced from the natal Lagna, Lagna Pad, the 7th from Upa Pad and the Lords thereof.

ects on Sons from the 9th of Upa Pad: If Sani, Chandra and Buddha are together in the 9th from one of the said places, there will be no son at all, while Surya, Guru and Rahu so placed will give a number of sons. Chandra so placed will give a son, while a mixture of Graha will delay in obtaining of a son. The son, caused by the Yuti of Surya, Guru and Rahu, will be strong, valorous, and greatly successful and will destroy enemies. If Mangal and Sani are in the said 9th, there will be no son, or a son will be obtained by adoption, or brother's son will come in adoption. In all these cases odd Rashi will yield many sons, while even Rashi will cause only a few.

Many Sons and Many Daughters: If Simha happens to be Upa Pad and receives a Drishti from Chandra, there will be a limited number of children. Similarly Kanya will cause many daughters.

Co-born form Lagna Pad: Rahu and Sani in the 3rd, or the 11th from Lagna Pad will destroy his/her co-born. Rahu and Sani in the 11th will indicate the destruction of elder brothers and/or sisters and in the 3rd younger ones. If Sukra is in the 3rd, or the 11th from Lagna Pad, there would have been an abortion to the mother earlier. Same is the effect, if Sukra is in the 8th from natal Lagna, or from Lagna Pad. Should Chandra, Guru, Buddha and Mangal be in the 3rd or the 11th from Lagna Pad, there will many valorous co-born. Should Sani and Mangal be in the 3rd, or the 11th from Lagna Pad, or give Drishti thereto, younger and elder co-born will, respectively, be destroyed. If Sani is alone in one of the said Bhava, he/she will be spared, while the co-born will die. Ketu in the 3rd or the 11th will give abundant happiness from one's sisters.

Other Matters from Lagna Pad: If the 6th from Lagna Pad is occupied by a malefic and is bereft of a Yuti with, or a Drishti from a benefic, he/she will be a thief. If Rahu is in the 7th, or the 12th from Lagna Pad, or gives a Drishti to one of the said Bhava, he/she will be endowed with spiritual knowledge and be very fortunate. If Buddha is in Lagna Pad, he/she will Lord over a whole country, while Guru will make him a knower of all things. Sukra in this context denotes a poet/speaker. If benefic occupy the 2nd from Upa Pad, or from Lagna Pad, he/she will be endowed with all kinds of wealth and be intelligent. He/She will surely become a thief, if the Lord of the 2nd from Upa Pad is in Dhan Bhava and is there yuti with a malefic Graha. If Rahu is in the 2nd from the Lord of the 7th, counted from Upa Pad, he/she will have long and projected teeth. Ketu in the 2nd from the Lord of the 7th, counted from Upa Pad, will cause stammering and Sani in the 2nd from the Lord of the 7th, counted from Upa Pad, will make one look ugly. Mixed will be the effects, if there are mixed Graha.

12.3 Predictions by Dara Pad

Dara Pad is nothing but the Arudha Pad of 7th house (A7). The Dara Pad is the external manifestation of the matters of the 7th House. A7 will indicate details of the native sexual life. If the Upa Pad (UL) is well placed from the AL but the A7 is in a Dustasthana, the couple may live together but sexual happiness may be lacking in the relationship. If the Lord of the UL is exalted in Rashi and Navamsa, it will indicate a spouse from a family more elevated conventionally than that of the native and vice versa. If the 2nd from the UL is afflicted by malefic or debilitated Graha, the length of the marriage is suspect. The Dhan Pad / A2 are the physical manifestation of assets of a movable nature.

12.4 Predictions by Pranapad

Calculation of Pranapad - method 1: Multiply Ishtaghati by 4 and we get 9 x 4 = 36 Ghati (i) 31 Pala ÷ 15 give 2 s quotients and 1 as remainder. We add 2 to 36 and divide the sum by 12; we get 3 as the quotient and 2 as the remainder. Leaving out 3 as so many cycles, we accept 2 as the Rashi number and doubling the remainder Pala, we get 2^0 as the degree number. The mean Pranapad is, therefore, 2^0 of Mithuna.

Method 2: Convert the given Birth Time into Vighati and divide the same by 15. Now, convert the result to the Rashi, degree, minutes and seconds. The resultant Rashi, degrees be added to Surya longitude, if he is in a Movable Rashi, it will yield Pranapad. If Surya is in a Fixed Rashi, add 240 degrees additionally, it will yield Pranapad and, if in a Dual Rashi, add 120 degrees further to get Pranapad. The birth will be auspicious, if Pranapad falls in the 2nd, 5th, 9th, 4th, 10th, or 11th from the Natal Lagna. In other Bhava, Pranapad indicates an inauspicious birth.

Predictions: If Pranapad is in Thanu Bhava, he/she will be weak, sickly, dumb, lunatic, dull witted, defective-limbed, miserable and emaciated. If Pranapad is in Dhan Bhava, he/she will be endowed with abundant grains, abundant wealth, abundant attendants, abundant children and be fortunate. If Pranapad is in Sahaj Bhava, he/she will be injurious (or mischievous), proud, hard-hearted, very dirty and be devoid of respect for elders. If Pranapad is in Bandhu Bhava, he/she will be happy, friendly, attached to females and elders, soft and truthful. If Pranapad is in Putra Bhava, he/she will be happy, will do good acts, and be kind and very affectionate. If Pranapad is in Ari Bhava, he/she will be subdued by his relatives and enemies, be sharp, will have defective digestive fire, be wicked, sickly, affluent and short-lived. If Pranapad is in Yuvati Bhava; he/she will be green-eyed, ever libidinous, and fierce in appearance, be not worth respect and be ill-disposed. If Pranapad is in Randhra Bhava; he/she will be afflicted by diseases, be troubled and will incur misery on account of the king, relatives, servants and sons. If Pranapad is in Dharma Bhava, he/she will be endowed with sons, be very rich, fortunate, charming, will serve others and be not wicked, but be skilful. If Pranapad is in Karma Bhava; he/she will be heroic, intelligent, and skilful, is an expert in carrying out royal orders and will worship gods. If Pranapad is in Labh Bhava; he/she will be famous, virtuous, learned, wealthy, and fair-complexioned and attached to mother. If Pranapad is in Vyaya Bhava, he/she will be mean, wicked, and defective-limbed, will hate Brahmins and relatives and suffer from eye diseases, or be one-eyed.

13

Divisional (Varga) Charts

Vargas (Divisional) Charts:

The term varga is Sanskrit name, in Indian astrology, of the division of a Zodiacal Sign (Rashi) into divisions. Each such fractional division of a Sign is also known as an aṃsa. We choose and study the divisional chart for detail analysis of the horoscope for two purposes, such as, (i) firstly, to see the strength of a planet by observing its placement in various divisions like in its own sign, mooltrikona sign or exaltation sign, where it is treated as strong. If it is placed in its sign of debilitation, it is treated as weak. A planet which is weak in the natal chart does not improve its position by occupying benefic, exaltation or own sign in the divisions. However, wearing a gemstone or a yantra (Kavach) strengthen those planets, which are weak but functionally favourable for a person. We treat Rahu, Ketu and those planets whose mooltrikona signs fall in the sixth, eighth or twelfth houses from the natal ascendant as functional malefic planets. The sign Cancer is treated as the mooltrikona sign of the Moon; and (ii) secondly, the divisional charts are studied for identifying the results of a particular sector of life by studying the corresponding divisional chart along with the concerned house in the natal (birth) chart. Nata Chart is given more importance than amsa in any context or area of life predictions. What so ever is not promised by the Natal chart cannot be given by an amsa chart. The amsa chart has to support the Natal chart for full fructification or the full potential of the Natal chart predictions.

Divisional Dignities:

Bṛhat Parasara Horashastra Slokas - 42-51 and Sarvartha Chintamani St. - 32-35 speak that if any planet occupies a friendly sign or its own sign or its exaltation sign then that planet gives good results in full and makes one wealthy, successful and renowned. Planets become more and more auspicious or dignified by gaining more and more of their Own Sign, Exaltation Sign, Moolatrikona Sign, or Friendly Sign or a Sign that is owned by a planet ruling an angle (1, 4, 7, 10) from the Arudha. Accordingly the status thus acquired by planets stands upgraded. The corresponding count of number of good Vargas (divisional charts) except the Rashi Chart (D-1) is known as Varga-Dignity. When a planet acquires such own sign or exalted sign etc in two out of sixteen vargas or divisional charts then it is known to have gained the status called the Parijatamsa or Bhedakamsa; when it acquires three vargas then it is known to have gained the Uttamamsa or Kusumamsa or Vyanjanamsa; acquirement of four is known as the Gopuramsa or Naagpushpamsa or Kimshukamsa or Chaamaramsa; if five then the Simhasanamsa or Kundakamsa or Chhatramsa , if six – the Parvatamsa or Keralamsa or Kundalamsa, seven – the Devalokamsa or Kalpavrkshamsa or Mukatamsa, eight – the Kumkumamsa or Brahmalokamsa or Chandanvanamsa, nine – the Iravatamsa or Poornachandramsa, ten – the Vyshnavamsa or Shridham or Ucchaishrvamsa, eleven – the Saivamsa Dhanvantriamsa, twelve – the Bhaswadamsa or Suryakantamsa, thirteen – the Vaisheshikamsa or Vidrumamsa, fourteen – the Indrasanamsa, fifteen – the Golokamsa, and for sixteen good Vargas, the planet is known to have gained the status called the Shrivallabhamsa.

Divisional Chart Making:

First Method: Convert the longitude into minutes of arc and multiply by the figure of Divisional Chart concerned. Divide the product by 1800. We get some quotient and some fraction in the form of decimals. Consider a digit 1 for the fractions. We add 1 for every fraction with the quotient to get the total number. If the total number is less than twelve (12), then count the number from Aries, i.e. the first Sign of the Lagna Chart. Then place that Ascendant or Planets or Points in that house where it falls. If the total number is more than twelve (12), we divide the total number by 12 and we get some remainder digit from 1 to 11 or some time the remainder will be zero. If the remainder is zero, we consider it as 12. Then, we count the remainder digit number from Aries, i.e. the first Sign of the Lagna Chart and place that Ascendant or Planets or Points in that house where it falls.

Second Method: We divide the total measure of the Longitude, i. e. Angle in degree including their minutes of arc of position from Aries by one Division Angle Measures. We get some quotient and some fraction in the form of decimals. Consider a digit 1 for the fractions. We add 1 for every fraction with the quotient to get the total number. If the total number is less than twelve (12), then count the number from Aries, i.e. the first Sign of the Lagna - Chart. Then place that Ascendant or Planets or Points in the house where it falls. If the total number is more than twelve (12), we divide the total number by 12 and we get some remainder digit from 1 to 11 or some time the remainder will be zero. If the remainder is zero, we consider it as 12. Then, we count the remainder digit number from Aries, of the Lagna - Chart and place that Ascendant or Planets or Points in that house where it falls.

Shodasa varga: "Shodasa Varga" means "sixteen divisions". Shodasa Varga is a group of the divisional charts: (1) (D-1), (2) D-2, (3) D-3, (4) D-4, (5) D-7, (6) D-9, (7) D-10, (8) D-12, (9) D-16, (10) D-20, (11) D-24, (12) D-27, (13) D-30, (14) D-40, (15) D-45, and, (16) D-60. In Shodasa Varga scheme, the dignity-designations commence from Bhedakaamsa for acquiring 2 (two) good Vargaas; Kusumaamsa for acquiring 3 (three) good Vargaas; Nagapurushaamsa for 4, Kandukaamsa for 5, Keralaamsa for 6, Kalpavrikshaamsa for 7, Chandanavanaamsa for 8, Poornachandraamsa for 9, Uchchaisravaamsa for 10, Dhanvantaryamsa for 11, Sooryakaantaamsa for 12, Vidrumaamsa for 13, Indraasanaamsa for 14, Golokaamsa for 15, Sree Vallabhaamsa for acquiring 16 (sixteen) good vargas.

13.1 Significations of Shodasa varga (Divisional Charts):

Divisional charts are also called Varga charts, Amsha charts, sub-charts, and D-charts.

1) D-1 (Lagna Chart- Deha, physique, body): This is the Lagna / Natal Chart. This is main Varga showing the body, material life, physical matters, social-maters, circumstances and actualization. It provides all about our general attitude, fortune and its overall picture of all situations. Rasi is like Head and it describes perspective of individual and resources for living.

2) D-2 (Hora Chart- Sampati, Liquid Wealth, money): This division is about dividing the Rasi (sign) into two equal parts. Hora Chart is used to judge the Financial Status & Wealth, the liquid assets such as cash money, collections of art, music, food, wine, or other valuables, the knowledge-treasury and Hoarded Wealth. It is seen for ascertaining the quantum of wealth and the way the person is going to make his earning. The first half of an odd Rashi of Hora is ruled by Sun. The second half of an odd Rashi of Hora is ruled by Moon. There is reverse in the case of an even Rashi, i.e. the first half of an even Rashi of Hora is ruled by Moon and the second half of an even Rashi of Hora is ruled by Sun. Male planets in the Sun's Hora and female planets in the Moon's Hora are beneficial. If maximum planets are in the Sun's Hora it is said that the native earns money as a result of hard work and the quantum of wealth may be very good. If Moons Hora predominates, the earning will be relatively easy. Followings are the special Rules for Hora: There are special rules for bala calculation in Hora. See table below.

Planet	Cancer	Leo
Sun, Mars, Jupiter	0	20
Mercury	20	20
Moon, Venus, Saturn	20	0

3) D-3 (Drekkana Chart- Sukha, happiness from co-born, siblings): This division is about dividing the Rasi (sign) into three equal parts. One third of a Rashi is called Drekana. It is used to judge the siblings, happiness through co-born and their ties with you, success through own efforts, mental health, nature of death and longevity of the native. Third Bhava lord for odd Lagna or eleventh lord for even Lagna of D-3 will show fortune and work of sibling. Rahu and Ketu will show amount of brothers/sisters. The planets falling in Sarpa Dreshkona and Phasha Dreshkona are not considered good for the native.

4) D-4 (Chaturthamsa- Bhagya, fortune, immoveable houses/Land properties): This division is about dividing the Rasi (sign) into four equal parts. Each Chaturthamsa is one fourth of a Rashi. 4th Bhava in D-4 will signify fixed assets such as wealth in the form of owned properties, landed properties, luck and Fortunes, childhood, Well-being, Pleasure & Happiness, foreign immigration or change of residence, Marriage of a girl child, Emotion, and prosperit. Lagnesh will show our attitude. Eighth lord in D4 shows us blockage to our fortune. The Lords of the 4 Kendra from a Rashi are the rulers of respective Chaturthamsa of a Rashi, commencing from Mesh. The deities, respectively, are Sanak, Sanand, Kumar and Sanatan

5) D-5 (Panchamsa- Fame, authority and power): This division is about dividing the Rasi (sign) into five equal parts. It is One-Fifth Sign. It is used to predict the fame, authority and power as well as the spiritual inclinations. It also reveals the spiritual evolution of the native and whether one has a leaning for religion, philosophy or even atheism or moral fibre. The Panchamsha D5 shows the inner spiritual nature and ethics. This Vargas is attributed by Jaimini.

6) D-6 (Shashtiamsa- Health & Diseases): This division is about dividing the Rasi (sign) into six equal parts. It is One-Sixth Sign. It is used to predict the diseases and hereditary diseases, if any, and general health of the native. The indications about the diseases help to take the remedial measures as suggested by the chart. The Shashtamsha D6 indicates health issues specifically, more than other divisional charts. This Vargas is attributed by Jaimini.

7) D-7 (Saptamsa- Putra, sons, matters with one's children and grandsons, and the prosperity of the spouse): This division is about dividing the Rasi (sign) into seven equal parts. It is One-Seventh Sign. Saptamsa is used for study of progeny or childlessness; own Children, grand children, prosperity of issues and childlessness and happiness through children. The Saptamsa counting commences from the same Rashi in the case of an odd Rashi. It is from the seventh Rashi for an even Rashi. In Saptamsa Varga scheme, the designations of the seven divisions in odd Rashi are Kshaar Ksheer, Dadhi, Ghrith, Ikshu, Ras, Madhya and Suddh Jal. These designations are reversed for an even Rashi. Saptamsa is an interesting chart. Fifth Bhava (ninth for even Rasi of Lagna, or odd Rasi with Ketu) will show character, sex and fortune of first child. Relation between Lagnesh of D7 and this lord will show the relations with our kid. The Karaka and the avashtha of the fifth lord of the birth chart and the placement and influences on the fifth house and the fifth lord of the said divisional chart have to be observed.

8) D-8 (Ashtamsa- Sudden and unexpected troubles, litigation, major challenges, and crises, including the deaths of that close): It is One-Eighth of Sign. It is used to predict the accidents and the longevity of individual with a fair accuracy. It is used to predict the unexpected troubles, sudden unexpected changes in services / professions, cycle of death and rebirth, the accidents and the longevity of individual with a fair accuracy. This Vargas is attributed by Jaimini.

9) D-9 (Navamsa- Kalatra, spouse, Marriage): It is One-Ninth of Sign. It is used for study mainly about the spouse, the couple compatibility by matching the two Navamsa Charts of the couple instead of Natal Chart, the quality of spouse and the time of marriage, longevity of the marriage life. It reveals the temperament, mental and moral character of spouse, the number of the marriages and whether one's married life will be happy or otherwise, the quality of spouse and the time of marriage, longevity of the marriage life This is used to see the actual strength of a planet. An exalted planet in the birth chart (if it debilitates in the Navamsha) may not give very good results or its effects can get diluted. However, a planet getting empowered in Navamsha by getting vargottama or by getting exalted will give good results. The lord of 64th Navamsha from the Moon becomes highly important for the calculation of Longevity and the Maraka dasa.

10) D-10 (Dasamsa- Mahat Phalam, Power, Position, Leadership, Profession, Career, and achievements): It is One-Tenth of Sign. Dasamsa is used for study the profession, carrier & allied matters and success, educational, vocation and power and position, of the native. Dasamsa gives the destiny and the field in which native's good fortune resides. This also gives indication of wealth. Second, sixth and tenth Bhava of D-10 will show type of work (profession). Venus will show private sector, while Luminaries will give work in politics. Opposite produce inauspicious effects. The placement and avashtha of the tenth lord of the birth chart is important here. Apart from that the rashi rising in the 10th house of this divisional chart and the influence on the 10th house has also to be considered, to judge the profession and any professional growth or fall of the native.

11) D-11 (Ekadasamsa- Rudramsa- Flow of income in a person's life, Death and destruction): It is One-Eleventh Sign. It is about gain, cure, inheritance, the legacies, and sudden inflow of money in chance games, speculations and gambling, gains, success and progress, success in speculation, inheritance, lottery, gifts and prizes, Death and Destruction, Joining, Connection, Gainful Association, inheritance, the legacies, and sudden inflow of money in chance games, speculations and gambling. This Vargas is attributed by Jaimini.

12) D-12 (Dwadasamsa- parents, uncles, aunts, grand-parents): It is One-Twelfth Sign. It is about Parents, details about parents, their length of life and higher consciousness and links the past and future lives. Dvadasamsa can also predict skill in astrology and philosophy and in depth knowledge, wealth and longevity of parents and inheritance from them, and showering His Blessings on you as per your past Karma, past karma, Subconscious mind, Heredity and higher consciousness and links the past and future lives of the native correctly. Dvadasamsa can also predict skill in astrology and philosophy and in depth knowledge of the native correctly. here, (as in all divisional charts) the Avashtha of the natural Karaka has to be seen. Here, the Sun represents the father and the Moon represents the mother. Here, the influence on the ninth and the tenth house has to be understood. The Lord of the 88th Dwadashamsha is important in longevity analysis. This chart also is said to reveal whether a person inherits a disease from his parents.

13) D-14 (Chaturadasamsa- Vehicles/Vahana): It is One-fourteenth Sign. Chaturadasamsa helps in study of Conveyance and general happiness from the conveyance. 4th house of D-14 will indicate the Conveyance and general happiness from the conveyance.

14) D-16 (Shodasamsa- Sukha/Asukham-happiness/distress from Vahana-Vehicles): It is One-Sixteenth Sign. It indicates the General happiness with conveyance (Vehicle), the Movable assets such as 'moving shelters', vehicles such as cars, trucks, airplanes, ships, rail-cars, and Traveling, Comforts, Home, luxuries and happiness, accident of vehicle, luck for conveyance, General happiness with conveyance (Vehicle) and timings of potential accidents in the native's life. The House, whose Lord is in a benefic Shodashamsa will flourish. For this the Karaka for conveyance, Venus, has to be seen. Affliction to the 4th Lord and the 4th house has to be seen along with the affliction of Venus. This can lead to violent accidents if otherwise also promised in the Birth Chart.

15) D-20 (Vimsamsa- Upasana, worship, Sadhna, spirituality): It is One-Twentieth Sign. Lagnesh in that Varga will show philosophical-spiritual development, Permanent wisdom, our approach to spirituality. Surya here is the real source/ and core of spirituality while Jupiter shows our Jnana. Mangal in Lagna can show focus on Ahimsa. Twelfth and Forth lords show meditation/dhyana while Kona's are connected to Jnana Marg (fifth is bhakti). The planet placed in the 5th and the 9th houses are important here. The overall exaltation or debilitation of planets holds a clue to understand the overall growth of the native spiritually.

16) D-24 (Chaturvimshamsa- Vidya- Learning, knowledge, education and training): It is One-Twenty fourth Sign. It signifies higher education, intellect and academic achievements, grooming oneself educationally in the professional arena, post higher secondary level education & training, achievements, Learning, Knowledge and Spiritual knowledge. Moon will show how we study. If it is connected to night-planets then we like to study in night. Tenth lord shows how we teach. The house to be considered is the 5th house. Further, influences on the fifth house have to be seen along with the avashtha of the 5th Lord of the birth chart. The connection of 5th lord with the 6th, 8th and the 12th lord is supposed to bring disturbances and educational change for the native. If however, such planets are under benefic influence the change in education will be for the betterment of the native. Example of this can be that a person does Engineering and then does Management.

17) D-27 (Saptavimshamsa/Nakshatramsa-General vitality, ability to raise, strengths and weaknesses, inherent nature): It is One-Twenty seventh Sign. It is about strength of native. The Saptavimshamsa Lords are, respectively, the presiding deities of the 27 Nakshatra. These are for an odd Rashi. Count these deities in a reverse order for an even Rashi. It signifies stamina, endurance and physical strength of an individuall.

18) D-30 (Trimsamsa- Vistha Phala- Evils effects, punishment, diseases): One thirty of a Rashi is called Trimsamsa. Trimsamsa shows knowledge about vices and diseases. It is also used to time death and punishment. It is about Malefic Effects and Bad luck. These are totally 360. Trimsamsa indicates the Evil effects. Trimsamsa signifies misfortunes and sufferings, painful and incurable diseases, sufferings of illness, Dangers, Health problems, Enmity (Aarishta), General Mischief, Evils, Failure, Bad Luck and the character of a woman native knowledge about vices and diseases. It is also used to time death and punishment. It is about Malefic Effects and Bad luck. These are totally 360. This chart in women horoscopy is also said to show the character of the native. It is said that in case of women if Trimshamsha of Mars is rising she is of questionable character. However, this chart can also be read to verify the character in case of a male native.

19) D-40 (Khavedamsa- Subha/Asubha-auspicious/inauspicious effects): One forty of a Rashi is called Khavedamsa. It is about Auspicious and inauspicious effects. It signifies Maternal Legacy, Auspicious / Inauspicious effects and Habits. But, it has to be noted that in prediction this chart cannot be used much as even few seconds of time difference can change the lagna. This should be used only on thoroughly rectified horoscopes.

20) D-45 (Akshavedamsa- All indications): It is One-forty-fifth Sign. It is a general indication about Paternal Legacy, General indications, Ethical nature, good and bad effects and Conditions of the native, character of the native and for general auspicious and inauspicious effect. Here, as in above case this chart becomes too time-sensitive, making it difficult to take it into account while giving prediction.

21) D-60 (Shashtiamsa- Karma of past life and its effects on current life): It is One-Sixtieth Sign or half a degree each. It is a general indication about good and bad effects in all areas of life. Shashtiamsa indicates the All Conditions. Restrictions on Present's life fruits, Debt owned from past lives, good and bad effects in all areas of life. The House, whose Lord is in a malefic Shashtiamsa will diminish. Graha in Benefic Shashtiamsa produces auspicious, while the opposite produce inauspicious effects. Maharishi Parashara has given lots of importance to this chart. This should be seen for every prediction. This chart makes the reading interesting and makes the analysis accurate. However, again the limitation of this is that it can be used on an astrologically rectified chart.

Methods of Evaluating Divisional Charts:

Many Vedic astrologers interpret divisional charts using the same methods as they use in evaluating birth charts. I have been taught to evaluate divisional charts using the method introduced by Sheshadri Iyer and explicated by Hart de Fouw. In this method, the primary emphasis is upon dynamic analysis, i.e. evaluating the placement of the dasa lord in each of the D-charts, and thereby determining how the theme of each D-chart will go during the planet's dasa. There are a few static analysis principles used in this approach too.

Static Analysis Methods of D-charts:

The placement of the 'significator of a person' in the First House of the D-chart represents that person. **Example:** Venus is the Significator of the Souse and D-9 deals with the Spouse. So, Venus in the First House in the D9 represents Spouse. Similarly, Jupiter in the First House in the D7 represents Child; Either of Mars or Jupiter in the First House of the D3 represents Younger or older sibling respectively; either of the Sun or the Moon in the First House of the D12 represents Dad or Mom respectively. If the planet is weak, afflicted or is malefic in the relevant D-chart, there will be more accentuated difficulties to that person indicated by that planet.

Dynamic Analysis of Placement: Whether the themes of a D-chart will go well or poorly depends primarily on the placement of the Dasa Lord in each D-chart.

The 1, 4, 5, 7, 9, and 10 houses are positive. The 2 and 11 houses are neutral (mildly positive). The 3, 6, 8, and 12 houses are negative. The Moon, Mercury, Venus and Jupiter are benefic. The Sun, Mars, Saturn, Rahu and Ketu are malefic. There are six pairs of companion houses: (1 + 7); (2 + 12); (3 + 11); (4 + 10); (5 + 9); and (6 + 8)

The four primary rules for assessing placement in each D-chart:

1. If a benefic is in a positive house, its dasa goes well. (If there is any planet in its companion house, it goes even better.).
2. If a benefic is in a negative house, its dasa goes poorly -- unless there is a planet in its companion house, in which case it goes well.
3. If a malefic is in a negative house, its dasa goes poorly. (If there is any planet in it's companion house, it goes even worse.)

4. If a malefic is in a positive house, its dasa goes well. Unless there is a planet in its companion house, in which case it goes poorly.
5. If Rahu is in one angle opposite to Ketu, and no other planets are in either house, then both Rahu and Ketu dasa go well for the themes of the D-chart.

Dynamic Analysis by Strength:

The strength of the Dasa Lord indicates how well one deals with the themes. There are four possible extremes:

1. Strong Lord + Positive Placement: Themes thrive and are made the most of them.
2. Strong Lord + Negative Placement: Challenges arise but are well met.
3. Weak Lord + Positive Placement: Themes are good but hard to make most of them.
4. Weak Lord + Negative Placement: Challenges arise and are difficult to handle.

Divisional Charts: Yogas in the Birth Chart:

When one or more yogas, representing a specific area of life or life theme (e.g. success in career) are present in the birth chart and are formed by one or more strong planets then the yogas can supercede divisional chart evaluations for that theme. E.g. If someone has one or more strong yogas in the 10 house, then even when they run the dasa of a planet poorly placed in their D10 chart, the career may in fact prosper.

In the above example, the Sun dasa goes well for career. Note: in the birth chart, the Sun is in its own sign and with dig bala and Moon is Full and has dig bala, and together they form a great Raja Yoga of the lord of the 9 house with the lord of the 10 house, and the Sun is in the 10 th house and also a karaka of success in career.

4. The influence of a debilitated planet upon the dasa lord. If the dasa lord is in the same house or aspected by a planet in its sign of debilitation, then the themes of the D-chart during that dasa become disrupted, especially during the bhukti of the debilitated planet.

Divisional Charts Page 115. The influence of multiple malefics on the dasa lord.

If two or more strong malefics aspect the dasa lord, the themes of the D-chart become

more challenged. If the sole enemy also participates, even more difficulties occur.
Note: in the above example, Mars is retrograde and aspect Jupiter with its 8 aspect.

Tequniques of Reading the Varga Charts:

Selection of correct Divisional chart, House, Reference and others are more important as mentioned below:
(i) Divisional Chart: We should choose the correct Divisional Chart, such as, for 'happiness from a vehicle select D-16'; for 'criminal's psychology select D-30; for 'marriage select D-9 and for one's 'religious activities D-20'.
(ii) House: We should choose the correct house, such as, for analyzing someone's 'learning & education' selecrt the 4th house in D-24; for 'intelligence, scholarship, academic reputation, academic distinctions / awards' select the 5th house in D-24 and for pursuit of 'knowledge and interacts with others' select the 7th house in D-24.
(iii) Arudha: Sometimes, an Arudha pada is more appropriate to see a matter than a house. For example, we can see Darapada (A7) in D-24 to figure out what kind of people one typically interacts with one's learning related activities. We can see one's academic distinctions/awards in A5, because they are maya (illusion) related to intelligence and scholarship. The world forms an impression about one's intelligence and scholarship based on one's scores, ranks, grades, distinctions and awards.
(iv) Influence: Planets influence the event with its position and its drishti. We should judge the meaning of each planet's influence. Planets in the quadrants from a Lagna sustain it. Planets in trines from a Lagna let it prosper. Planets in upachayas let it grow. Planets in dusthanas bring obstacles. Example: Suppose we are analyzing A3 in D-10 and 3rd house for an author. The 3rd house shows one's writing skills and A3 shows one's books. If a planet is in a quadrant from A3, its periods may result in book writing. If a planet is in the 8th house from A3, its periods may bring obstacles in book writing. If a planet is a badhaka house from A3, it can create troubles in book-writing. The Trines indicate prosperity and flourishing. Quadrants indicate sustenance and vital activity. Upachayas

indicate gains and growth. Dusthanas indicate setbacks and obstacles. Argala Sthanas indicate decisive influences.

(v) D-Charts Common References for Houses: It may be noted from the list above that each house shows many matters. We have to note the area of life to be predicted in the divisional chart under examination. We have to choose the correct houses with reference to area of life. **(i) Natal Lagna:** Lagna is the most commonly used reference when finding houses. **(ii) Chandra Lagna:** Chandra lagna or Rashi means Moon house is taken as a reference. **(iii) Ravi Lagna (Sun lagna):** Ravi lagna means Sun house is taken as a reference. Sun is an important reference for things related to physical vitality. **(iv) Arudha Lagna:** Arudha lagna shows how a native is perceived in the world. It also shows the status of a native. **(v) Paaka Lagna:** Paaka Lagna is nothing but the house where Lagna Lord is positioned. Example: If someone with Pisces lagna has Jupiter in Cancer, then Cancer becomes Paaka Lagna for the native. Houses counted from Paaka lagna throw light on matters related to the physical body of a native. **(vi) Karakamsa Lagna:** Atma karaka stands for the soul of the person. Atma karaka is an important reference point in a chart, because the soul is an important factor in deciding the nature of inner self than the physical existence. The Rasi occupied by Atma karaka in it is called “Karakamsa”. We can analyze navamsa chart with respect to Karakamsa. **(vii) Ghati Lagna:** Ghati lagna (GL) shows self, from the point of view of power, authority and fame. When we analyze promotions in career or political power of politicians, this reference is very important. **(viii) Hora Lagna:** Hora lagna (HL) shows self, from the point of view of wealth. Hora is important when analyzing one’s wealth. **Example:** (i) The 4th house from Lagna shows education, vehicle, house Property, pleasure and mother. A list of the areas of life is given above. The 4th house from lagna, the 4th house from Arudha lagna and the 4th house from Paaka lagna can mean different things, depending on the matters shown by Lagna, Arudha Lagna and Paaka Lagna. (ii) Suppose, we are looking for “Success in Competition”, the most appropriate reference is Arudha lagna, the Lagna of perceiption of the native. So, the 5th house from Arudha lagna will show ‘success in competition’. But, ‘Intelligence and scholarship’, are seen from 5th from Radix lagna, the lagna of self. We also use Karaka as references instead of lagna and accordingly, 'scholarship' can be seen from the 5th from Mercury. 'Intelligence' can be seen from the 5th from Jupiter (Karaka). 'Academic Reputation' can be seen from the 5th from Sun (Karaka of reputation) in D-24. 'Memory' is seen from the 5th house from Paaka lagna (Lagna lord position). Saturn's transit over the Rasi containing one's lagna may throw obstructions and hamper one's activities. Saturn's transit over Chandra may create frustration and mental depression. Saturn's transit over one's Paaka lagna may show sickness all the time and attacking the physical vitality.

Tips on Examples: We see the following factors related to that event in the Varga:

Career: If Dashamsha lord (Lagna lord) is strong in radix, the native is choosing career, public service, and earthly power as a vehicle for self-knowledge on the road to God-realization. The dashamsha may also be read in the same method as reading in the radix to very finely tune the profile of profession. Radix L-10 should be strong in dashamsha to confirm maximum public leadership. Example 1: Suppose, someone wants to know about his career in medical profession. A common astrologer would look at his 'Lagna Chart', "Dasha Period" and may be 'Chalit Chart'. But, a good astrologer will also look at the divisional chart called Dasamsa Chart, or the D-10 Chart. This chart is the heart of the 10th house in the main chart "Lagna Chart". If career is shown from the 10th house, then Dasamsa chart will open up that house surgically and will show the real story. D-10 will tell he or she should go into the medical profession, but it will also tell if he/she will become 'Nurse', 'Doctor', 'Surgeon', 'Dentist', or should he or she should open up a medical business and hire doctors to work under them; this is why divisional charts are so important. No other astrology system in the world will have divisional charts; not even Egyptian. Mayan are very close, as they do provide 'some' divisional charts in their astrology, but their history and text of astrology are 90% lost. Example 2: The professional matter is normally seen from the 10th house of a Natal Chart (D-1). Suppose, the ruler of 10th house is exalted in a Natal Chart (D-1) and a common astrologer without consulting the relevant Varga Chart has predicted that the native has a great luck with regard to his profession, such prediction in all probability is likely to fail and misguiding for him if the same planet is debilitated in D-10 chart (Dasamasa chart). World leaders in Politics almost uniformly have a powerful 10th dashamsha.

Children: Radix L-5 should be strong in saptamsha (D-7) to confirm benefits from children.

Education: Radix L-4, radix L-7 and Radix Budha are is the primary significations for education. L-4 is usually the strongest of these, and L-4's Radix character will set the paradigm for education in his life time while L-7 and karaka Budha contribute the mastic variations within that paradigm. Example: L-4 = Shani, for Thula lagna and Vrischika lagna. We must check their performance within D-24. If confirmation of radix/navamsha strength does not occur in D-24, education expectations drawn from D-1 and D-9 Charts should be reduced.

Leadership: Radix L-10 should be strong in Dashamsha (D-10) to confirm maximum public leadership (e.g. G.W. Bush radix L-10, Kuja, is exalted in 7th dashamsha). The World leadership in Politics almost uniformly has a powerful 10th Dashamsha.

Marriage and Rlationships: D-5, D-7, and D-9 are seen for Marriage and Rlationships. Marriage is important area in the life. The marriage is seen from the 7th house of a horoscope. In case of life partner, 7th house in D-1 shows approach towards relationship, 7th house in D-9 indicates traits and character of partner & 7th from Venus distance from the location of partner, Upapada gives hints of type of marriage, Upapada Lord stands for status of wife, 7th from Upapada may reveal impediments in marriage, AI from UL meeting place of partners, 2nd and 8th from UL determines the longevity of marriage, Sexual desire is expressed by Saptamarudha. Suppose, the lord of the 7th house is exalted in the Natal Chart (D-1), but afflicted or debilitated or in evil houses in D-9, D-4 and D-30 charts, the native will have delayed marriage and his married life would be a troublesome. It will lead to wrong predictions without divisional charts analysis. The astrological predictions given merely on the basis of only Natal chart cannot be considered to be complete and foolproof. Venus gives native an opportunity to experience the relationship. Radix L-7 should be strong in Navamsha (D-9) to confirm a happy marital situation. The lord of the 6th in Navamsha (D-9) is a trouble-maker in marriage. Periods of the L-6 in navamsha causes the spouse to create or experience a loss of marital agreement. If, within the navamsha, the native has L-6 occupying the D-9's lagna, then this native is probably an instigator of one's own marital conflict. The nature of the trouble-making behavior is more vividly described by the lord of the 6th navamsha's behavior in radix. If the mahadasha pati occupies a good house in Navamsha and is also a natural benefic, all is well. A natural malefic in a good house is pretty. A natural malefic in a Navamsha's Dushthamsha is not favourable. If circumstances in the navamsha are bad enough, it would be wise to delay the marriage or, be exceptionally cautious in all marriage communications, whilst the inauspicious mahadasha remains in effect.

Marriage timing: Marriage timing is predicted to occur during the bhukti of L-7 or L-1 in Navamsha (D-9) or Shukra or Rahu, simultaneous with gochara Rahu-Ketu visiting a key partnership axis = lagna of Chandra, lagna of R-K, or 1st bhava lagna of the D-1 or D-9.

Parents Character: The 12th in dwadashamsha (D-12) reveals character of parents.

Promotion: Suppose we want to see when one would get a promotion. We should analyze D-10 as it shows one's career and achievements. GL (Ghati Lagna) shows power and authority, planets or Sign connected with GL will give a promotion usually. If they are in GL or aspect GL, it is favourable for promotion. AL, planets associating with AL or the 5th or the 10th from it inn D-10 are favourable for promotion. When the lord of AL is in the 10th from it and aspects GL, one will get a promotion.

Sibling's marriage & Business contracts: The 7th house of the D-3 [Drekkana] confirms sibling's marriage, business contracts and capacity to make agreements. The 10th house of the D-9 [Navamsha] confirms the spouse's career and public standing. Also read the role which the radix house lords assume within the varga.

Spiritual Path: If the lord (Lagna lord) of D-20 is strong, spiritual practice is a site of profound self-revelation to the native. The spiritual path in this case may be smooth if it is a benefic or ragged, if it is a violent malefic.

Spouse Career & Personality: The 7th navamsha is the key to spouse's personality. The internal 10th house of the D-9 [navamsha] confirms your spouse's career and public standing.

Teamwork, Siblings, Competition, and Self-made capital wealth: If the lord (Lagna lord) of Dreshkhamsha-3 is strong in radix, then teamwork, siblings, group mental process, competition, and self-made capital wealth are powerful venues for self-realization. Mahadasha of the Varga lord and its bhukti periods will emphasize matters of the varga whose lord is strong in radix.

13.2 Predictions by Hora (Half Sign-D/2) Chart

Method –1: Hora in Aries: If the birth be in the first half of Aries, he/she will roll his eyeballs unnecessarily, be cruel, rich, and very bright in appearance, will have a fierce spouse, be tall, irascible and be a leader of thieves. If the birth be in the second half of Aries, he/she will be a thief, be careless, will have afflicted foot, fingers, large eyes and tall stature and be very intelligent.

Hora in Taurus: If the birth be in the first half of Taurus, he/she will be dark in complexion, will have broad eyes, fore face and chest, be eminent, disposed to carnal pleasures and will have strong bones and strong physique. If the birth is in the second half of Taurus, he/she will have broad, long and round (i.e. muscular) limbs, be liberal, will have attractive hair, thin waist and beautiful eyes.

Hora in Gemini: If the first Hora of Gemini be the Ascendant, he/she will have a long waist, be very skilful, will have a moderate physique and soft hair and feet, be courageous and fond of sexual union, and be wealthy and learned. If the birth be in the second half of Gemini, he/she will have beautiful and large eyes, be sensuous, soft, is a gifted speaker and be attached to others' housewives.

Hora in Cancer: If the birth be in the first half of Cancer, he/she will be endowed with elevated body, (i.e. be prominent in appearance), a beautiful head, eminence and intelligence, be weak sighted, will have limbs moving off and on, be crafty, dark in complexion, ungrateful and will have broken teeth. If the birth be in the second half of Cancer, he/she will be fond of gambling and travels will have a broad chest, be quite truthful, irascible and will have a harsh physique.

Hora in Leo: He/She born in the first Hora of Leo will have blood-red eyes, be bold, be endowed with a large and long physique, be crooked, happy and decidedly firm in his/her undertakings. He/She born in the latter half of Leo will be fond of opposite sex, sweets, foods, drinks and robes, will move frequently, will have a harsh physique, be liberal, fond of travels, will have a few sons, will enjoy pleasures and be firm in friendship.

Hora in Virgo: If the birth be in the initial half of Virgo, he/she concerned will have a soft and beautiful body, be a sweet speaker, be fond of music, females and sexual union, and be sweet, fortunate and excellent. Born in the second half of Virgo, he/she will be short-stature, will have knowledge of Hath Yoga and Vedas, will possess a big head, be agreeable, argumentative, learned in service, drawing and writing, will have chequered prosperity and (yet) be happy.

Hora in Libra: If the first Hora of Libra ascends at birth he/she will have a round face, elevated nose and back and large eyes, be sportive, will have a strong body and strong bones, be wealthy and will be affectionate to his/her men. If the second Hora of Libra ascends at birth, he/she will be very wealthy, will have firm wealth, and will be endowed with black and curly hair, be crafty, round-eyed, will have a strong lower body and attractive skin and will have defective feet.

Hora in Scorpio: He/She born in the first half of Scorpio will have blood-red eyes, be adventurous in acts, be valorous in battle, wicked, fond of females and be rich. If born in the second half of Scorpio, he/she will have broad, well grown, long and fleshy limbs, will serve a king, will have many debts and many friends and will possess blown eyes (expanded, as a flower).

Hora in Sagittarius: He/She born in the first half of Sagittarius will possess a wide and split face and chest, with eyes and neck contracted (i.e. small), will be discarded by elders in boyhood and be pious. If the latter half of Sagittarius rises, he/she will possess eyes akin to the petals of a lotus and long arms, will have knowledge of Shashtra, be good in appearance, will speak sweetly, and be fortunate and famous.

Hora in Capricorn: If the first half of Capricorn is on the Ascendant, he/she will be dark in complexion, will possess eyes, like that of a deer, be fortunate, will conquer women, be beautiful, crafty, opulent, will eat purified food, will do good acts and will have elevated nose. If born in the second half of Capricorn he/she will have blood-red eyes, be indolent, be interested in long travels, be foolish, dark in complexion, will have a hairy physique, be clever and fierce.

Hora in Aquarius: He/She born in the first half of Aquarius will be endowed with a female and with friends, be a person of tastes, or feelings (or be an alchemist), be soft in disposition, will have a few sons, be virtuous, valorous, will be endowed with physical complexion akin to copper, be radiant and interested in travels. If born in the second half of Aquarius, he/she will have copper-brilliant eyes, be emaciated, firm, very insignificant in appearance, indolent, not outspoken, very dejected, miserly and very crafty.

Hora in Pisces: He/She born in the first 15 degrees of Pisces will be short-stature, be endowed with broad and beautiful body, a large fore face, large face and broad chest, be dear to women, and be very famous, skilful and valorous. If born in the second Hora of Pisces, he/she will be liberal, will have an elevated nose, be skilful, intelligent, will possess charming eyes be dear to king and will speak affably. 25. The (good effects) due to a Hora will come to pass in full measure, if either the Sun, or the Moon is strong in aspect to the Ascendant Lord, or, if the Ascendant Lord is himself in an Angle.

13.3 Predictions by Drekkana Chart (D-3)

A decanate (Drekkana in Sanskrit) is 1/3 rd of a house. The first Drekkana is, therefore, 0 to 10 degrees, the second decanate is 10 to 20 degrees & the third decanate is 20 to 30 degrees. Since each Sign has 3 decanates, there are 36 decanates in all. Check out which Drekkana your Ascendant falls in. Look at the Ascending Degree and check where it falls. The first decanate of any sign is the same Sign and is ruled by the same owning planet, second decanate is the 5th house Sign and is ruled by the 5th house owning planet or by the 5th lord and the third decanate is the 9th house Sign and is ruled by the 9th house owning planet or by the 9th lord. For example, the first decanate of Aries is ruled by Mars, the second decanate is owned by Sun and the third decanate is owned by Jupiter.
22nd Drekkana is considered fatal. 22nd Drekkana falls in the 8th house from the natal ascendant. The 1st, 5th and the 9th Rashi from any Rashi are the three Drekkana of that Rashi and are, respectively, lorded by Narada, Agasthya and Durvasha. In this harmonic, the first 1/3 or first ten degrees (00° 00' to 10° 00') of any sign is ruled by itself. The middle 1/3, (10° 00'-20° 00'), is ruled by the subsequent sign of the same element and finally the last 1/3, (20° 00'-30° 00'), is ruled by the third sign of the same element. Example: The first 1/3 of Sagittarius is in the harmonic third of Sagittarius, the second 1/3 in the harmonic third of Aries, and the last 1/3 in the harmonic third of Leo as they all belong to same Fire Element (Aries, Leo & Sagittarius). Hence, if Saturn is located in a birth chart at 15° 20' Sagittarius, it would be in the harmonic third of Aries. D-3 is a key to longevity prediction, because the lord of the 22nd dreshkhamsha is associated with physical death. Lord of the 22nd dreshkhamsha is easily seen by graha ruling the 8th rashi from D-3 lagna (lord of the 8th Dreshkona). There are also predictive uses for the Sarpa dreshkhamsha and the Phasha dreshkhamsha. "Sarpa" Dreshkona (serpent decanate) is the 1st and 2nd decanate of Karka; 1st and 2nd decanate of

Vrischika; 2nd and 3rd decanate of Meena. Each Decanates in a sign gets the traits or qualities of the sign ruling it. The Decanates are said to have psychological and physical significances and are related to the mind, body and soul. We get much more precise information with Decanates. It helps further refine exactly what type of a personality the natives have. So for example, all Aries are not the same. Depending on what Decanates the native has in, he is a certain type of Aries.

Prediction by Drekkana Chart: If the Sign owning the Drekkana is afflicted by malefic, he/she will undergo imprisonment in any offence. If the Moon at the time of birth is in its own Drekkana or in that of a friendly, he/she will be very beautiful appearance and will be endowed with all good qualities. If the Moon at the time of birth is in that of a non-friendly Decanates, he/she will have the stature and quality pertaining to that planet. If the Moon at the time of birth is in a Serpent Drekkana, he/she will be very wrathful. If the Moon at the time of birth is in Weapon or Armed Drekkana, he/she will do evil deeds or even he can murder. If the Moon at the time of birth is in a Quadruped Drekkana, he/she will take women who are not of his equal dignity. If the Moon at the time of birth is in Bird (Pakshi) Drekkana, he/she will be wanderer.

Example: The Sun at 18 degrees 25' Virgo is in the Decanates of Capricorn. So we may add some Capricorn traits to the Virgo Sun sign. This person expresses his Virgo traits of needing to feel useful in a Capricorn way, such as a strong sense of duty and responsibility. Further, it may also be refined or blended the delineation in the sign and house placement of the ruler of Capricorn which is Saturn. Suppose, Saturn is in Leo sign in the seventh house. With Saturn in Leo, the sense of responsibility found in Capricorn might manifest in a take charge attitude (Leo) towards ones committed relationships as is in seventh house. From Leo keywords, he is something like flamboyant or generous, but because of the Capricorn energy, it suggests him authority and ambition.

The first 0 degrees to 9 degrees 59' 59" is called as the First Decanates: This Decanates corresponds to the physical side and relates to the body. The first Decanates of any sign corresponds to the sign being studied so that the first Decanates of Aries will relate to Aries. The first Decanates is

the strongest and it corresponds to the double strength of the sign.
The second 10 degrees to 19 degrees 59' 59" is called as the second Decanates: This Decanates corresponds to the mental or the Mind side. Degree 10 represents new pursuits and degree 19 represents the peak of mental activity. The second Decanates of any sign takes on the characteristics of the next sign in the same element and hence the second Decanates for Aries is Leo.
The third 20 degrees to 30 is called as the Third Decanates: This Decanates corresponds to the spiritual aspects of a person. When many planets or in this Decanates, the native would be spiritually inclined. Degree 29 relates to the spiritual peak. The third Decanates of any sign is the last sequential sign within the same element and so for Aries it would be Sagittarius. A person born during the third Decanates will be less typical of the sign under consideration.
Let us look further into the Decanates of Aries. We can describe the first ten degrees of Aries as Aries/Aries or Mars/Mars giving an impulsive and pioneering nature to the planet that is there. The second ten degrees of Aries can be described as Aries/Leo or Mars/Sun giving a less aggressive and sunnier nature to the planet there. The third ten degrees of Aries can be described as Aries/Sagittarius or Mars/Jupiter giving a mutable, outgoing nature to the planet there. We find out the correct Drekkana Division from the birth Nakshatra Pad with the help of following table. Find out the type and nature of Drekkana with the help of Drekkana Division for predictions of horoscope.

13.4 Predictions by D-4 (Chaturthamsha) Chart:

Chaturthamsha (D-4) indicates Bhagya, luck, fortune, happiness, welfare, reward, defensibility; owned Properties, school, fixed abode and security. "The Lords of the four Kendra from a Rashi are the rulers of respective Chaturthamsha of a Rashi commencing from Mesha. The deities, respectively, are Sanaka, Sananda, Kumara and Sanatana. D-4 can confirm the character of graha which produce the foundational physical and cultural securities. Look within the D-4. The character of graha, such as, karaka Chandra, the radix L-4, and L-4 from Chandra confirm the actualization of its potency. Look for confirmation of the fruits of the 4th house in D-4 (Chaturthamsha). Chaturthamsha reveals fine detail of how the results of mother's early training will manifest. Predictions can be partially confirmed via analysis of Chaturthamsha. D-4 assessment can assist the query of whether the native will change residence. Transits through the D-4 and bandhu bhava or transits affecting the radix L-4 within the D-4 will specify the timing of change residence. L-3 of radix (12th-from-4th) or L-12 of radix (sudden change of home, 8th-from-4th) should be affected as a first indication of bhava-move, whether across town or as international emigration to another continent.

13.5 Predictions by D-6 (Shastamsha) Chart

Shastamsha (D-6) defines the illness, accident, or imbalance in the physical body. D-6 confirms the existing character of radix L-6 and L-6 from Chandra. Budha, karaka for 'argumentation', should be confirmed in D-6. Budha can reveal the manifests in chronic illness, injury, poverty, and disagreement. The D-6 lagnesha has only minor powers to stimulate the subconscious patterns which in turn generate physical illness. However, if the D-6 lagnesha is malefic in radix, either a natural malefic or a dushthamsha lord, then the D-6 lagnesha becomes that much more pernicious during its bhukti. Additional confirmation of a range of misfortunes which include sickness, and recurring environmental influences, may be confirmed from D-30 Trimshamsha Chart. D-6, D-30, and D-60 all denote inauspicious effects arising from suppressed imbalanced conditions in the subconscious, which are carry-overs from unresolved conflict in past lives. D-6, D-30, and D-60 should all be consulted when predicting the outcome of enemies, debts-poverty, or disease.

13.6 Predictions by D-7 (Saptamsha / Saptans / putra) Chart:

Saptamsha (D-7) represents Children - Fertility - Fruitfulness of Partnership. Each sign is divided into seven equal parts of 04° 17' 09". The first 1/7 of odd signs is governed by the sign itself, with the rest following the signs in order through the zodiac. For even signs, the first 1/7 corresponds to the sign seventh from it, and the rest follow in order from that. Example: Hence, the first seventh of Aries of an odd sign is Aries; the second, Taurus; the third Gemini; and so on to Libra as its last seventh. The first seventh of Taurus (an even sign) would be Scorpio, the sign seventh from it. The second seventh division would be Sagittarius; the third, Capricorn; and so on to Taurus as its last seventh. In this way we are merely going through the signs in order. The following are the harmonic 1/7 divisions of the sign: 1. 00° 0' 0" to 04° 17' 09"; 2. 04° 17' 09" to 08° 34' 16"; 3. 08° 34'17" to 12° 51'26"; 4. 12° 51'26" to 17° 08'34"; 5. 17° 08'34" to 21° 25'43"; 6. 21° 25'43" to 25° 42'52" and 7th as 25° 42'52" to 30° 00'00". The names of the seven divisions in odd Rashi are: Kshaara (Ksheera), Dadhi, Ghritha, Ikshu, Rasa, Madhya, Suddha Jala. These designations are reversed for an even Rashi. Example: Thus if the Moon is located at 25° 10' Taurus, it would be in the sixth division, which would be Aries (sixth from Scorpio). Look first to all the effects upon Putra Bhava [the 5th house]; status of child-karaka (Guru); Dharma-bhava (5th-from-5th); 5th-from-Guru; and 5th-from-Chandra. Now, check the D-7 = Saptamsha: The house which defines and energizes "children or childlessness" sits in 3/11 angle to the children-house: it is yuvati bhava, the 7th house of marriage. It is considered that the relationship between the two parents (bhava-7) is the basis (lagna) from which the prospects for children should be measured. Thus children, a fruit (11) of marriage, are seen from D-7. Once the L-5 and L-9 are confirmed positive for children, look for confirmation of the "fruits" of the 7th house in the seventh divisional chart. Prepare the list of Jaimini Karakas, and find the MP (Matru-Putrakarka).

The position and strength of MP with the Saptamsa Lagna is noted. If the "Matru-Putrakarka" is found in the 6th, 8th, or 12th from Saptamsa Lagna, it is considered as denial of the child or childlessness. However, this single rule is not sufficient to show denial of children. One must not use this rule alone for prediction of childbirth. If the Natal Chart and the Saptamsa, both, clearly point out the denial of children or trouble in matters pertaining to children, then only this rule can be used for confirmation of the same.

13.7 Predictions by D-9 (Navamsha/Navans) Chart:

Aries Lagna (Rising) Navamsa: Following are the effects of births in the nine Navamsa, such as,

Effects of Aries Ascendant, First Navamsa (Up to 3^0 20'): He/She will have a face, resembling that of a he-goat, with nose and shoulders not being very prominent. He/She will have a fierce voice, ugly appearance and narrow eyes. His/Her body will be thin, but free from defects.

Effects of Aries Ascendant, Second Navamsa (3^0 20' to 6^0 40'): He/She will be dark in complexion; will have broad shoulders and long arms, small forehead, strong collar bones, sharp sight and prominent face and nose. He/She will be an affable speaker and will possess weak legs.

Effects of Aries Ascendant, Third Navamsa (6^0 40' to 10^0 00'): He/She will suffer loss of hair, be fair in complexion, will have irregular (defective) arms, charming eyes and nose, will be a scholar in direct poetic ability and will have weak thighs.

Effects of Aries Ascendant, Fourth Navamsa (10^0 00' to 13^0 20'): He/She will have an erratic sight, be irascible, short-nosed, and wandering-natured, will have rough legs and coarse hair, and be bereft of co-born and emaciated.

Effects of Aries Ascendant, Fifth Navamsa (13^0 20' to 16^0 40'): He/She will be fierce and will have eyes, resembling that of a supreme elephant, a fat nose, and thick eye brows, wide fore face, fat body and coarse hair.

Effects of Aries Ascendant, Sixth Navamsa (16^0 40' to 20^0 00'): He/She will be dark in complexion, soft in disposition, will have eyes, akin to that of a deer, be tall in stature, will have irregular (defective) stomach and hands, be a eunuch, and be timid and garrulous.
Effects of Aries Ascendant, Seventh Navamsa (20^0 00' to 23^0 20'): He/She will have complexion akin to green sprout, be fickle-minded, will possess white eyes, will marry an unchaste lady, be malicious and will have a broad physique.
Effects of Aries Ascendant, Eighth Navamsa (23^0 20' to 26^0 40'): He/She will have a face, akin to that of a monkey, be a good speaker, will have an afflicted and tawny body, will suffer from secret diseases, be torturous, be a liar, be fond of friends and be fierce.
Effects of Aries Ascendant, Ninth Navamsa (26^0 40' to 30^0 00'): He/She will be tall, emaciated, wandering, will have defective fore face and ears, will have a face, akin to that of a horse, will possess many names and be crooked.
Taurus Lagna (Rising) Navamsa: Following are the effects due to births in the Navamsa belonging to Taurus Ascendant:
Effects of Taurus Ascendant, First Navamsa (Up to 3^0 20'): He/She will have an even and dark coloured physique, be hard-hearted, will perform obsequies in the beginning and ending parts of his life, be base, will indulge in unnatural acts and will have crooked sight. **Effects of Taurus Ascendant, Second Navamsa (3^0 20' to 6^0 40')**: He/She will be endowed with majestic looks, be indolent, will have a bent body, be not very intelligent, will indulge in hostile acts and be a great liar. Effects of Taurus Ascendant, Third Navamsa **(6^0 40' to 10^0 00')**: He/She will be erratic in sight, be fierce, will have a short nose, be wandering natured, will have coarse legs and hair will have soft limbs, be beautiful, will have broad eyes and big limbs, be interested in Sacrifices and will have stiff legs and hands. Effects of Taurus Ascendant, Fourth Navamsa **(10^0 00' to 13^0 20')**: He/She will be short in stature, will be wandering natured, be easily irritable, will have eyes, akin to that of a he-goat, is tawny in complexion, be poor and will steal others' wealth. Effects of Taurus Ascendant, Fifth Navamsa **(13^0 20' to 16^0 40')**: He/She will be vicious, will have a well-elevated nose, and will appear, like a giant ox, will have crooked hair, be sportive, will have large shoulders and hips. Effects of Taurus Ascendant, Sixth Navamsa **(16^0 40' to 20^0 00')**: He/She will

have beautiful eyes and hair, be firm, be endowed with a fair complexioned physique, will speak sweetly, be pre-eminent, be fond of amusements, and be emaciated and skilful. Effects of Taurus Ascendant, Seventh Navamsa **(20^0 00' to 23^0 20')**: He/She will be interested in females, who lost their sons, will have somewhat elevated nose and prominent eyes, will possess a strong physique, will hate his own men and will have stout feet and exquisite hair. Effects of Taurus Ascendant, Eighth Navamsa **(23^0 20' to 26^0 40')**: He/She will have eyes, akin to that of a tiger and charming teeth, be unconquerable, will possess a full-blown nose, will work sparingly, will have curly and bluish hair and sharp nails and be garrulous. Effects of Taurus Ascendant, Ninth Navamsa **(26^0 40' to 30^0 00')**: He/She will be honourable, will not be very strong, and be timid, given to anger, will possess an even and charming body, be a rogue (or cheat), will gather money, be famous, will have a thin lower body.

Gemini Lagna (Rising) Navamsa: Following are the effects of births in the all Navamsa of Gemini Ascendant. Effects of Gemini Ascendant, First Navamsa **(Up to 3^0 20')**: He/She will have hair on the shoulders, will possess charming and dark eyes and an elevated nose, be akin to green (Durva) grass in complexion and will possess thin legs and thin hands. Effects of Gemini Ascendant, Second Navamsa **((3^0 20' to 6^0 40')**: He/She will have a pot-like head, will do dirty acts, be fond of torturous deeds, will have depressed nose, will speak much, will work much and will lead in strife and quarrels. Effects of Gemini Ascendant, Third Navamsa **(6^0 40' to 10^0 00')**: He/She will be fair in complexion, will possess blood-red eyes, charming nose and even physique be very intelligent, will have a long face and dark eye-brows and is a skilful speaker. Effects of Gemini Ascendant, Fourth Navamsa **(10^0 00' to 13^0 20')**: He/She will possess charming eye brows and forehead, be lustful, will possess a physique with the splendour of a blue lotus, will be broad-chest and white teethed, be soft in speech and will have attractive hair. Effects of Gemini Ascendant, Fifth Navamsa **(13^0 20' to 16^0 40')**: He/She will have a broad face, strong chest and big head, be wicked, cunning and will possess charming and friendly looks. Effects of Gemini Ascendant, Sixth Navamsa **(16^0 40' to 20^0 00')**: He/She will possess eyes with the hue of honey, be garrulous, will possess a broad fore face, even body and charming lips, be a rogue, be fickle-minded and be strong. Effects of Gemini Ascendant, Seventh Navamsa **(20^0 00' to 23^0 20')**: He/She will possess a copper coloured physique, copper red and prominent eyes and a broad chest, be skilful in teaching and arts and be jocular in disposition. Effects of Gemini Ascendant, Eighth Navamsa **(23^0 20' to 26^0 40')**: He/She will be dark in complexion, be great, intelligent, soft in disposition, sweet in speech, will have a broad and tall physique and large and black eyes and will be an expert in arts. Effects of Gemini Ascendant, Ninth Navamsa **(26^0 40' to 30^0 00')**: He/She will have round and dark coloured eyes and charming body, be successful, very intelligent and be fond of sexual cohabitation, poetry and worldly knowledge.

Cancer Lagna (Rising) Navamsa: Following are the effects for births in various Navamsa related to Cancer Ascendant: Effects of Cancer Ascendant, First Navamsa **(Up to 3^0 20')**: He/She will have a clean, charming and fair coloured physique, beautiful hair, broad belly, impressive face, prominent eyes, thin body and thin shoulders. Effects of Cancer Ascendant, Second Navamsa **(3^0 20' to 6^0 40')**: He/She will be blood red in complexion, be fierce in quarrels, will like fine arts, will possess face and eyes, akin to that of a cat, will be well disposed to scarify for others and will have weak knees and shanks. Effects of Cancer Ascendant, Third Navamsa **(6^0 40' to 10^0 00')**: He/She will be fair in complexion, will possess beautiful eyes, be an eloquent speaker, will have a soft body, akin to that of a female, be intelligent be a sparing and light worker and be indolent. Effects of Cancer Ascendant, Fourth Navamsa **(10^0 00' to 13^0 20')**: He/She will be black in complexion, will have pressed eyebrows, be graceful in appearance, will have charming eyes and nose, be courageous, liberal, will perform acts prescribed for superior caste-men and be crafty. Effects of Cancer Ascendant, Fifth Navamsa **(13^0 20' to 16^0 40')**: He/She will possess voice akin to the sound of bell, crooked, or stooping face, allied eyebrows and very long arms, be interested in worship, be bereft of dutifulness, will injure others and be not very intelligent. Effects of Cancer Ascendant, Sixth Navamsa **(16^0 40' to 20^0 00')**: He/She will have a long and broad physique, charming eyes and great courage, be fair in complexion, be a good speaker and will possess beautiful nose and big teeth. Effects of Cancer Ascendant, Seventh Navamsa **(20^0 00' to 23^0 20')**: He/She will have scattered hair, big body and sinewy knees, will be disposed to protect others' families and will be akin to a crow in appearance. Effects of Cancer Ascendant, Eighth Navamsa **(23^0 20' to 26^0 40')**: He/She will have a head with bell shape, charming face, shoulders and limbs, will be a degraded artisan (i.e. an infamous worker), will have the gait of a tortoise and crooked nose and be dark in complexion. Effects of Cancer Ascendant, Ninth Navamsa **(26^0 40' to 30^0 00')**: He/She will be fair complexioned, will possess eyes resembling fish, be great, be soft-bellied, broad-chest, and will have prominent chins and lips, large, but weak knees and similar ankles.

Leo Lagna (Rising) Navamsa: Following effects will mature in the various Navamsa prevailing in Leo Ascendant at birth: Effects of Leo Ascendant, First Navamsa **(Up to 3^0 20')**: He/She will have an even belly (like a lion), be fierce, will have sharp and blood-red nose and a big head, be valorous and will possess a prominent and fleshy chest. Effects of Leo Ascendant, Second Navamsa **(3^0 20' to 6^0 40')**: He/She will have a prominent and broad fore face, a square body, broad eyes, broad chest, long arms and big nose. Effects of Leo Ascendant, Third Navamsa **(6^0 40' to 10^0 00')**: He/She will have hairy and broad arms, eyes, akin to that of (Greek) partridge (said to live on moon-beams), be fickle-minded, charitable, will have an elevated nose, pure white physique and round neck. Effects of Leo Ascendant, Fourth Navamsa **(10^0 00' to 13^0 20')**: He/She e will have (ash-coloured) body akin to ghee, large and black eyes, soft hair, peculiar voice, big hands and legs and stomach, resembling that of a frog. Effects of Leo Ascendant, Fifth Navamsa **(13^0 20' to 16^0 40')**: He/She will have a bell-shaped head with limited hair, charming nose and eyes, hairy body and long belly, be fierce and will have unsightly teeth and strong and broad cheek. Effects of Leo Ascendant, Sixth Navamsa **(16^0 40' to 20^0 00')**: He/She will have limited, but soft hair on his physique, white and large eyes, be tall in stature, dark in complexion, will have proven skill (only) among females, be swaggering and be learned. Effects of Leo Ascendant, Seventh Navamsa **(20^0 00' to 23^0 20')**: He/She will have a long face, be sinewy, will have a prominent physique, be unfortunate in the matter of wife (or females in general), dark-complexioned, fierce, hairy-bodied and be cunning and harsh in speech. Effects of Leo Ascendant, Eighth Navamsa **(23^0 20' to 26^0 40')**: He/She will be endowed with excellent speech, firm limbs, charming and majestic looks, be undutiful, poor and crafty. Effects of Leo Ascendant, Ninth Navamsa **(26^0 40' to 30^0 00')**: He/She will have a face akin to a donkey's, will possess dark eyes, long arms, and charming legs and be troubled by breathing disorders.

Virgo Lagna (Rising) Navamsa: The various Navamsa emanating from Virgo Ascendant at birth will give the following effects: Effects of Virgo Ascendant, First Navamsa **(Up to 3^0 20')**: He/She will possess eyes, akin to that of an antelope, be a good speaker, be charitable, will enjoy sexual pleasures, be very rich, dark in complexion and large hearted. Effects of Virgo Ascendant, Second Navamsa **(3^0 20' to 6^0 40')**: He/She will have a charming face, charming eyes and fair complexion, be soft, argumentative, fickle minded and long-bellied. Effects of Virgo Ascendant, Third Navamsa **(6^0 40' to 10^0 00')**: He/She will have blown nose, prominent feet, long arms, pure speech, and fair complexion and be friendly. Effects of Virgo Ascendant, Fourth Navamsa **(10^0 00' to 13^0 20')**: He/She will be learned, will be sportive with the fair sex, is beautiful, sweet, blood red in complexion, sharp, intelligent, emaciated and will have charming eyes and face. Effects of Virgo Ascendant, Fifth Navamsa **(13^0 20' to 16^0 40')**: He/She will have large lips and hands, big body, broad chest, strong ankles and will depend on others. Effects of Virgo Ascendant, Sixth Navamsa **(16^0 40' to 20^0 00')**: He/She will have charming appearance, impressive speech, and splendour body and is an exponent of Shashtra, be very intelligent, skilful in writing and fine arts, be good hearted and will take pleasure in walking, or roaming. Effects of Virgo Ascendant, Seventh Navamsa **(20^0 00' to 23^0 20')**: He/She will have a small face, elevated nose, compact arms, very fair complexion, prominent belly, hands and legs and will have fear for water. Effects of Virgo Ascendant, Eighth Navamsa **(23^0 20' to 26^0 40')**: He/She will be very beautiful, fair in complexion, tall in stature, will have charming eyes, be fierce, honourable and will have long and stout arms and brown hair. Effects of Virgo Ascendant, Ninth Navamsa **(26^0 40' to 30^0 00')**: He/She will be famous, will have charming physique, broad eyes, incomparable vigour, be skilful, be with stooping shoulders and be learned a writer.

Libra Lagna (Rising) Navamsa: The various Navamsa out of Libra Ascendant at birth will produce following effects: Effects of Libra Ascendant, First Navamsa **(Up to 3^0 20')**: He/She will be fair complexioned, broad eyed, praiseworthy, long-faced, skilful in business, happy and famous. Effects of Libra Ascendant, Second Navamsa **(3^0 20' to 6^0 40')**: He/She will have squint and round eyes, elevated (ill formed) teeth, depressed waist, charming neck, large (physical) heart, ugly body and compact brows. Effects of Libra Ascendant, Third Navamsa **(6^0 40' to 10^0 00')**: He/She will be fair in complexion, will have a face, akin to that of a horse, be thin-bodied, famous, long-haired and long-nosed and will have beautiful legs. Effects of Libra Ascendant, Fourth Navamsa **(10^0 00' to 13^0 20')**: He/She will have weak hands, be timid, will have ill-formed teeth (some placed over others), weak body, rolling eyes, small nails, dark complexion and be devoid of virtues and be miserable. Effects of Libra Ascendant, Fifth Navamsa **(13^0 20' to 16^0 40')**: He/She will have majestic looks, be firm disposition, be not proud, rough haired, even-eyed and will possess a beautiful nose. Effects of Libra Ascendant, Sixth Navamsa **(16^0 40' to 20^0 00')**: He/She will have fleshy limbs, be fair in complexion, will have broad eyes, beautiful nose and white nails, be diplomatic and be learned in Shashtra. Effects of Libra Ascendant, Seventh Navamsa **(20^0 00' to 23^0 20')**: He/She will be blood-red in complexion, be intelligent, will have long physique and long arms and a big head, and be miserly, fierce and intelligent. Effects of Libra Ascendant, Eighth Navamsa **(23^0 20' to 26^0 40')**: He/She will have elevated shoulders and prominent neck, will enjoy pleasures, will have a coarse physique, long and dark brows, is a polite speaker and will have a beautiful chest and bruised head. Effects of Libra Ascendant, Ninth Navamsa **(26^0 40' to 30^0 00')**: He/She will have charming eyes, pleased mind, be fair complexioned, even and beautiful bodied, be skilful, fond of arts, and be charitable and jocular.

Scorpio Lagna (Rising) Navamsa: The Ascendant Scorpio will produce various following effects according to the nine Navamsa thereof. Effects of Scorpio Ascendant, First Navamsa **(Up to 3^0 20')**: He/She will be short in stature, will have prominent lips and nose, charming forehead, strong and fair complexioned body with belly, akin to that of a frog and will act, as a marriage broker (ascertaining genealogies and negotiating marital alliances). Effects of Scorpio Ascendant, Second Navamsa **(3^0 20' to 6^0 40')**: He/She will be fair in complexion, will possess a strong and broad chest and shoulders and reddish eyes, and will conquer his enemies be valorous and will have abundant hair. Effects of Scorpio Ascendant, Third Navamsa **(6^0 40' to 10^0 00')**: He/She will be learned, will have strong shoulders and arms and beautiful hair, be endowed with clear speech, fair complexion and charming lips. He is born of a virgin. Effects of Scorpio Ascendant, Fourth Navamsa **(10^0 00' to 13^0 20')**: He/She will be intent upon joining others' wives, will induce others to be active, be valorous, tall, dark in complexion, dark-haired and dark eyed. Effects of Scorpio Ascendant, Fifth Navamsa **(13^0 20' to 16^0 40')**: He/She will be majestic, will possess coppered eyes, depressed nose, be courageous, proud, will perform fearful acts, be famous and will have a strong physique. Effects of Scorpio Ascendant, Sixth Navamsa **(16^0 40' to 20^0 00')**: He/She will be impudent, intelligent of a high order, will have elevated nose and great strength, and will be endowed with knowledge of justice (or be diplomatic), be skilful, will possess less hair and compact brows. Effects of Scorpio Ascendant, Seventh Navamsa **(20^0 00' to 23^0 20')**: He/She will have a split face, strong body, and teeth in various sizes, depressed belly, and squint sight and be very splendour. Effects of Scorpio Ascendant, Eighth Navamsa **(23^0 20' to 26^0 40')**: He/She will have blown nose, be dark in complexion, devoid of virtues, dirty in appearance, will possess stiff hair and be foolhardy. Effects of Scorpio Ascendant, Ninth Navamsa **(26^0 40' to 30^0 00')**: He/She will have a fair coloured physique, be beautiful, like a deer, will possess calm and tawny eyes and similar hair and strong body and be amiable to elders.

Sagittarius Lagna (Rising) Navamsa: The native born in Sagittarius Ascendant, but in its various Navamsa will obtain following effects: Effects of Sagittarius Ascendant, First Navamsa **(Up to 3^0 20')**: He/She will have charming big nose, will possess sight, akin to that of a goat, be a gifted speaker, will have charming teeth and hair, be fair in complexion, will have inset testicles and be fierce. Effects of Sagittarius Ascendant, Second Navamsa **(3^0 20' to 6^0 40')**: He/She will have a prominent head, be firm in disposition, will have large eyes, strong waist and knees, ugly nose, tall stature and firm cheeks. Effects of Sagittarius Ascendant, Third Navamsa **(6^0 40' to 10^0 00')**: He/She will have skill in educating and in fine arts, be majestic, just, fond of females, intelligent and jocular.

Effects of Sagittarius Ascendant, Fourth Navamsa **(10^0 00' to 13^0 20')**: He/She will be skilful, be tawny in complexion, will possess round eyes, fair-coloured physique and a belly akin to tortoise, be intelligent, wandering-natured, will have charming hair and charming appearance. Effects of Sagittarius Ascendant, Fifth Navamsa **(13^0 20' to 16^0 40')**: He/She will have large ears, eyes and face, will possess a (majestic) physique, like a lion, widely spread eye brows, strong shoulders and arms, hairless physique and firm disposition. Effects of Sagittarius Ascendant, Sixth Navamsa **(16^0 40' to 20^0 00')**: He/She will have beautiful and large eyes, broad forehead and broad face, be a poet, and be mean and interested in scholarly discussions. Effects of Sagittarius Ascendant, Seventh Navamsa **(20^0 00' to 23^0 20')**: He/She will be dark in complexion, soft in disposition, will keep up his promise, will have a prominent head, will be interested in accumulating savings, be tall in stature, will possess broad eyes and be liberal. Effects of Sagittarius Ascendant, Eighth Navamsa **(23^0 20' to 26^0 40')**: He/She will be flat-nosed, broad headed, inimical, and erratic sighted, garrulous and be dear to elders. Effects of Sagittarius Ascendant, Ninth Navamsa **(26^0 40' to 30^0 00')**: He/She will be fair in complexion, will have a face, akin to that of a horse and broad and dark eyes, will speak sparingly, be truthful, miserable and will possess crooked walking limbs.

Capricorn Lagna (Rising) Navamsa: The various Navamsa resulting in the Ascendant Capricorn will emanate the following effects: Effects of Capricorn Ascendant, First Navamsa **(Up to 3^0 20')**: He/She will have interstice and outwardly visible teeth, be dark in complexion, will speak with broken words, or be stammering, will have coarse hair, be famous, be interested in music and amusement, be emaciated and will have fluctuating wealth. Effects of Capricorn Ascendant, Second Navamsa **(3^0 20' to 6^0 40')**: He/She will be indolent, crafty, crooked nosed, fond of music, broad bodied, be interested in many females, will prattle much and be skilful. Effects of Capricorn Ascendant, Third Navamsa **(6^0 40' to 10^0 00')**: He/She will have lust for music, be famous, fair complexioned, be endowed with superior looks and charming nose, will be fond of many friends and relatives and achieves fulfilment of desires. Effects of Capricorn Ascendant, Fourth Navamsa **(10^0 00' to 13^0 20')**: He/She will have round eyes with a mix of blood-red and black hue, large forehead, emaciated body and thin arms, scattered hair, interstice teeth and broken speech. Effects of Capricorn Ascendant, Fifth Navamsa **(13^0 20' to 16^0 40')**: He/She will have prominent neck, nose and belly, will enjoy pleasures, be attached to women, be dark in complexion, will have round knees and arms and will attain successful beginnings in his undertakings. Effects of Capricorn Ascendant, Sixth Navamsa **(16^0 40' to 20^0 00')**: He/She will possess a splendorous body, will attire charmingly, be libidinous, will possess small and even teeth, is a good speaker and will have big cheeks and large forehead. Effects of Capricorn Ascendant, Seventh Navamsa **(20^0 00' to 23^0 20')**: He/She will be dark in complexion, be indolent, be an eloquent speaker, be short-haired, big-bodied, be harsh in disposition, will possess soft hands and legs, and be intelligent and very virtuous. Effects of Capricorn Ascendant, Eighth Navamsa **(23^0 20' to 26^0 40')**: He/She will be endowed with majestic sight and charming nose, reddish face, uneven nails and hair, grotesque body and forehead protruding, like a pot. Effects of Capricorn Ascendant, Ninth Navamsa **(26^0 40' to 30^0 00')**: He/She will have broad chest and large eyes, high intelligence, fully developed face, interest in musical studies, be endowed with sweetness and strength, be gentle and diplomatic.

Aquarius Lagna (Rising) Navamsa: The various Navamsa ascending, while Aquarius is on the East will produce following effects: Effects of Aquarius Ascendant, First Navamsa **(Up to 3^0 20')**: He/She will be dark in complexion, be soft in disposition, will have an emaciated body and prominent cheeks, be learned in poetry and Shashtra, be libidinous, interested in carnal pleasures and be splendour. Effects of Aquarius Ascendant, Second Navamsa **(3^0 20' to 6^0 40')**: He/She will possess coarse skin, nails, sight and hair, be kind to the helpless, be gentle, be tall in stature, foolish and will have a 'distinct' head. Effects of Aquarius Ascendant, Third Navamsa **(6^0 40' to 10^0 00')**: He/She will have a compact body (limbs sticking close to each other), be fond of females, will possess the splendour of lapis lazuli, be learned in the meanings of Shashtra and will act accordingly. Effects of Aquarius Ascendant, Fourth Navamsa **(10^0 00' to 13^0 20'):** He/She will be fond of women, be fair in complexion, will have a split face, will destroy enemies, be majestic, courageous and be fond of pleasures and sexual enjoyments. Effects of

Aquarius Ascendant, Fifth Navamsa **(13^0 20' to 16^0 40')**: He/She will be learned in Shashtra and in fine arts and dark in complexion and will have coarse hair on legs, concealed (not prominent) neck and ears. Effects of Aquarius Ascendant, Sixth Navamsa **(16^0 40' to 20^0 00')**: He/She will have a face, resembling that of a tiger, be bold, short-haired, will have unchanging aims, will kill living beings viz. tiger, deer, snake etc. and be dear to king. Effects of Aquarius Ascendant, Seventh Navamsa **(20^0 00' to 23^0 20')**: He/She will have eyes and face, resembling that of a goat, be fierce in disposition, be delighted in village life, insulted by females, will suffer diseases of bilious imbalances and be endowed with strength and courage. Effects of Aquarius Ascendant, Eighth Navamsa **(23^0 20' to 26^0 40')**: He/She will possess a no diminishing strength, firm be disposition and affection, be a warrior with the king, or be a king himself, be beautiful and will have strong teeth and broad eyes. Effects of Aquarius Ascendant, Ninth Navamsa **(26^0 40' to 30^0 00')**: He/She will be dark in complexion, will possess unclean and elevated teeth, be disunited from his wife, children and wealth, be an affable speaker and be famous and skilful.

Pisces Lagna (Rising) Navamsa: Should Pisces ascend at birth, the various Navamsa thereof will yield following specific effects: Effects of Pisces Ascendant, First Navamsa **(Up to 3^0 20')**: Though he/she may be white in complexion, his body will reveal the splendour of blood-red hue; he will be soft in disposition, be akin to a female in mental makeup, (i.e. will act, like a female), be fickle-minded and will have a short neck and emaciated waist. Effects of Pisces Ascendant, Second Navamsa **(3^0 20' to 6^0 40')**: He/She will have a big nose, be skilful in his assignments, will eat meat and the like, be endowed with a charming physique, will wander in forests and hills and will have a big head. Effects of Pisces Ascendant, Third Navamsa **((6^0 40' to 10^0 00')**: He/She will be white in complexion, be crafty, will possess beautiful eyes, and be beautiful, righteous, learned, courteous, modest and charming in appearance. Effects of Pisces Ascendant, Fourth Navamsa **(10^0 00' to 13^0 20')**: He/She will have praiseworthy attributes, will fall into adversity, will serve aged people, be skilful in his assignments, learned, very strong versed in justice and will have elevated nose. Effects of Pisces Ascendant, Fifth Navamsa **(13^0 20' to 16^0 40')**: He/She will be tall in stature, dark in complexion, be valorous, be not peaceful, will have a small nose and charming eyes, be fond of torturing others, be impatient, and will possess beautiful teeth and prattle. Effects of Pisces Ascendant, Sixth Navamsa **(16^0 40' to 20^0 00')**: He/She will be self-respected, righteous, excellent, strong, miserable, crafty, and unsteady and be a minister. Effects of Pisces Ascendant, Seventh Navamsa **(20^0 00' to 23^0 20')**: He/She will be self-respected, will show interest in other religions, be excellent, and be a minister, strong, miserable, cruel and unsteady. Effects of Pisces Ascendant, Eighth Navamsa **(23^0 20' to 26^0 40')**: He/She will be tall in stature, will have a big head, be emaciated, indolent, will have uneven (or dirtied) eyes and hair, will have foolish children (or a few children), be interested in earning money and be skilful in war (or quarrels). Effects of Pisces Ascendant, Ninth Navamsa **(26^0 40' to 30^0 00')**: He/She will be short, soft in disposition, courageous, broad chest, broad eyed, big nosed, be bright, will have a broad physique, be intelligent, virtuous and famous.

13.8 Predictions by D-10 (Dashamsha/Dashans) Chart:

This is the 1/10th Portion or Tenth Divisional Chart of a sign or a division of 3 degrees of sign. The first 1/10th Portion in the D-10 (Dashamsha) start from the same Rashi for an odd Rashi and from the 9th Rashi for an even Rashi. In case of an odd Rashi, these are presided over by the ten rulers of the cardinal directions, Indra, Agni, Yama, Rakshasa, Varuna, Vayu, Kubera, Isana, Brahma and Anantha. In case of an even Rashi, these are presided over by in the reverse order of these presiding deities. The 10th house from lagna in D-10 shows one's Career, Professional Ascendance, Public Dignity, Social Responsibilities, Executive or Leadership roles in the society, Promotions, Recognition in the status hierarchy or Matters of public opinion and Respect. If confirmation of radix/navamsha strength does not occur in dashamsha, it reduces the career expectations derived from D-1 and D-9. The 10th house from

AL [Arudha lagna] in D-10 shows perceptions about one's conduct in society and his career. If L-10 is uchcha in D-1 but nichha in dashamsha, there is a struggle in the career. If L-10 radix is strong in both D-1 and D-10, while the lord of the 10th dashamsha is auspicious, this is a sure sign of a vibrant and successful career.

The dashamsha 10th lord and the dashamsha 1st lord in the radix assess power. Examples: If there are significant graha in the 4th or 10th from lagna in D-10 chart, it provides strong positive or negative public acknowledgement for conduct in the parent and educator roles depending on the nature of the graha(s). If there are significant graha in the D-10's 6/12 axis, it provides strong public acknowledgement for conduct in the conflict-policing and sanctuary roles. When bhava-6 is strong in D-10 the native should be professionally involved with conflict, either creating or managing it. Also strong graha in bhava-4 of the D-10 suggest that the public may give special attention to facts regarding the native's childhood home or upbringing. If the graha is Rahu, read 'unusual, unstable, or foreign' elements in the upbringing. If the graha is uttama-Guru, read wealth and privilege, or a magnificent education. Significant graha in any Kendra of D-10 gives the native an unusually strong public presence. As in the D-1, the more graha in Kendra, the more the native is well-known. No graha in D-10 or D-1 Kendra suggest that the native is not known outside of their own neighbourhood or village. For determining career, by far the most important Graha in D-10 are, the graha which is the lord of radix bhava-10; the graha which is the lord of dashamshalagna; the graha which is the lord of bhava-10 in dashamsha; Shani - for position in large hierarchies; and Budha - for public communication skills. Evaluate radix L-10 & its role in D-10; dashamsha lagnesha & its role in D-1; navamsha lagnesha; lord of radix L-10 in D-9; lord of radix L-7 in D-9; lord of radix L-7 in D-10; rashi of dashamsha lagna; rashi of navamsha lagna; Arudha lagna of radix karmaa bhava and the current Vimshottari periods lords. "The 5th house in D-10 shows the power of a political leader in a position. So A5 [arudha-5] can show the trappings of power enjoyed by one. It can also show awards.

Various planets govern various traits of a dynamic personality. Example: A strong Sun in one's horoscope provides ability of organising, acquiring and cultivating a diligent team. Jupiter rules dependability, honesty and knowledge. Mars rules enterprising and aggressive persuasion. The Moon rules the public relations and imagination. Venus rules interactivity, teamwork and tolerance. The Sun and Jupiter rule the integrity while the planet Mercury rules the analytical and communicative capabilities. If any of the planets is weak in the horoscope, the traits ruled by the said planet remain weak endangering success. Astrology offers strengthening of concerned weak planets for choosing the right profession, averting setbacks in profession and deriving prosperous results from one's profession for the ultimate peaceful life with the help of the astrological science. The lords of the tenth and the second houses are considered as the primary determinants for analysing the professional affairs. In case there is no mooltrikona sign in any of these houses, then the lord of the first house is considered the prime determinant. The planets influencing the tenth and the second houses or the ascendant, if there is no mooltrikona in the tenth and the second houses, or their lords by way of close aspects/conjunctions as well as the planets placed in these houses act as secondary determinants for analysing the professional affairs. The lord of the ascendant of dasamsa containing a mooltrikona sign becomes an additional primary determinant for profession. The operating planetary sub-periods between the ages of 16 to 24 years become supplementary primary determinants for one's educational/professional pursuits. There are a number of factors that help us in arriving at a decision on this vital issue and these are as under:

(i) The strength and placement of the planet(s) of prime and secondary determinant(s) of profession;
(ii) The prime and secondary determinant(s) planetary influences on the houses ruling professional matters;
(iii) The strength of the operating planetary periods and
(iv) The nature of planets exerting influence on these planets.

The weak planets and/or afflictions to them cause the problems in life. Therefore, a two way application of astral remedies is administered, after diagnosing the problematic planetary influences in a chart. Firstly, the strength is provided to the weak functional benefic planets. The strength can be provided by various methods, such as, gemstones, colour therapy, a Kavach (Zodiac Pendant) containing mystical numbers of the planets in an auspicious time. Secondly, the malevolence of the functional malefic planets is reduced by offering propitiatory charities concerning these planets. The two-way application helps in reducing the impact of malefic planetary influences to a large extent. The preventive use of astral remedies is much more useful than the curative astral remedies.

Predictions of rise and fall in one's career by Dasamsa:

Factors influencing career, profession and life are, the planets in the tenth house; lord of the tenth; planets aspecting the tenth; karakas, Saturn and Sun; 10th lord dispositor in D-9; Dasamsa or D-10; strong planet in tenth; Yoga formed by the conjunction of the planets 10th Bhava; and each of the planets in the tenth gives its effects in its periods and sub-periods more strongly, while the influence of the other planet(s)

The following planetary position has both positive and negative implications to interpret:

• Tenth lord in 1,4,7,10: Good generally (Angles are Vishnu sthanas and supportive)

• Tenth Lord in 1,5,9: Good (Trines are lakshmi sthanas and lead to prosperity)

• Tenth lord in 2 or 11: Wealth giving (2H: finance/wealth & 11H: fulfilment/gains), though other problems may be there

• Tenth lord in 3H: Hard working and undaunted, both qualities being good for career (3H is the house of efforts and valour)

• Tenth lord in 6,8,12 (trik): Ups and downs, obstacles, changes (not necessarily always bad though rise and fall are dramatic at times); of these the 6th is probably better since it is an Upachaya house (house of improvement/growth)

With the above principles in mind let us move on to the Dasamsa where I will expand the same principles. Is the 9th house (being the house of luck and fortune) good for career? With respect to career it may not prove to be so good often. Because, the 9th being the 12th from the 10th negates the

10th. Dharma (9th) negates Svadharma (10th). So the 9th may not be very conducive to career matters depending on some factors. But the same 9th is good for gains from father, hereditary vocations (again linked to father), or gains from higher knowledge or spiritual knowledge.

Excellent Career:

The following factors indicate a good career, public life, recognition of one's work etc, (i) D-1Lagna Lord in D-10 lagna; or (ii) D-1 Lagna Lord in good houses in D-10 (Kendras, Trikona, 2 or 11); or (iii) D-1 10th lord or planets in the 10th of natal well placed in D-10 (well placedmeans placed in Kendra, Trikona, 2, or 11 and no malefic conjunctions or aspects); or (iv) D-10 lagna lord well placed in D-1; or (v) D-10 10th lord well placed in D-1 or (vi) D-10 9L or 11L or 2L in benefic places in D-1. In other words, if a planet is well placed in both D-1 and D-10, it indicates rise in career. But, if a planet is well placed in one chart, and badly placed in the other, like, (i) D1 lords of 3, 6, 8, 12 in benefic places in D-10, then maleficience gets reduced; (ii) D-1 benefic lords (lords of 1, 5, 9) badly placed in D-10, then it will give results but only after initial struggle and difficulties. D-10 trik lords (6, 8, 12) connected to D-1 Lagna or 10th house provides obstacles or difficulties with respect to career. Strong 10th House in D-10, or planets in the 1st or 1st lord in D-10, or planets in Kendras and Trikonas indicates status, recognition and great achievements. Al results indicated by a planet in the rasi and amsa chart come to fructification in their dasa, antardasa and sensitizing transits.

Planets in Kendras or Trikona are auspicious. Dasa lord or antardasa lord getting connected to the D-10 Lagna lord or D-10 tenth lord leads to a rise in career. Planets in trikonas in D-10 are good. Planets in or lords of 1, 5, 9 are good for career. Planets in Trika houses 6, 8 and 12 signify struggles, obstacles and difficulties. Planets in 6, 8, 12 in D-10 indicate difficulties, tensions and worries with respect to career especially if they are unaspected by lagna lord or 10th lord. If aspected by the LL or 10L, initial difficulties followed by relief. When the Trika lords of D-10 are related to the 10th house or 10th lord, the dasa or antardasa may be tough. Here again relations are by placement as well as aspects. Malefics in Trika houses are better than benefics here (in the dasamsa only). Strong Moon (Sukla paksa or bright half, especially fewdays before and after full moon), Jupiter, Venus and Mercury without malefic company or influence are considered benefic. The Sun, Saturn, Rahu, Ketu and weak Moon are considered malefic. If the Antardasa lord is in 6, 8, 12 from Dasa lord in D-10, it indicates difficulties and tensions. If the antardasa lord is not in 6, 8, 12 from dasa lord in the rasi chart, it is considered well placed. : Planets in Upachaya houses i.e. 3, 6, 10, 11 generally give good results eventually but with hard work and struggle. The benefics are good in the 10th and 11th, they are fine in the 3rd, and may not be good in the 6th. Malefics in the Upachayas are good generally. • Planets in the 10th and 10th lord in D-10; or Lagna lord and lagna in D-10; or 10th house and lord in rasi become the promittors for career matters. Planets in the 10th in D-10 can trigger a rise in career while transiting over the D-10 lagna or or lagna lord. D-10 tenth lord's transit over the rasi 10th house or rasi 10th lord can also trigger some favorable event or rise in career, especially during the dasa or antardasa of such lords.

13.7.1 Predictions by D-12 (Dwadshamsha) Chart:

This is 1/12 Portion or or 2½ degrees each or a Twelfth Divisional Chart and contains 144 portions or divisions of a sign. D-12 indicates the Parents. The reckoning of the Dvadashamsha (one twelfth of a Rashi) commences from the same Rashi. In each Rashi the presidency repeats thrice in the order of Ganesha, Azvini Kumara, Yama and Sarpa for the 12 Dvadashamsha. The 9th-from-lagna or 9th-from-Surya in the D-12 (whichever is stronger) reveals the details of the father and 4th-from-lagna or 4th-from-Chandra in the D-12- (whichever is stronger) reveals details of the mother. See also the bhratri-karaka (graha holding the 3rd-highest degree in any rashi within the radix) for father and for mother, see also the matri-karaka (graha holding the 4th-highest degree in any rashi within the radix). For father, see Surya and Guru. See

Graha in dharma bhava, and L-9, for the material relationship with Father and his moral belief system. See 9th-from-Chandra for the emotional relationship with father. For mother, see Chandra.

See Graha in Bandhu bhava, and L-4, for the material relationship with Mother and her cultural manners and roots.

13.7.2 Predictions by D-16 (Shodashamsha) Chart:

This is 1/16 Portion or Sixteenth Divisional [Harmonic] or 1 degree 52 minute 30 second of a Rashi and indicates Vehicles or conveyance and its pleasure, convenience of travel by land, sea, and air; the potential for accident and injury through vehicular travel, which is a consequence of imbalance due to 6th in D-16. D-16 covers all modes of non-pedestrian travel from camels and ox-carts to F-1 fighter planes and zephyrs. First check Shukra (Karaka for the luxury of vehicular conveyances), bhava-4 radix, and L-4 radix before looking into the D-16. Example 1: Beatle George Harrison has uttama-Shukra yuti Rahu who enjoyed many luxury automobiles. Harrison was a driver and collector of speed-racing cars. He owned fleets of these super-expensive, exquisitely designed (Shukra) vehicles, and enjoyed much pleasure through them.

The native may not own a car, or even a donkey, if both Shukra and bhava-4 are weak. However one will enjoy the pleasures (Shukra) of world travel (9). Therefore, the D-16's prognosis for vehicles is mainly in context of world-travel (outside one's region) which is nearly always by public conveyance - plane, train, bus, ship. L-4 will help to decide whether one should buy a luxury car with pricey embellishments (which Shukra might crave) or stick with the basic model (e.g., a strict and austere Shani (L-4) or rather put the car money into savings and take the bus. Shodashamsha is specific to vehicles and their good or bad effects. D-16 indicates whether your vehicle is a lumbering elephant, a jewelled palanquin, a creaking oxcart or a luxury automobile. If Radix bhava-4 looks good; lord Kuja strong in bhava-5 in good company; Shukra (Luxmi/luxury) occupies bhava-4 receiving drishti of blessed uttama-Guru from the house of Private Imagination (bhava-12); the native would be able to manifest his imaginative dream of owning and operating a beautiful luxury automobile. The presence of L-4 (Kuja, Budha or Chandra) in Kumbha lagna of the D-16 indicates the resistance in owning a car. If Shani is the satkona-pati (L-6), it indicates poverty, illness, and exploitation as Shan is the King of Fear. Shani has way too much negative, scarcity-thinking power in the D-16.

13.7.3 Predictions by D-24 (Vimshamsha) Chart:

This is 1/20 Portion or Twentieth Divisional Chart. The calculations of Vimshamsha start from Mesha for a chara Rashi, from Dhanushya for a sthira Rashi and from Simha for a Dwisva-bhava Rashi. D-20 Analysis protocol: Check the conditions of bhava- 4, 5 and 9 and planets positioned in D-20 and the swabhava of Surya for Individual genius and Creative intelligence, celebration of life, brilliance, fashion, and royal glamour. According to Pt Narasimha Rao, "The 5th house in D-20, the chart of religious activities, shows one's devotion (bhakti) in religious activities. A5 [arudha-5] in D-20 shows the things based on which the world forms an impression about one's devotion.

13.7.4 Predictions by D-20 (Chaturvimshamsha) Chart:

This is 1/24 Portion or Twenty-Fourth Divisional or 1 degree 15 minutes each Chart. It indicates Fruits of schooling, Education (Vidya), knowledge, Science, learning, scholarship, philosophy and siddhi (abilities, skills). It indicates the matters of 2nd, 3rd and 4th house of Radix Lagna. The Chaturvimshamsha distribution commences from Simha and Karka, respectively, for an odd and an even Rashi. In D-24, the chart of learning, the 5th house may show intelligence. A5 [arudha-5] in D-24 shows the things based on which the world forms an impression about one's intelligence in learning. The 3rd house in D-24 (chart of learning and knowledge) and D-

10, (achievements in society) may show one's writing skills. A3 shows the Maya associated with one's writing skills. One may have excellent writing skills, but not write any book. In such a case, people may not really appreciate one's writing skills. One may be an average writer, but end up writing 20 books. Thus the exact books written by one decide the perceptions of the world about one's writing skills. So A3 in D-24 and D-10 may show the books or articles written by him. Example: In D-10, A3 is in Pisces mean that he will be known for some books on saattwik and traditional subjects and astrology.

13.7.5 Predictions by D-27 (Saptavimshamsha / Nakshatramsha / Bhamsha) Chart:

This is 1/27 Portion or Twenty-seventh Divisional Chart. "The Saptavimshamsha Lords for an odd Rashi are, respectively, the presiding deities of the 27 Nakshatra starting from Dastra (Azvini Kumar) and ending with Pushya. Count these deities in a reverse order for an even Rashi. The Saptavimshamsha distribution commences from Mesha and other Movable Rashi for all the 12 Rashi. Look for Bhamsha D-27 Lagna.

13.7.6 Predictions by D-30 (Trimshamsha) Chart:

This 1/30th Portion or Thirtieth Divisional [Harmonic] Chart. The Trimshamsha Lords for an odd Rashi are Mangala, Shani, Guru, Budha and Shukra. Each of them in order rules 5, 5, 8, 7 and 5 degrees. The deities, ruling over the Trimshamsha, are, respectively, Agni, Vayu, Indra, Kubera and Varuna.

13.8. Predictions by Ashtakavarga:

Saravali and Brihat Parashara Hora Sashtra provide the most important method for predictions of the auspicious and inauspicious effects in life of the native by Ashtakavarga of planets during transit in Signs. The purposes of Ashtakavarga are also to assess the strength of planets and houses in natal charts. When we look at the house occupied by a transit planet with respect to natal Moon and Transit Lagna Chart, we can look at the houses occupied by a transit planet with respect to all the eight references used in Ashtakavarga. Mainly 3 types of Ashtakavarga charts are used to study the effects of the planets, such as Binnashtakavarga, Sarvashtakavarga and Samudaya Ashtakavarga.

Bhinnashtakavarga (BAV Charts):

The Bhinnashtakavarga charts are prepared for the Lagna and the 7 Planets. Rahu & Ketu are omitted in Asthakavarga. The Asthakavarga is made by dividing each of the 12 Rasi into eight sub divisions called Kaksha based on the Lagna and the 7 main Planets. Each Kaksha has a span of 3:45′.

Binnashtakavarga Chart for Sun

Sun Sign	**2**	**3**	**4**	**5**	**6**	**7**	**8**	**9**	**10**	**11**	**12**	**1**	**Total**
Saturn	1	0	1	0	0	1	1	1	1	1	0	1	8
Jupiter	0	1	0	1	0	0	0	0	0	1	1	0	4
Mars	0	1	1	0	1	0	0	1	1	1	1	1	8
Sun	1	1	0	1	0	0	1	1	1	1	1	0	8
Venus	1	0	0	0	0	0	1	1	0	0	0	0	3
Mercury	0	1	0	1	1	0	0	1	1	1	1	0	7
Moon	0	0	1	0	0	0	1	1	0	0	0	1	4
Lagna	0	0	0	1	1	1	0	0	1	1	0	1	6
Total	**3**	**4**	**3**	**4**	**3**	**2**	**4**	**6**	**5**	**6**	**4**	**4**	**48**

The AshtakaVargaa chart is prepared with 10 horizontal lines and 14 vertical lines. The chart so prepared will consist of 117 apartments having the figures. In the chart, incorporate the names of the seven Planets and Lagna in the first left Column. Write the numbers of all the 12 Rashi in the first top Line. Based on their individual positions in a horoscope, the Lagna and the 7 main Planets contribute certain positive (auspicious) called Sthan and negative (inauspicious) influences called Karan on the 12 signs of the horoscope. These influences like Karan (Negative / inauspicious) are signified by Karanprad or a dot or (0) significators and Sthan (Positive / auspicious) is signified by Rekhapradas or a Rekha or a line or Bindu or (1) significators. The 1st Kaksha of each Rasi is ruled by Saturn, the 2nd by Jupiter, 3rd by Mars, 4th by Sun, 5th by Venus, 6th by Mercury, the 7th by Moon and the 8th by the Lagna. The meaning of AshtakaVargaa is literally the group of eight divisions. In other words it is the combination of the good and bad positions of the 7 Planets and Lagna or combination of the benefic (the Rekhas) and malefic marks (the Bindus) with reference to the position of the eight Planets (Lagna is treated as a Planet). Ashtakavarga give us predictive clues with respect to the planets nature results. Ashtakavarga is also employed to study longevity of a native. The Ashtakavarga is an advanced application. There will be no contradictions in judging the effects of happiness and sorrows and determination of the longevity.

In the Bhinnashtakavarga chart for Sun, the Rasi are marked horizontally by the respective Rasi numbers, i.e. 1 for Aries, 2 for Taurus etc. The 8 Kaksha are to be seen vertically. A Bhinnashtakavarga Kaksha of any Rasi can contain only one point. If a positive point is contributed by any planet or Lagna, it is marked by 1 in a Kaksha. If a negative point is contributed by any planet or Lagna, it is marked by 0 in a Kaksha. So we can see that in the chart of the Bhinnashtakavarga of Sun, in the sign Aries (under column 1), Saturn has contributed a positive point marked by 1 in his Kaksha, Jupiter has contributed a negative point marked by 0 in his Kaksha below 1 column and so on. The total positive points contributed in Aries are 4. We can see the other signs similarly. The Bhinnashtakavarga charts of planets are very useful to study the transit effects of planets.

Sarvasthakavarga Charts:

The Sarvasthakavarga chart is the combined detailed summary of how many positive points each of the 7 Planets and Lagna are contributing in all the Rasi of their BAVA, such as, it is filled up here for 12 Signs and is highlighted for Sun Ashtakavargathe highlighted figures .

Table: Sarvashtakavarga Chart

Sun Sign / House No.	1	2	3	4	5	6	7	Sub Total
Lagna	7	1	4	6	5	4	3	30
Sun	**4**	**3**	**4**	**3**	**4**	**3**	**2**	**23**
Moon	6	1	2	5	7	4	5	30
Mars	5	1	2	5	3	5	2	23
Mercury	6	3	4	4	4	7	3	31
Jupiter	5	4	4	6	6	4	3	32
Venus	4	3	6	4	5	4	3	29
Saturn	4	3	2	1	5	4	1	20
Total	**34**	**18**	**24**	**28**	**34**	**31**	**19**	**218**

Sarvashtakavarga Chart

Sun Sign / House No.	8	9	10	11	12	Sub Total	Total
Lagna	4	1	6	5	3	19	49
Sun	**4**	**6**	**5**	**6**	**4**	**25**	**48**
Moon	4	3	3	6	3	19	49
Mars	3	3	4	3	3	16	39
Mercury	3	5	5	5	5	23	54
Jupiter	4	6	4	4	6	24	56
Venus	3	5	4	7	4	23	52
Saturn	5	3	2	5	4	19	39
Total	**26**	**31**	**27**	**36**	**29**	**168**	**337**

Example: Sun transits a Rasi for one month. But the results he gives are not uniform throughout the month. He produces good results while transiting the Kaksha where there is a positive point and he gives bad results if negative point is there. In this case while Sun transiting the 1st Kaksha of Aries, from 0 degrees to 3:45′, he gives good results as Saturn has contributed a positive point there. In the next Kaksha, from 3:45 to 7:30′, where there is a negative point contributed by Jupiter, he gives bad results. As Aries has 4 positive and 4 negative points in Suns BAV chart, Suns transit results in Aries for the native of this chart are 50% good and 50% bad. When the Sun transits Libra (7) for one month its results are very bad as it has only 2 positive points. Suns transit of Sagittarius (9) is the best with 6 Positive Points. So a Rasi with 4 good points gives medium results, better with more and worse with fewer points. Benefic results will be predominant, if these planets happen to be exaltation, own or friendly Sign, the malefic effects will be ordinary. But, if these planets happen to be in debilitation or inimical Signs, the malefic results will be more and benefic effects will be ordinary. The good or bad effects will have a relation with the planet contributing the point, either positive or negative. They will have a link to its nature, house ownership. In similar fashion we can study the transit effects of the other planets with the help of the BAV charts.

Samudaya Ashtakavarga (SAV) / Aggregational Ashtakavarga:

Samudaya Ashtakavarga (SAV) is used to judge good or bad effects of the Rasi Kundali.

29	34	18	24
36	**Samudaya Ashtakavarga**		28
27			34
31	26	19	31

Write down a Rasi Kundali with 12 Houses including Lagna. Then insert the total of the Rekhas (Positive Points) obtained in Sarvashtakavarga Chart under the Rashis Column in all the Ashtakavargas of the Planets in the Rasi concerned. The Ashtaka Varga with such Rekhas is called the Samudaya Ashtaka Varga, or the Aggregational Ashtaka Varga. The Rasi, which has more than 30 Rekhas in the Samudaya Ashtakavarga, gives favourable effects, that having between 25 and 30 Rekhas, produces medium effects; and that Rasi having less than 25 Rekhas, yields adverse effects. Auspicious functions like marriage should be performed when the Planet moves into the Rasi with favourable effects. The Rasi, which is productive of adverse effects, should be avoided for these purposes.

Example: The strength of Moon is generally acceptable for all auspicious functions. Therefore auspicious functions should be performed or started when Moon is in the Rasi with maximum number of Rekhas. The Planet in the Rasi with favourable number of Rekhas produces auspicious effects and the Planet in the Rasi with unfavourable number of Rekhas yields evil results. Amongst the 12 Houses more than 30 Rekhas advance the effects of a House, between 25 and 30 Rekhas produce medium effects and the effects of the House, which contains less than 25 Rekhas, get damaged.

There will be all-round increase in wealth, happiness in respect of children and enjoyments in the Samvat, month and Nakshatra of the Rasi, which has more than 30 Rekhas and increase in wealth, property, children and good reputation, if the Rasi has more than 40 Rekhas. The Rasi, which is auspicious in Samudaya Ashtakavarga, is considered auspicious for all auspicious functions. Consequently the auspiciousness of Ashtaka Varga should be got checked before performing any function, like marriage. If a Rasi is not auspicious in Ashtaka Varga, then its auspiciousness should be checked from transit effects. It is not necessary to check transit effects, if a Rasi is auspicious in Ashtaka Varga. Thus the auspiciousness of the Rasi in the Samudaya Ashtakavarga should be considered, as paramount. The Samudaya Ashtakavarga chart gives a brief, at a glance, idea of how many total good points are there in each house. Houses with 25 or more produce good results when planets transit them. If the points are below 25 then the results are negative. If there are less than 25 Rekhas in 6th, 8th and 12th House, their effects become favourable. The effects will become adverse, if these Houses contain more than 25 Rekhas. If in a Rasi Kundali there is larger number of Rekhas in the 11th than those in 10th and there is smaller number of Rekhas in the 12th than those in the 11th and Lagna contains largest number of Rekhas, the native will be wealthy and will enjoy all kinds of comforts and luxury.

Sections of Life Period:

Divide the 12 Houses in 3 sections of containing 4 Houses each. There will be sufferings and distress in that part of the life, which is represented by the section of the Rasi Kundali with more malefic. There will be happiness in the part of the life, represented by the section of the Rasi Kundali, containing more benefic. There will be mixed results in that area of life, when the relative section of the Rasi Kundali has equal number of benefic and malefic. The Houses from Lagna up to the 4th signify childhood, those from 5th to 8th youth and those from the 9th to 12th represent old age. There will be danger of death in the month of the Rasi, while Sun's transit in that Rasi, which has 7 or less than 7 Rekhas in the Samudaya Ashtaka Varga.

Sun's AshtakaVargaa: In Sun's AshtakaVargaa, Lagna, Moon, Jupiter, Venus and Mercury in the 1st, 2nd and 8th from Sun; Sun, Mars, Saturn, Moon and Jupiter in the 12th; Mercury, Moon, Venus, Jupiter in the 4th; Lagna, Moon, Venus in the 9th; Sun, Saturn and Mars in the 6th; Lagna, Mercury, Jupiter and Moon in the 7th; Venus in the 11th; Sun, Saturn, Venus, Jupiter and Mars in the 3rd; Jupiter and Venus in the 10th; Sun, Saturn, Moon, Lagna Mars and Venus in the 5th are Karanprad, or dot significators.

Saturn, Mars and Sun in the 2nd, 8th and 1st; Jupiter and Mercury in the 5th; Mercury, Moon and Lagna in the 3rd; Lagna, Sun, Saturn and Mars in the 4th; Lagna, Sun, Saturn, Mars, Mercury and Moon in the 10th; Sun, Moon, Mars, Mercury, Jupiter, Saturn and Lagna in the 11th; Lagna, Venus and Mercury in the 12th; Lagna, Venus, Mercury, Jupiter and Moon in the 6th; Sun, Mars, Saturn and Venus in the 7th; Sun, Mars, Saturn, Mercury and Jupiter in the 9th from their own places are Rekhapradas (line significators).

Moon's AshtakaVargaa: In Moon's AshtakaVargaa six Planets in the 9th and the 2nd, 5 Planets in the 4th, 8th and the 1st, one Planet in the 10th and the 3rd, 4 Planets in the 5th, 3 Planets in the 6th and the 7th and 1st, eight in the 12th are Karanprad. Thus Lagna, Sun, Mars, Saturn and Venus, these five in the 1st; Lagna, Mercury, Sun, Moon, Saturn and Venus, these 6 in the 2nd; Jupiter in the 3rd; Sun, Saturn, Moon, Lagna and Mars, these 5 in the 4th; Venus, Mercury, Jupiter, these 3 in the 6th; Mars, Lagna and Saturn, these 3 in the 7th; Mars, Lagna, Saturn, Venus and Moon, these five in the 8th; Lagna, Sun, Mars, Saturn, Mercury and Jupiter, these 6 in the 9th; Saturn only in the 10th; none in the 11th; all the eight in the 12th from their own places are Karanprad, or dot significators. These Planets in the other Houses are Rekhaprad (line significators).

Mercury, Moon and Jupiter in the 1st; Jupiter and Mars in the 2nd; Mercury, Sun, Moon, Mars, Saturn, Venus and Lagna in the 3rd; Jupiter, Venus and Mercury in the 4th; Mars, Mercury, Venus and Saturn in the 5th; Sun, Moon, Mars, Saturn and

Lagna in the 6th; Sun, Moon, Jupiter, Mercury and Venus in the 7th; Sun, Mercury and Jupiter in the 8th; Venus and Moon in the 9th; Sun, Mercury, Jupiter, Venus, Moon, Lagna and Mars in the 10th and all the 8 Planets in the 11th from their own places are Rekhaprad (line significators). No Planet is Rekhaprad in the 12th.

AshtakaVargaa of Mars: In the AshtakaVargaa of Mars, Planets in the 12th, 4th and 7th, 5 Planets in the 5th, 6 Planets in the 2nd, 7 Planets in the 9th, 5 Planets in the 1st and 8th, 4 Planets in the 3rd, 3 Planets in the 10th and 2 Planets in the 6th are Karanprad. In the 11th no Planet is Karanprad. In other words all the Planets in the 11th from their own places are Rekhaprad. Thus Sun, Moon, Mercury, Jupiter and Venus, these five in the 1st; Lagna, Sun, Moon, Mercury, Jupiter and Saturn, these 6 in the 2nd; Venus, Mars, Jupiter and Saturn, these 4 in the 3rd; Sun, Moon, Mercury, Jupiter, Venus and Lagna, these 6 in the 4th; Moon, Mars, Jupiter, Venus and Lagna, these 5 in the 5th; Mars and Saturn, these 2 in the 6th; Mercury, Moon, Sun, Venus, Jupiter and Lagna, these 6 in the 7th; Mercury, Moon, Sun, Lagna and Jupiter, these five in the 8th; Sun, Moon, Mars, Mercury, Jupiter, Venus and Lagna, these 7 in the 9th; Venus, Moon and Mercury, these 3 in the 10th; none in the 11th; Sun, Saturn, Mercury, Moon, Lagna and Mars, these 6 in the 12th from their own places are Karanprad, or dot significators.

Lagna, Saturn and Mars in the 1st; Mars in the 2nd; Lagna, Mercury, Moon and Sun in the 3rd; Saturn and Mars in the 4th; Mercury and Sun in the 5th; Mercury, Moon, Jupiter, Sun, Lagna and Venus in the 6th; Saturn and Mars in the 7th; Saturn, Mars and Venus in the 8th; Saturn in the 9th; Mars, Sun, Jupiter, Saturn and Lagna in 10th; all in the 11th and Jupiter and Venus in the 12th from their own places are Rekhaprad (line significators).

AshtakaVargaa of Mercury: In the AshtakaVargaa of Mercury, 3 Planets in the 1st, 2nd, 4th, 10th, 6th and 9th, 2 Planets in the 8th, 6 Planets in the 3rd and the 7th, none in the 11th, 5 Planets in the 5th and 12th are Karanprad. Thus Sun, Moon and Jupiter, these 3 in the 1st; Jupiter, Sun and Mercury, these 3 in the 2nd; Lagna, Sun, Mars, Saturn, Moon and Jupiter, these 6 in the 3rd; Mercury, Sun and Jupiter, these 3 in the 4th; Jupiter, Mars, Moon, Saturn and Lagna, these 5 in the 5th; Venus, Saturn and Mars, these 3 in the 6th; Mercury,

Moon, Lagna, Sun, Venus and Jupiter, these 6 in the 7th; Mercury and Sun, these 2 in the 8th; Jupiter, Moon and Lagna, these 3 in the 9th; Sun, Jupiter and Venus, these 3 in the 10th; none in the 11th; Lagna, Moon, Mars, Saturn and Venus these 5 in the 12th from their own places are Karanprad, or dot significators.

Lagna, Saturn, Mars, Venus and Mercury in the 1st; Lagna, Mars, Moon, Venus and Saturn in the 2nd; Venus and Mercury in the 3rd; Lagna, Moon, Saturn, Venus and Mars in the 4th; Mercury, Saturn and Venus in the 5th; Jupiter, Mercury, Sun, Moon and Lagna in the 6th; Mars and Saturn in the 7th; Mars, Saturn, Lagna, Moon, Venus and Jupiter in the 8th; Saturn, Mars, Sun, Mercury and Venus in the 9th; Lagna, Saturn, Mars, Mercury and Moon in the 10th; all in the 11th and Jupiter, Mercury and Sun in the 12th from their own places are Rekhaprad.

AshtakaVargaa of Jupiter: In the AshtakaVargaa of Jupiter, one Planet in the 2nd and 11th, 2 Planets in the 10th, 7 Planets in the 12th, 4 Planets in the 6th, 5 Planets in the 8th and 3rd, 3 Planets in the remaining Houses are Karanprad. Thus Venus, Moon and Saturn, these 3 in the 1st; Saturn in the 2nd and 11th; Lagna, Mars, Moon, Mercury and Venus, these 5 in the 3rd; Sun, Jupiter and Mars, these 3 in the 5th; Venus, Saturn and Moon, these 3 in the 4th; Mercury, Venus and Saturn, these 3 in the 7th; Jupiter, Mars, Sun and Moon, these 4 in the 6th; all except Saturn, these 7 in the 12th; Moon and Saturn, these 2 in the 10th; Saturn, Mars and Jupiter, these 3 in the 9th; Lagna, Saturn, Venus, Moon and Mercury, these 5 in the 8th from their own places are Karanprad, or dot significators.

Lagna, Mars, Sun and Mercury in the 1st and 4th; Jupiter, Lagna, Mars, Sun, Mercury, Moon and Venus in the 2nd; Saturn, Jupiter and Sun in the 3rd; Venus, Moon, Lagna, Mercury and Saturn in the 5th; Venus, Lagna, Mercury and Saturn in the 6th; Lagna, Mars, Jupiter, Sun and Moon in the 7th; Jupiter, Sun and Mars in the 8th; Venus, Sun, Lagna, Moon and Mercury in the 9th; Jupiter, Mercury, Mars, Sun, Venus and Lagna in the 10th; all except Saturn in the 11th and Saturn in the 12th from their own places are Rekhaprad.

AshtakaVargaa of Venus: In the AshtakaVargaa of Venus, 2 Planets in the 5th, 8th and 3rd, 5 Planets in the 1st, 2nd, 12th, 10th, 8 Planets in the 7th, 6 Planets in the 6th, one in the 9th, 3 in the 4th, none in the 11th are Karanprad. Thus Sun, Mars, Mercury, Jupiter and Saturn, these 5 in the 1st and the 2nd; all the 8 Planets in the 7th; Jupiter and Sun, these 2 in the 3rd; Sun and Mars, these 2 in the 5th; Sun in 9th; Sun, Mercury and Jupiter, these 3 in the 4th; Mars and Mercury, these 2 in the 8th; Venus, Sun, Moon, Saturn, Lagna and Jupiter, these 6 in the 6th; none in the 11th; Lagna, Saturn, Mercury, Venus and Jupiter, these 5 in the 12th; Lagna, Mars, Mercury, Moon, Sun, these 5 in the 10th from their own places are Karanprad, or dot significators.

Lagna, Venus and Moon in the 1st; Lagna, Venus and Moon in the 2nd; Lagna, Venus, Moon, Mercury, Saturn and Mars in the 3rd; Lagna, Venus, Moon, Saturn and Mars in the 4th; Lagna, Mercury, Moon, Jupiter, Saturn and Venus in the 5th; Mercury and Mars in the 6th; none in the 7th; Venus, Sun, Moon, Jupiter, Lagna and Saturn in the 8th; all except Sun in the 9th; Venus, Jupiter and Saturn in the 10th; all in the 11th; Mars, Moon and Sun in the 12th from their own places are Rekhaprad.

AshtakaVargaa of Saturn: In the AshtakaVargaa of Saturn, 7 Planets in the 2nd, 7th, 9th, 6 Planets in the 8th, Lagna and 4th, 4 Planets in the 10th, 3rd and 12th, one Planet in the 6th, 5 Planets in the 5th, none in the 11th are Karanprad. Thus Moon, Mars, Mercury, Jupiter, Venus and Saturn, these 6 in the 4th and the 1st; Moon, Mars, Mercury, Jupiter, Venus, Saturn and Lagna, these 7 in the 2nd and the 7th; Sun, Moon, Mars, Jupiter, Venus, Saturn and Lagna, these 7 in the 9th; Moon, Jupiter, Venus and Saturn, these 4 in the 10th; Jupiter, Sun, Mercury and Venus, these 4 in the 3rd; Sun in the 6th; Lagna, Moon, Saturn, Sun, these 4 in the 12th; Venus, Sun, Moon, Mercury and Lagna, these 5 in the 5th; Moon, Mars, Jupiter, Venus, Saturn and Lagna, these 6 in the 8th; none in the 11th from their own places are Karanprad, or dot significators. The remaining places are auspicious and are Rekhaprad (line significators).

Sun and Lagna in the 1st; Sun in the 2nd; Lagna, Moon, Mars and Saturn in the 3rd; Lagna and Sun in the 4th; Jupiter, Saturn and Mars in the 5th; all except Sun in the 6th; Sun in the 7th; Sun and Mercury in the 8th; Mercury in the 9th; Sun, Mars, Lagna and Mercury in the 10th; all in the 11th; Mars, Mercury, Jupiter and Venus in the 12th from their own places are Rekhaprad.

AshtakaVargaa of Lagna: In the AshtakaVargaa of Lagna, 3 Planets in the 1st and 4th, 2 Planets in the 3rd, 5 Planets in the 2nd, 6 Planets in the 5th, 8th, 9th and 12th, one Planet in the 10th, 11th and 6th and all except Jupiter in the 7th are Karanprad. Thus Lagna, Sun and Moon in the 1st; Lagna, Mars, Moon, Sun and Saturn in the 2nd; Jupiter and Mercury in the 3rd; Lagna, Moon, Mars, Mercury, Saturn and Sun in the 5th; Lagna, Moon and Mars in the 4th; Venus in the 6th; all except Jupiter in the 7th; Lagna, Sun, Moon, Mars, Jupiter and Saturn in the 8th; Lagna, Sun, Moon, Mars, Mercury and Saturn in the 9th; Venus in the 10th and 11th; Lagna, Mars, Mercury, Jupiter, Venus and Saturn in the 12th from their own places are Karanprad.

Saturn, Mercury, Venus, Jupiter and Mars in the 1st; Mercury, Jupiter and Venus in the 2nd; Lagna, Sun, Moon, Mars, Venus and Saturn in 3rd; Sun, Mercury, Jupiter, Venus and Saturn in the 4th; Jupiter and Venus in the 5th; all except Venus in the 6th; Jupiter in the 7th; Mercury and Venus in the 8th; Jupiter and Venus in the 9th; all except Venus in the 10th; all except Venus in the 11th and Sun and Moon in the 12th from their own places are Rekhaprad.

Shodhana (Rectification) of the AshtakaVargaa: Shodhana (Rectification) of the AshtakaVargaa of a Planet is done to see in which Rasi the Planet is posited. Beginning from that Rasi, the names of the 12 Rasi should is written and then the names of the Planets, posited in them, is mentioned against them. Thereafter the Rekhas, gained by that Rasi, are written below them and the number, achieved after Shodhana, below it.

Trikona Shodhana in the AshtakaVargaa: After preparing the AshtakaVargaa of all the Planets and Lagna, Trikona Shodhana (Rectification) has to be done for each Rasi. A Trikona is made of three Rasi equidistant from each other. Thus Aries, Leo and Sagittarius; Taurus, Virgo and Capricorn; Gemini, Libra and Aquarius; Cancer, Scorpio and Pisces; form the Trikona of the Rasi. Trikona Shodhana (Rectification) is done by writing the Rekhas or 1 in the AshtakaVargaa of Sun under the Rasi Aries. Amongst the Trikona Rasi, the Rasi having lesser number of Rekhas should be allotted Rekhas, arrived at by deducting its number of Rekhas from the greater number of Rekhas of the three Trikona Rasi. No Trikona Shodhana is necessary, if any of the Trikona Rasi has no

Rekha. Shodhana should be done, if all the three of them have equal number of Rekhas that is a zero should be written against all of them. After doing Trikona Shodhana, Ekadhipatya Shodhana should be taken in hand.

Ekadhipatya Shodhana in the AshtakaVargaa: Ekadhipatya Shodhana (Rectification) is done after writing the numbers for Rasi arrived at by Trikona Shodhana. Ekadhipatya Shodhana is done, if both the two Rasi, owned by a Planet, have gained a number after Trikona Shodhana. Ekadhipatya Shodhana is not to be done, if one Rasi has got a number and the other is bereft of any number. The following are the rules for Ekadhipatya Shodhana. If both the Rasi are without a Planet and the Trikona Shodhana numbers are different, both should be given the smaller number. If both the Rasi are with Planets, no Shodhana is to be done. If amongst the two Rasi one is with a Planet and a smaller Trikona rectified number and the other is without Planet with a bigger number, deduct the smaller number from the bigger number and the number of the Rasi with Planet should be kept unchanged. If the Rasi with the Planet has a bigger number than that of the Rasi without Planet, the Shodhana should be done of the number of the Rasi without Planet and the number of the Rasi with Planet should be kept unchanged. If both the Rasi are without Planets and possess the same numbers, Shodhana of both the numbers should be done and the rectified numbers should be reduced to zero. If one Rasi is with Planet and the other is without any Planet, the number of the latter should be reduced to zero. Sun and Moon own one Rasi only; their numbers should be kept unchanged. After doing Ekadhipatya Shodana, Pinda Shodhana should be taken in hand.

Pinda Shodhana (Rectification) in the AshtakaVargaa: After completing the Trikona and Ekadhipatya Shodhana in the Ashtakavargaas of all the Planets, the rectified number should be multiplied by the measure of the Rasi. If there is any Planet in any Rasi, the rectified number should be multiplied by the measure of the Planet also. Then, after multiplying the rectified number of each Rasi, the products should be added up. The total so arrived at will be Pinda of that Planet. The multiples of Rasi are 10 for Taurus and Leo, 8 for Gemini and Scorpio, 7 for Aries and Libra, 6 for Capricorn and Virgo. The multipliers of the remaining Rasi are the same, as their numbers. (**Rasiman Chakra:** Aries 7, Vrishab 10, Mithun 8, Cancer 4,

Leo 10, Kanya 6, Libra 7, Vrischik 8, Sagittarius 9, Makar 5, Kumbh 11, Meena 12) The multipliers of Planets are 10 for Jupiter, 3 for Mars, 7 for Venus, 6 for Mercury, Sun, Moon and Saturn. (Planet man Chakra. Sun 5, Moon 5, Mars 8, Mercury 5, Jupiter 10, Venus 7, Saturn 5)

Calculation: The following procedure is adopted to ascertain the effects of a house. Multiply the number of Rekhas with the Yoga Pinda (Rasi Pinda plus Planet Pinda), connected with the AshtakaVargaa of that Planet and divide the product by 27. The remainder will denote the number of the Nakshatra. During the transit of Saturn in that Nakshatra the House concerned will be harmed.

Yoga Pinda: Yoga Pinda is the sum of Rasi Pinda and Planet Pinda.

Predictions of Ashtakavarga Effects:

General Effects: If a planet is associated with 8 Bindus (8 points) in the Sign occupied by it, a person, even an ordinary, will become a king; with 7 Bindus will have fulfilment of all desires; with 6 Bindus will have fame and financial gains; with 5 Bindus will have happiness and friendship; with 4 Bindus will have eradication and misery; with 3 Bindus will have loss of money; with 2 Bindus will have worries; with 1 Bindu will have bodily emaciation and with the complete absence of Bindus will have evils at all times.

Effects on Father: Sun is Pitra Karaka (significator of father). Therefore all about father is ascertained from the 9th house from the Sun's AshtakaVargaa as the 9th house from Sun at the time of birth deals with father. The Rekhas of that Rasi, as marked in Sun's AshtakaVargaa, is multiplied by the Yoga Pinda (Rasi Pinda plus Planet Pinda), and the product be divided by 27. The remainder will denote the number of Nakshatra. The father will be in distress, or he will otherwise suffer, when Saturn in transit passes through the Nakshatra. Even, when Saturn passes in transit the Trikona Nakshatras, father, or relatives, like father, may die, or suffer. If the AshtakaVargaa Rekha number is multiplied by the Yoga Pinda and the product is divided by 12, the remainder will denote the Rasi, through which, or through the Rasi in Trikona to it, the transit of Saturn will cause harm, or unfavourable effects to father. Death of the father may occur, if the Dasa prevailing at that time be unfavourable. If the Dasa be favourable, father will face only adverse effects. The death of the father may be expected, if Rahu, Saturn, or Mars are in the 4th from Sun at the time of transit of Saturn through any of the above three Rasi (Trikona Rasi). The death of the father will come to pass by such transit, if at that time Saturn, associated, or aspect by a malefic, be in the 9th from Lagna, or Moon and/or the Dasa of the Lord of the 4th from Lagna be in operation. The death does not take place, if a favourable Dasa be in force at the time of Saturn's transit. If the Rasi of Lagna of the native be the 8th Rasi from Lagna of the father, or, if the Lord of the 8th from fathers Lagna be in Lagna of the native, he takes over all the responsibilities of his father after the latter's death. The father enjoys happiness in the Dasa of the Lord of the 4th from Lagna. The native is obedient to his father, if the Lord of the 4th be in Lagna, or the 11th; or in the 11th, or 10th from Moon. If the birth be in the 3rd Rasi from Lagna, or Moon of the father, the native makes proper use of the wealth, inherited from his father. If the birth be in the 10th Rasi from Lagna, or Moon of the father, the native will inherit all the good qualities of his father. If the Lord of the 10th be in Lagna, the native will be more distinguished than his father.

Example: Suppose we want to time the good and bad periods of a native's father. Father is seen from Sun and the 9th house. We take Sun's BAV and find the number of Rekhas in the 9th house from Sun. Suppose Sun is in Aquarius its BAV contains 5 Rekhas in Libra (the 9th from Aquarius). Suppose Sun's Sodhya Pinda is 86. Multiplying 86 with 5, we get 430. If we divide 430 by 27, the quotient is 15 and the remainder is 25. The 25th Nakshatra is Poorva Bhadrapada. So Saturn's transit in Poorva Bhadrapada is bad for father and Jupiter's transit in the same nakshatra is good. Now we find the Rasi by dividing 430 by 12, which gives a quotient of 35 and a remainder of 10. So Saturn's transit in Capricorn (the 10th Rasi of Aries) is bad for father and Jupiter's transit in Capricorn is good timing for his father.

Effects on Mother: Consideration, regarding mother, house and village, should be done by the fourth House from the Moon's Ashtakavarga as the 4th house from Moon at the time of birth deals with mother. Therefore multiply the number of Rekhas in the AshtakaVargaa of Moon by the Yoga Pinda of that AshtakaVargaa and divide the product by 27. The death of, or distress to mother may be expected, when Saturn passes in transit through the Nakshatra, denoted by the remainder. Then divide the product by 12. The death of the mother may occur, when Saturn transits the Rasi, denoted by the remainder. Distress to mother may be predicted, when Saturn transits the Nakshatras, or Rasi in Trikona to Nakshatra and Rasi, indicated above.

Effects on brothers (co-born): Consideration of brothers (co-born), valour and patience is done by the third house from Mars' Ashtaka Varga as the 3rd house from Mars at the time of birth deals with brother. If the number of Rekha is larger in any Rasi after Trikona Shodhana, there will be gains of land, happiness from wife and great happiness to brother, when Mars passes through that Rasi in transit. If Mars be weak, the brothers will be short lived. There will be distress to brothers, when Mars transits a Rasi without Rekhas. Here also the Yoga Pinda of Mars should be multiplied by the number of Rekhas in the Ashtakavargas of Mars and the product be divided separately by 27 and the remainders will denote the Nakshatra and Rasi. The brother will suffer, whenever Saturn transits that Nakshatra, or Rasi (or the Trikona Nakshatras, or Rasi).

Family, maternal uncle and friends: Consideration in regard to family, maternal uncle and friends should be done by the 4th House from Mercury's Ashtakavarga as the 4th house from Mercury at the time of birth deals with family, maternal uncle and friends. The family will enjoy happiness during the transit of Mercury's Ashtaka Vargas. After performing Trikona and Ekadhipatya Shodhana in Mercury's Ashtaka Varga, the happiness, or distress of the family should be predicted from the transit of Saturn through the resultant Nakshatra and Rasi (and those in Trikona to them).

Knowledge, son (progeny), religious inclinations: All about knowledge, religious inclinations of the native and son (progeny) is to be ascertained by the 5th House from Jupiter. If the Rekhas in the 5th House from Jupiter's Ashtaka Varga are larger in number in the Ashtaka Varga, there will be great happiness in respect of progeny. If the dots are larger in number, the happiness in respect of progeny will be meagre. The number of children are equal to the number of Rekhas in the 5th House (from Jupiter), provided it is not the Rasi of debilitation of Jupiter, or his enemies Rasi. In that case the number of children will be very limited. The number of children is also equal to the number of Navamsa, in which the Lord of the 5th from Jupiter is posited. Multiply the Yoga Pinda of Jupiter by the number of Rekhas in the AshtakaVargaa and divide the product separately by 27 and 12. The remainders will denote the Nakshatra and Ras. Transit of Saturn through that Nakshatra and its Trikona Nakshatras and of that Rasi and its Trikona Rasi will be inauspicious. During that period the knowledge, learning and religious activities of the native will also be adversely affected.

Wealth, land, happiness and marriage: Consideration of wealth, land, happiness and marriage is done by the 7th house from Venus Ashtakavarga as the 7th house from Venus at the time of birth deals with marriage. The effects should be judged in the manner, already explained earlier, after multiplying the Rekhas in the 7th House from Venus by the Yoga Pinda. There will be gain of wealth, land and happiness and marriage, whenever Venus passes in transit through the Rasi, which have larger number of Rekhas in the Ashtakavargaa of Venus. These gains will be from the directions of the 7th Rasi from Venus and of its Trikona Rasi.

Effects of the Sun's A.V: Note the Bindus in the Sign occupied by the Sun. Particular number of Bindus (auspicious points) in the Sign occupied by him will lead to respective effects, such as, with 8 Bindus wealth from the king; with 7 Bindus transcendental beauty, happiness and riches; with 6 Bindus increase of valour and fame; with 5 Bindus influx of money; with 4 Bindus neither gainful nor losing; with 3 Bindus fatigue due to (frequent) travels (or difficulties on account of travels, proneness to accidents and the like); with 2 Bindus fear of diseases; with 1 Bindu difficulties; and with absence of Bindus death or troubles of a very severe nature.

Effects of the Moon's A.V: Note the Bindus in the Sign occupied by the Moon. Particular number of Bindus (auspicious points) in the Sign occupied by it will lead to respective effects, such as, with 8 Bindus, the native will enjoy affluence and pleasures; 7 Bindus financial gains through robes, food, scented articles etc., 6 Bindus acquaintance with the virtuous; 5 Bindus acquisition of courage and intelligence due to association with Brahmins (or the learned); 4 Bindus equilibrium of happiness and grief; 3 Bindus enmity with relatives; 2 Bindus separation from kith and kin and deprival of wealth; 1 Bindu evils and absence of auspicious points grief and difficulties following excitement.

Effects of the Mars's A.V: Mars be associated with various benefic points in the Sign tenanted by him. Particular number of Bindus (auspicious points) in the Sign occupied by it will lead to respective effects, such as, with 8 Bindus acquisition of wealth and land and victory over foes; 7 Bindus increase of fortunes and splendour; 6 Bindus royal favours; 5 Bindus increase of fame; 4 Bindus equality of wealth and calamity; 3 Bindus separation from co-born and conjugal partner; 2 Bindus troubles from king, fire and bilious disorders; 1 Bindu ulcer and stomach diseases and absence of Bindus diseases of the eye and difficulties comparable to death.

Effects of the Mercury's A.V: The effects of Mercury's association with various auspicious points in the Sign occupied by him will be as given. Particular number of Bindus (auspicious points) in the Sign occupied by it will lead to respective effects, such as, with 8 Bindus honour from the ruler; 7 Bindus wealth, knowledge and happiness; 6 Bindus success in all undertakings; 5 Bindus new acquaintances; 4 Bindus unemployment; 3 Bindus mental worries on account of financial losses; 2 Bindus enmity with wife, children and friends causing loss of courage and wisdom; 1 Bindu evils of all kinds and absence of Bindus will bring about death.

Effects of the Jupiter's A.V: Based on the number of auspicious points, the Sign occupied by Jupiter will produce the following effects. Particular number of Bindus (auspicious points) in the Sign occupied by it will lead to respective effects, such as, with 8 Bindus spotless fame, happiness and growth of wealth; 7 Bindus fortunes and happiness; 6 Bindus acquisition of robes, conveyances, gold etc.; 5 Bindus destruction of enemies and success in undertakings; 4 Bindus no loss, no gain; 3 Bindus loss of hearing, sight and masculine vigour; 2 Bindus incurring royal wrath; 1 Bindu distress due to diseases and absence of Bindus destruction of relatives, wealth and progeny.

Effects of the Venus's A.V: The association of Venus with various number of Bindus will generate the under mentioned effects. Particular number of Bindus (auspicious points) in the Sign occupied by it will lead to respective effects, such as, with 8 Bindus enjoyments of all kinds and acquisition of robes, wife, scents, food, drinks etc.; 7 Bindus plentiful of ornaments and pearls; 6 Bindus acquisition of a girl of choice; 5 Bindus association with friends; 4 Bindus equality of inauspicious and auspicious results; 3 Bindus enmity with people of his community and senior members of the village; 2 Bindus displacement; 1 Bindu phlegmatic disorders and absence of Bindus becoming a synonym of all kinds of evils.

Effects of the Saturn's A.V: If Saturn is associated with a certain number of Bindus, the effects will be as given. Particular number of Bindus (auspicious points) in the Sign occupied by it will lead to respective effects, such as, with 8 Bindus lordship over villages, towns and people; 7 Bindus acquisition of female servants, asses and elephants; 6 Bindus gains from thieves, hunters and Army heads; 5 Bindus advent of wealth, corn and happiness; 4 Bindus happiness caused by association with others; 3 Bindus destruction of progeny, wife, attendants and money; 2 Bindus imprisonment, emotions and diseases; 1 Bindu/absence of Bindus loss of wealth, family.

Timing with Ashtakvarga, Sodhya Pinda: Parasara taught some techniques of finding of timing of events based on Sodhya Pindas. When we want to time particular events, first fix the relevant planet related to it and then the relevant house. Then we find the number of Rekhas in that house from that planet in that planet's Binnashtaka Varga (BAV). Multiply it by the Sodhya Pinda (also called Yoga Pinda) of the planet. By dividing the product with 27 or 12 and taking the remainder, we find the associated Nakshatra or Rasi counting it from Aswini Nakshatra or Aries. Then we can time key events based on the transits in that Nakshatra and Rasi. Based on the Benefic and malefic transiting in this Particular Nakshatra, it will give good or bad results (respectively) relating to the original house of the event. Especially Saturn's transit is important. Saturn's transit in the Nakshatra corresponding to a particular house makes the signified matters suffer. Jupiter's transit is beneficial. Some people also take the 10th or 19th nakshatra from the Nakshtra found above. That Nakshatra, the 10th and 19th Nakshatras from it are owned by the same planet under Vimsottari dasa scheme and they go together. We can also find a Rasi/House by dividing the product with 12 instead of 27 and counting Rasi from Aries. Saturn's transit in the resultant Rasi brings misfortune relating to the original house. However, predictions by Nakshatras are more important.

Marriage: Auspicious functions like marriage is performed in the month of Rasi (when Sun transits that Rasi), which has more number of dots in Sun's AshtakaVargaa. The same applies to the Samvatsar of that Rasi (when the mean Jupiter transits that Rasi). Auspicious functions should be performed, when Sun or the Jupiter transits the Rasi, which has more Rekhas in Sun's AshtakaVargaa. Auspicious functions shall not be performed during the transit of Moon in the Rasi, which has larger number of dots in Moon's AshtakaVargaa.

Longevity by Ashtakavarga: The 8th House from Saturn signifies death, as well as longevity. Assessment about longevity should be made by the 8th House from Saturn's Ashtakavarga as the 8th house from Saturn at the time of birth deals with Death and Longevity of Native. Multiply the Yoga Pinda (sum of Rasi Pinda and Planet Pinda) by the number of Rekhas in the Ashtakavargaa for Saturn and divide the product by 27. The death of the native will take place, when Saturn passes in transit through the Nakshatra, denoted by the remainder, or it's Trikona Nakshatras. Again, divide the product by 12. The native will face danger of death, when Saturn passes in transit through the Rasi denoted by the remainder, or through its Trikona Rasi. The results will be favourable, when Saturn passes in transit through Rasi, which have larger number of Rekhas in Saturn's AshtakaVargaa. Saturn's transit through Rasi, which have larger number of dots, will produce only evil effects.

13.9 Predictions by Karakans

Karakans Definition: Karakans is the Navamsa Sign occupied by the Atma Karaka by Jaimini Astrology.

Predictions by Karakans in Various Rashi:

Mesh Karakans: There will be nuisance from rats and cats at all times. A malefic joining to Karakans will further increase the nuisance.

Vrishabh Karakans: Happiness from quadrupeds will result.

Mithuna Karakans: He/She will be afflicted by itch.

Karka Karakans: There will be fear from water.

Simha Karakans: Fear will be from tiger.

Kanya Karakans: Itch, corpulence, fire will cause trouble.

Tula Karakans: He/She will make one a trader and skilful in making robes.
Vrischika Karakans: It will bring troubles from snakes and also affliction to mother's breasts.
Dhanu Karakans: There will be falls from height and conveyances
Makara Karakans: It denotes gains from water dwelling beings and conch, pearl, coral.
Kumbha Karakans: He/She will construct tanks.
Meena Karakans: It will grant final emancipation. The Drishti of a benefic in Karkans will remove evils, while that of a malefic will cause bad.

Predictions by Benefic or Malefic in the Karakans:

If there be only benefic in Karakans and the Navamsa of Lagna receives a Drishti from a benefic, he/she will undoubtedly become a king. If the Kendra/Kona from the Karakans be occupied by benefic or is devoid of malefic association, he/she will be endowed with wealth and learning. If the Upketu (Upagraha) is in its exaltation, or own, or friendly Rashi and is devoid of a Drishti from a malefic, he/she will go to heaven after death. If the Atma Karaka is in the divisions of Chandra, Mangal, or Sukra, he/she will go to others wives, otherwise will not go.
Predictions by Graha in the Karakans: If Surya is in the Karakans, he/she will be engaged in royal assignments. If the full Chandra is there, he/she will enjoy pleasures and be a scholar, more so, if Sukra gives a Drishti to the Karakans. If strong Mangal is in Karakans, he/she will use the weapon spear, will live through fire and be an alchemist. If strong Buddha be in Karakans, he/she will be skilful in arts and trading, be intelligent and educated. Guru in Karakans denotes him/her, doing good acts, endowed with spiritualism and Vedic learning. If Sukra is in Karakans, he/she will be endowed with longevity of 100 years, be sensuous and will look after state affairs. If Sani in Karakans, it will give such livelihood, as due to the natives family. Rahu in Karakans denotes a thief, a bowman, a machinery maker and a doctor, treating poisonous

afflictions. If Ketu be in Karakans, he/she will deal in elephants and be a thief.

Predictions by Rahu and Surya in Karakans: If Rahu and Surya be in Karakans, there will be fear from snakes. If a benefic gives a Drishti to Rahu-Surya in Karakans, there will be no fear, but a malefic Drishti will bring death (through serpents). If Rahu and Surya occupy benefic Shad Vargas, being in Karakans, he/she will be a doctor, treating poisonous afflictions, while the Drishti from Mangal on Rahu-Surya in Karakans denotes that the native will burn either his own house, or that of others. Buddha Drishti on Rahu-Surya in Karakans will not give the burning of his/her own house, but that of others. If Rahu and Surya happen to be in Karakans and are in a malefic Rashi, receiving a Drishti from Guru, he/she will burn a house in one's neighbourhood, while the Drishti of Sukra will not cause such an event.

Predictions by Gulika in Karakans: Should the full Chandra give a Drishti to Gulika, placed in the Karakans, he/she will lose his wealth to thieves, or will himself be a thief. If Gulika is in Karakans, but does not receive a Drishti from others, he/she will administer poison to others, or he will die of poisoning. Buddha Drishti in this context will give large testicles.

Predictions by Ketu in Karakans: If Ketu is in Karakans, receiving a Drishti from malefic, his/her ears will be severed, or one will suffer from diseases of the ears. Sukra, giving a Drishti to Ketu in Karakans, denotes him/her, initiated into religious order. He/she will be devoid of strength, if Buddha and Sani give a Drishti to Ketu in Karakans. If Buddha and Sukra give a Drishti to Ketu in Karakans, he/she will be the son of a female slave, or of a female remarried. With Sani Drishti on Ketu in Karakans one will perform penance, or is a servant, or will be a pseudo-ascetic. Sukra and Surya together, giving a Drishti to Ketu in Karakans, will make him/her serve the king.

Predictions by Planet in Houses from Karakans

2nd from Karakans: If the 2nd from Karakans falls in the divisions of Sukra, or Mangal, he/she will be addicted to others' wives and, if Sukra, or Mangal give a Drishti to the 2nd from Karakans, the tendency will last till death. If Ketu is the 2nd from Karakans in a division of Sukra, or Mangal, addiction to other's wives will not prevail, while the position of Guru will cause such an evil. Rahu in the 2nd from Karakans will destroy wealth.

3rd from Karakans: A malefic in the 3rd from Karakans will make him/he valorous, while a benefic in the 3rd from Karakans will make him/he timid.

4th from Karakans: If the 4th from Karakans happens to be occupied by Sukra and Chandra, he/she will own large buildings, like palaces etc. Similar is the effect of an exalted Graha in the said 4th. A house, made of stones, is denoted by the occupation of the 4th from Karakans by Rahu and Sani. Mangal and Ketu in the 4th from Karakans indicate a house, made of bricks, while Guru in the 4th from Karakans denotes a house, made of wood. Surya in the 4th from Karakans will give a house of grass. If Chandra is in the 4th from Karakans, he/she will have union with his wife in house without compound.

5th from Karakans: If Rahu and Mangal are in the 5th from Karakans, he/she will suffer from a pulmonary consumption, if Rahu and Gulika, he/she will fear from mean people and poison, if Buddha, he/she will be an ascetic of the highest order. If Surya, he/she will be using a knife. If Mangal, using a spear. Sani denotes a bowman. Rahu denotes a machinist. Ketu denotes a watch maker. Sukra in the 5th from Karakans will make him/he a poet and an eloquent speaker. Sukra denotes, he/she will be eloquent and a poet. Guru denotes, he/she is an exponent and be all knowing, but be unable to speak in an assembly. He/She will be further a grammarian and a scholar in Vedas and Upanishads. Sani will make one ineffective in an assembly. Buddha will make him skilful in Karma Mimansa. Mangal in Karakans or the 5th will make him/her justice. Chandra in Karakans, or the 5th denotes a Sankhya Yogi, a rhetoric, or a singer. Surya will make him/he learned in Vedanta and music. Ketu will make one a mathematician and skilful in Jyotish. If Guru be related to the said Ketu, these learning will be by inheritance. If Guru and Chandra are in Karakans, or the 5th thereof, he/she will be an author. Sukra will make one an ordinary writer, while Buddha indicates the writing skills is less than those of an ordinary writer. Guru is alone; he/she will be a writer and be versed in Vedas and Vedanta. Mangal denotes a logician. Buddha denotes a Mimamsaka. Sani indicates that he/she is dull-witted in the assembly. Surya denotes that he/she is a musician. Chandra denotes a follower of Sankhya philosophy and indicates, that he/she is versed in rhetoric and singing. Ketu, or Rahu denotes, that one is a Jyotish.

6th from Karakans: If the 6th from Karakans is occupied by a malefic, he/she will be an agriculturist, while he will be indolent, if a benefic is in the 6th from Karakans. The 3rd from Karakans should also be similarly considered.

7th from Karakans: If Chandra and Guru are in the 7th from Karakans, he/she will beget a very beautiful wife. Sukra in the 7th form Karakans denotes a sensuous wife. Buddha in the 7th from Karakans indicates a wife, versed in arts. Surya in the 7th from Karakans will give a wife, who will be confining domestic core. Sani in the 7th from Karakans denotes a wife of a higher age bracket, or a pious and / or sick wife. Rahu in the 7th from Karakans will bring a widow in marriage.

8th from Karakans: If a benefic, or the Graha, owning the 8th from Karakans, happens to be in the 8th from Karakans, he/she will be long-lived, while a malefic, placed in the 8th from Karakans, will reduce the life span. Drishti/Yuti of both benefic and malefic will yield a medium span of life.

9th from Karakans: If the 9th from Karakans receives a Drishti from, or is occupied by a benefic, he/she will be truthful, devoted to elders and attached to his own religion. If a malefic gives a Drishti to, or occupies the 9th from Karakans, he/she will be attached to his religion in boyhood, but will take to falsehood in old age. If Sani and Rahu give a Drishti to, or occupy the 9th from Karakans, he/she will betray his elders and be averse to ancient learning. If Guru and Surya give a Drishti to, or occupy the 9th from Karakans, he/she will betray his elders and will be disobedient to them. Should Mangal and Sukra give a Drishti to, or occupy the 9th from Karakans and are joining in six identical Vargas, a female, ill-related to the native, will die. Buddha and Chandra giving a Drishti to, or occupying the 9th from Karakans and joining in six identical Vargas will cause imprisonment of him/her, due to association with a female not of his own. If Guru is alone, related to the 9th from Karakans by Drishti, or by Yuti, he/she will be addicted to females and be devoted to sensual enjoyments.

10th from Karakans: If the 10th from Karakans receives a Drishti from, or is conjoined by a benefic, gives a Drishti to, or occupies the 9th from Karakans, will have firm riches; be sagacious, strong and intelligent. A malefic, giving a Drishti to the 10th from Karakans, or occupying this Bhava, will cause harm to his profession and deprive him of paternal bliss. Buddha and Sukra, giving a Drishti to the 10th from Karakans, or conjoining this Bhava, will confer many gains in business and will make him do many great deeds. Surya and Chandra, giving a Drishti to the 10th from Karakans, or conjoining this place and receiving a Drishti from, or be in Yuti with Guru, gives a Drishti to, or occupies the 9th from Karakans, will acquire a kingdom.

11th from Karakans: If the 11th from Karakans receives a Drishti from, or is yuti with a benefic, gives a Drishti to, or occupies the 9th from Karakans, will enjoy happiness from co-born apart from gaining in every undertaking of his. If a malefic is in the 11th from Karakans, gives a Drishti to, or occupies the 9th from Karakans, will gain by questionable means, be famous and valorous.

12th from Karakans: If the 12th from Karakans has a benefic, the expenses will be on good account, while a malefic in the 12th from Karakans will cause bad expenses. If the 12th from Karakans is vacant, then also good effects will follow. If there happens to be a benefic Graha in exaltation, or in own Bhava in the 12th from Karakans, or, if Ketu is so placed and receives a Drishti from, or is yuti with a benefic, he/she will attain heaven after death. He/she will attain full enlightenment, if Ketu is in the 12th identical with Mesh or Dhanu and receives a Drishti from a benefic. If Ketu is in the 12th from Karakans and receive a Drishti from a malefic, or there is yuti with a malefic, he/she will not attain full enlightenment. If Surya and Ketu are in the 12th from Karakans, he/she will worship Lord Shiva. Chandra and Ketu denote a worshiper of Gauri. Sukra and Ketu denote of Lakshmi and a wealthy person. Mangal and Ketu denote of Lord Subramanya. Rahu will make him/her to worship Durga, or some mean deity. Ketu alone denotes Subramanya's, or Ganesh's worshipper. If Sani is in the 12th from Karakans in a malefic Rashi, he/she will worship mean deities. Sukra and Sani in the 12th from Karakans in a malefic Rashi will also make him/her worship mean deities. Similar inferences can be drawn from the 6th Navamsa, counted from Amatya Karaka Navamsa.

Miscellaneous Other Effects of Karakans:

If there are two malefic in a Kona from Karakans, he/she will have knowledge of Mantras and Tantras (formulas for the attainment of super-human powers). If a malefic simultaneously gives a Drishti to two malefic in a Kona from Karakans, he/she will use his learning of Mantras and Tantras for malevolent purposes, while a Benefice's Drishti will make him/her use the learning for public good. If Chandra is in the Karakans, receiving a Drishti from Sukra, he/she will be an alchemist and, if receiving a Drishti from Buddha he/she will be a doctor capable of curing all diseases. If Chandra is in the 4th from Karakans and receives a Drishti from Sukra, he/she will be afflicted by white leprosy. If receiving a Drishti from Mangal, he/she will have blood and bilious disorders and, if receiving a Drishti from Ketu, he/she will suffer from black leprosy. If Rahu and Mangal be in the 4th, or 5th from Karakans, he/she will suffer from pulmonary consumption and, if simultaneously there happens to be Chandra Drishti on the 4th, or the 5th, this affliction will be certain. Mangal alone in the 4th or the 5th will cause ulcers. If Ketu is in the 4th, or the 5th, he/she will suffer from dysentery and afflictions, due to (impure) water. Rahu and Gulika will make one a doctor, curing poisonous afflictions, or will cause troubles through poison. If Sani be alone in the 4th, or 5th, he/she will be skilful in archery. Ketu lonely placed in the 4th, or the 5th will make one a maker of watches etc. Buddha lonely placed in the 4th, or the 5th will make him/her an ascetic of the highest order, or an ascetic, holding staff. Rahu, Surya and Mangal, respectively, in these places denote a machinist, a knife user and a spear, or arrow user. Chandra and Guru in the Karakans or in the 5th there from denote a writer well versed in all branches of learning. The grade of writer ship will comparatively descend in the case of Sukra and even further in the case of Buddha. If Ketu be in the 2nd, or 3rd from Karakans, he/she will be defective in speech, more so, if a malefic gives a Drishti to Ketu, as above. If malefic be in Karakans, Arudha Lagna and the 2nd and 8th from these places, there will be Kemadrum Yoga, the effects of which will be still severer, if Chandra's Drishti happens to be there. The effects, due for these Yoga, will come to pass in the Dasha periods of the Rashi, or Graha concerned. Kemadrum Yoga will operate additionally, if there are malefic in the 2nd and 8th from the Rashi, whose Dasha will be in currency. The results of such Yoga will also be inauspicious. If the 2nd and 8th in the Kundali, cast for the beginning of a Dasha, have malefic, then also Kemadrum prevails throughout the Dasha.

13.10 Predictions by Argala:

Support provided by one house to another is called Argala and the obstruction offered to supporting houses is called Virodha Argala. Planets in 2nd, 4th and 11th house cause Argala on a given house, whereas the planets in 12th, 10th and 3rd cause virodha Argala to 2nd, 4th and 11th respectively. Benefic generally give Subha Argala, malefic offer papa Argala. If however a malefic has an Argala on house of which it is a significator, such an Argala can be termed as Subha. For example a malefic in 10th house cast papa Argala to 9th house as 10th house is second from 9th. This may make him/her non religious and give bad relations with boss/teacher, provided there is no virodha Argala from 8th.

Argala is located to know the definite effects of Bhava and Graha. Argala is caused by the Graha in the 2nd, 4th, and 11th from a Bhava counted in opposite direction, i.e. Clockwise direction in North Indian Chart and anticlockwise in South Indian Chart. The concept of Argala can be applied to planets as well. Similarly, the 5th and the 8th Houses have Secondary Argala. Secondary Argala exerts less conspicuous than those of the Primary Argala. These Argala are the obstructions by planets and signs in the same place from the original sign or planet under consideration. Thus, the Argala caused on Lagna by the 2nd House is the obstructions by the 12th House (Signs or Planets). Similarly, the Argala caused by the 4th house on Lagna is obstructions by the 10th House. The Argala caused by natural malefic causes malefic intervention and that caused by natural benefic is benefic intervention. If the Argala causing planet is stronger than the caused one, the former will prevail. Or, if the number of Argala causing Graha is more than the caused Graha, then also the former will prevail. As Rahu and Ketu have retrograde motions, the Argala and obstructions are also counted accordingly in a reverse manner. Maharashi say, that the Argala, caused by one Graha, will yield limited effect, by two medium and by more than two, excellent effects. The Argala effects will be derived in the Dasha periods of the Rashi, or Graha concerned. Argala can be by a malefic to a benefic, so that the native does not enjoy good effects, due to the benefic. This is Pap (malefic) Argala.

Predictions: If there be Argala for the Arudha Pad, he/she will be famous and fortunate. A malefic, or a benefic, causing unobstructed Argala, giving a Drishti to Lagna will make him/her famous. Similarly a malefic, or a benefic, causing unobstructed Argala, giving a Drishti to Dhan Bhava denotes acquisition of wealth and grains, to Sahaj Bhava happiness from co-born, to Bandhu Bhava residences, quadrupeds and relatives, to Putra Bhava sons, grandsons and intelligence, to Ari Bhava fear from enemies, to Yuvati Bhava abundant wealth and marital happiness, to Randhra Bhava difficulties, to Dharma Bhava fortunes, to Karma Bhava royal honour, to Labh Bhava gains and to Vyaya Bhava expenses. The Argala by benefic will give various kinds of happiness, while benefic effects will be meddling with malefic Argala. Argala by both benefic and malefic will yield results. Should there be (unobstructed) Argala for Lagna, Putra and Dharma Bhava, he/she will doubtlessly become a king and fortunate.

Example: In a chart the Lagna and the 7th house are afflicted by Exalted Nodes. Jupiter is the Retrograde Lagna Lord in Pisces in the 4th House causing Hamsa Mahapurusha Yoga and Buddha is the Retrograde. 7th and 10th Lord in the 10th House causes Bhadra Mahapurusha Yoga. Buddha is the Badhakasthana but its exaltation in the 10th House is a very good placement. The Moon is exalted in a Vipreeta Raja Yoga in the 6th House from the Lagna as the 8th Lord of the chart. Mars, the 5th Lord of the chart is in the 11th House is placed in the 11th House. The Sadesati has been going on for the chart, but the terrible financial and other troubles are because of the Retrograde Saturn in the 2nd House. Saturn has unobstructed malefic Argala on the Lagna. As the lord of the 2nd in the 2nd House, Saturn, empowered as a retrograde malefic gives grave and controlled speech, especially since Mercury, the significator of speech, is exalted. The main point is that the malefic Argala of Saturn causes paucity of money and makes the personality emaciated and otherwise Saturnine. There is also a malefic Argala on the 4th House of family life, since

Saturn in the 2nd is in the 11th House from the 4th. This is obstructed by the benefic Argala of the exalted Moon. The native may want to psychologically transform the domestic environment and would probably succeed since the Moon is more powerful than Saturn in exaltation. But Saturn would also not be defeated easily due to its retrogression in Capricorn. The wisdom and optimism of Guru wants to assert itself through the Argala on the Lagna from the 4th House but the trade minded and fickle Mercury corrupts the purity of Jupiter by obstructing the Argala. Mars has an unobstructed malefic Argala on the Lagna, making for rage and violence in the personality. Thus Argala explains some surprising features of life that may not seem explicable merely by an analysis of some of the best Yoga in charts.

14

Planets Transit (Gochara)

14.1 Planets' Description

There are nine Graha, namely, Sun, Moon, Mars, Mercury, Jupiter, Venus, Saturn Rahu and Ketu. But after further studies, the Uranus, Neptune and Pluto have been added to the astrology to make it more fascinating subject. Moon, Mercury, Jupiter and Venus are benefic by nature and others are malefic.

Introduction of Grahas:

Surya: Surya's eyes are honey-coloured. He has a square body. He is of clean habits, bilious, intelligent and has limited hair (on his head).

Chandra: Chandra is very windy and phlegmatic. She is learned and has a round body. She has auspicious looks and sweet speech, is fickle-minded and very lustful.

Mangal: Mangal has blood-red eyes, is fickle-minded, liberal, and bilious, given to anger and has thin waist and thin physique.

Buddha: Buddha is endowed with an attractive physique and the capacity to use words with many meanings. He is fond of jokes. He has a mix of all the three humours.

Guru: Guru has a big body, tawny hair and tawny eyes, is phlegmatic, intelligent and learned in Shashtra.

Sukra: Sukra is charming, has a splendours physique, is excellent or great in disposition, has charming eyes, is a poet, is phlegmatic and windy and has curly hair.

Sani: Sani has an emaciated and long physique, has tawny eyes, is windy in temperament, has big teeth, is indolent and lame and has coarse hair.

Rahu (The Dragon Head/ North Node of Moon): The Moon's orbit and the earth's orbit intersect and these two intersecting

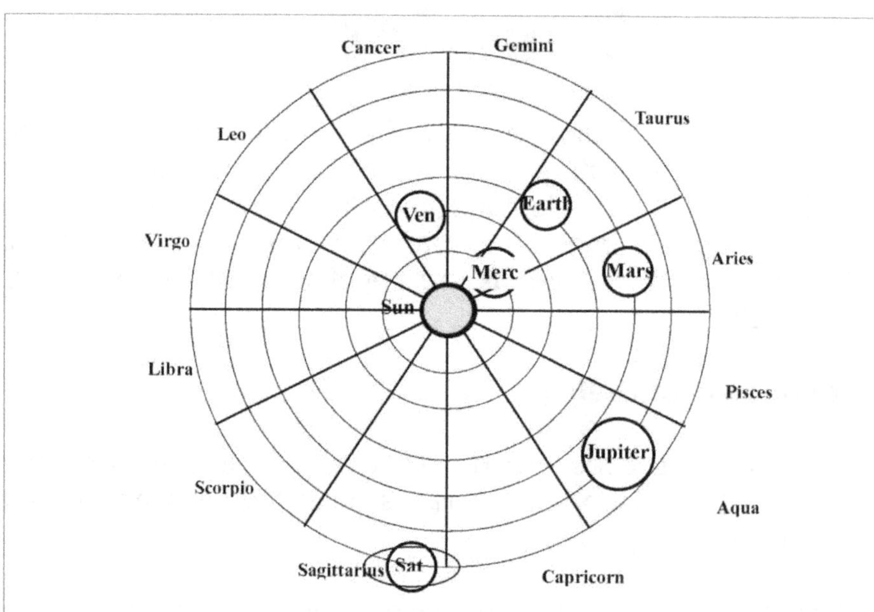

Fig 1: Actual Planet Movement in Zodiac

Points are known as North Node (Rahu) and the South Node (Ketu). These orbits differ by 8 degrees. They are mathematical points which influence human behaviour. Hence great importance has been assigned to these Nodes in Vedic Astrology. Rahu is also known as the dead body of the lusty demon, killed by Vishnu. So, Rahu is the head of the dragon and symbolizes the trouble. Rahu has smoky appearance with a blue mix physique. He resides in forests and is horrible. He is windy in temperament and is intelligent.

Ketu (Dragons Tail/South Node of Moon): Ketu is the dead body of the lusty demon, killed by Vishnu. So, Ketu is the tail of the dragon and losses and symbolizes the death along with Saturn. Ketu has smoky appearance with a blue mix physique. He resides in forests and is horrible. He is windy in temperament and is intelligent. Ketu is akin to Rahu.

Note: Each planet has been allotted some points, such as Sun - 48; Moon - 49; Mars - 39; Mercury - 54; Jupiter - 56; Venus - 52; Saturn - 39. Thus it is totalling 337 points all together.

Table 1: Astronomical Detail of Planets

Planet	Mean Distance from sun in '000 km	Sidereal (Orbiting) Period
Earth	14,94,56.180	365.25 days
Mars	22,77,21.610	687 days
Mercury	579,36.240	88 days
Jupiter	77,77,94.020	11.86 year
Venus	10,78,25.780	224.7 days
Saturn	142,60,36.100	29.46 year
Uranus	286,9453.000	84.01 year
Neptune	449,48,86.600	164.43 year
Pluto	589,98,40.4000	248.43 year

Table 2: Astronomical Detail of Planets

Planets	Axial Rotation Period	Equatorial Diameter in 000 km	Max. Surface Temperature (^{0}F)
Earth	23 hr 56 min.	12.756	(+) 140^0
Mars	24 hr 37 min.	6.759	(+) 85^0
Mercury	88 day	4.667	(+) 770^0
Jupiter	9 hr 51 min.	142.748	(-) 200^0
Venus	N. A	12.392	(+) 880^0
Saturn	10 hr 14 min.	120.861	(-) 240^0

Uranus	10 hr 48 min.	47.153	(-) 310^0
Neptune	14 hr	44.579	(-) 360^0
Pluto	6 days 9 hr	5.794	Not Available

4.2 Apparent Motion of Planet

Planets in Apparent Motion (Graha's Transit or Bhraman): Transit refers to the movement or revolution of planets across the Sun. Transit Chart is the position of the Planets at a particular time and place while orbiting the Sun because of their daily motion. The transits Chart show the effect of planets, the ups and the downs brought about by circumstances in our life that we face on daily or monthly basis. The planet's power increases when it is retrograde or stationary. The following table represents the average daily angular motion of the planets and the approximate time they spend in one sign and they need to complete a full circle of the Sun. The symbols in the brackets indicate the direction of their movement. In Primary Directions the apparent motion of the planets and the House-cusps is clockwise, resulting from the counter-clockwise motion of the Earth's periphery. Direct The true motion of the planets in the order of the Signs is counter-clockwise within the Zodiac.

Table 3: Astronomical Detail of Planets

Planet	**Average Daily Motion**	**Time spent in a Sign/House**	**Cycle Duration to cover 12 Signs**
Sun (D)	$0^0 59'$	30 days	1 year
Moon (D)	$13^0 10'$	2 1/4 days	1 month
Mars (Θ)	$0^0 31'$	1 1/2 months	18 months
Mercury(Θ)	$4^0 5'$	27 days	1 year
Jupiter (Θ)	$0^0 5'$	1 year	12 years
Venus (Θ)	$1^0 36'$	28 days	1 year
Saturn (Θ)	$0^0 2'$	2 1/2 years	30 years

Rahu (R)	0^0 3'	1 1/2 years	18 years
Ketu (R)	0^0 3'	1 1/2 years	18 years

Legend: D = Always Direct Motion; R = Always Retrograde Motion & Θ = Mostly Direct Motion but sometimes Retrograde Motion planets.

Motion of the Planets: There are eight kinds of motions to planets from Mangal to Sani. These are Vakra (Retrograde), Anuvakra (entering the previous Rashi in retrograde motion), Vikal (Stationary, i.e. fixed or devoid of motion,), Mand (slower motion than usual), Mandatar (slower than the previous motion), Sama (increasing in motion), Char (faster than Sama motion) and Atichar (entering next Rashi in Accelerated motion). The strengths, allotted due to such 8 motions are 60, 30, 15, 30, 15, 7.5, 45 and 30 respectively.

Daily Forecast Analysis (Moon Transit): Moon rotates once every 2.25 days. This makes daily forecast analysis based on the current positions of the Moon in the sky and how he aspects on other planets, i. e. the angles between the Transit Moon and Planet and the Signs in Natal Chart. For example, when, the transit Moon makes a Trine (a 120 degree angle) to the Moon in the native Natal Chart, he makes intuition strong. He/She identifies and act upon his/her gut instincts about things easily.

Table 4: Astronomical Details of Planets

Planet	**Angular Distance from Sun**		**No. Of Days Stationary**	**No. of Days Retrograde**
Mars	228^0	132^0	3	80
Mercury	$14\text{-}20^0$	$17\text{-}20^0$	1	24
Jupiter	245^0	115^0	5	120
Venus	29^0	26^0	2	42
Saturn	251^0	109^0	5	140

Weekly, Monthly & Yearly Transit & Forecast Analysis: The weekly and Monthly forecast analysis is based on the current positions of the planet in the sky and how they aspects between them, i. e. the angles between the Transit planet and Planet in Natal Chart and the Signs in that weak of predictions. Like that the different aspects of individual life are highlighted each week to get the most advantageous and beneficial results. This is how a weekly or monthly horoscope is forecast.

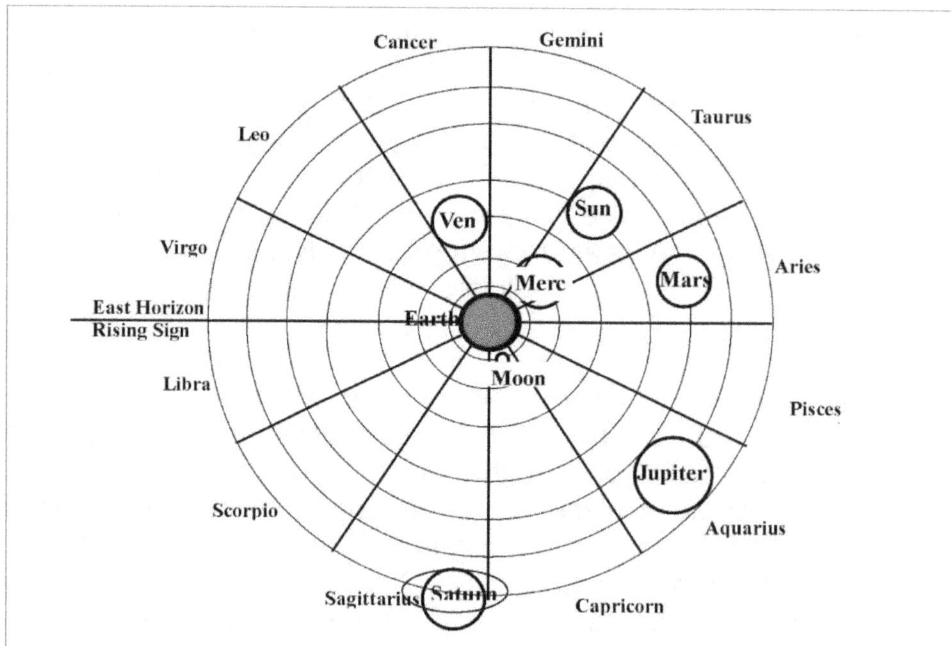

Figure 2: Imaginary Movement of Planets as viewed from the Earth

(i) Stationary Motion (Vikal): Some times, planets apparently seem to slow down or stand still or proceed backwards and then again stop, turn around and go forwards again. There is a point where for a short time, to its cresting on the edge of its orbit in relation to us; it appears to stand still or stationary, when we judge by looking at the sign (Aquarius) behind it in Zodiac. This is called 'Stationary Motion' (see Fgure-3). Except the Sun and the Moon, the other planets, such as, Mars, Mercury, Jupiter, Venus and Saturn change their proper motion in the Zodiac periodically and appear to move backwards for short period and are called "Retrograde" and after some time they resume their direct motion. Most of the time, they move in the "Direct" way, but at some times they fall into a Retrograde cycle. Before retrogression occurs, the planet gets stationary for a certain period of time. The same thing occurs when the planet ends its retrograde cycle. Sun and moon will never retrograde. The planet gives a very strong and steady effect while it is stationary.

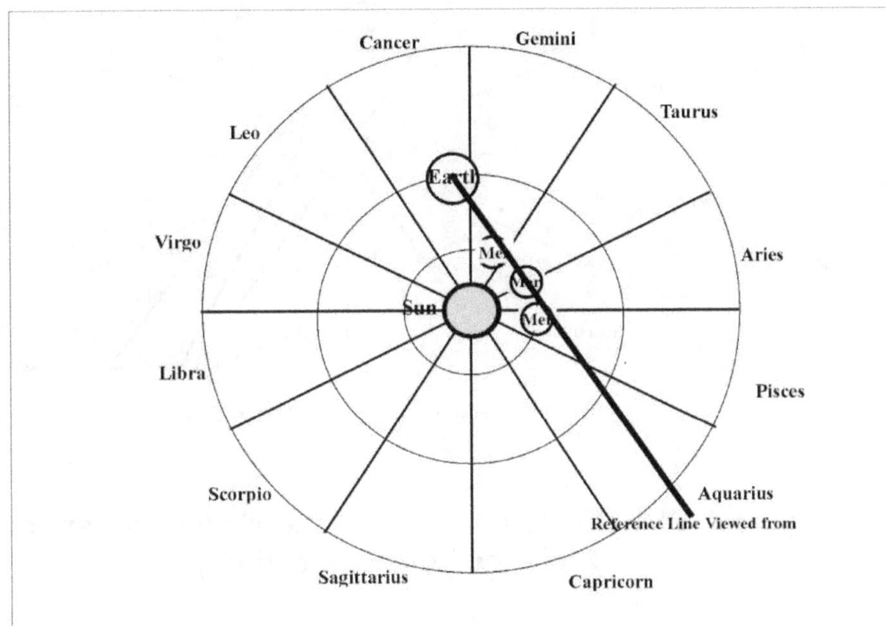

Figure 3: Mercury appears "stationary" or "standstill" as viewed from Earth

ii) Retrograde Motion (Vakra): The planets move within the belt of the zodiac with a different average of speed. When any planet appears moving apparently in the opposite direction to the Sun, as viewed from the earth, that motion is called Vakra or Saktha Avastha or Retrograde Motion. The two planets, i.e. the Sun and the Moon have steady and direct motion. The planet in retrograde is marked in the horoscope with the mark 'R'. A retrograde planet becomes more powerful. The lunar nodes, Rahu and Ketu always move in retrograde direction. In the figure shown below, while Mercury moves from position-I to the position-II (see Figure-4) around the Sun, it cast a shadow of position I and II in opposite direction in the Aquarius Sign, which appears the Mercury moving in Retrograde, i.e. in opposite direction. This is called Retrograde Motion.

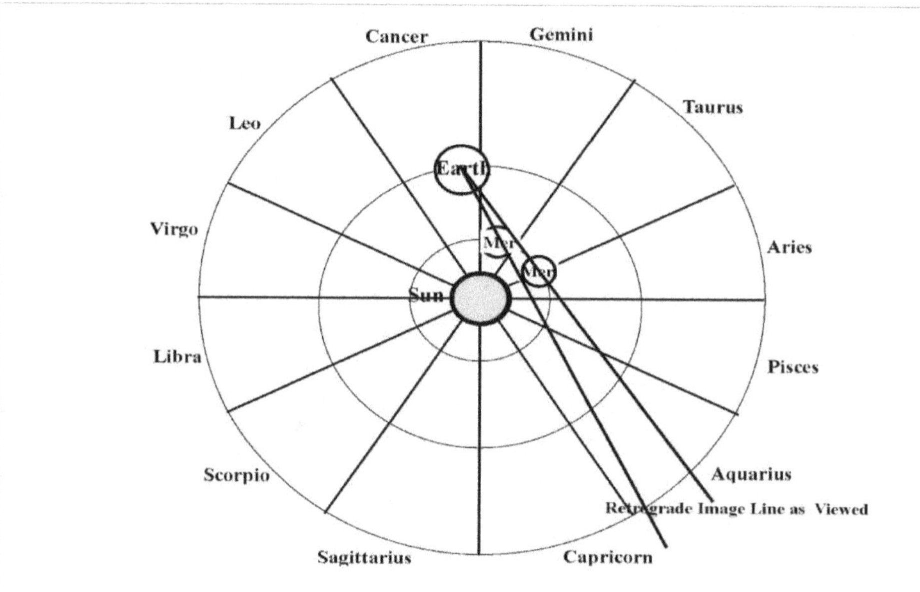

Figure 4: Planet in "Retrograde Motion" against Capricorn Sign (the Zodiac backdrop)

14.3 Planet Transit (Gochara)

The movement of the planets through twelve Rashi is called Gochara or transit. According to the position of a transit planet from natal moon (Rashi) or Lagna or a planet in the birth chart, the astrological effects on native will change. The position of the planets in Natal chart is fixed. The Gochara or Transits of the planets is the instrument used by the astrologers to time the event. At any given time, these 9 planets in transit (Gochara) have a major bearing on native destiny. It is said that the Natal Chart and the Vimshotari Dasa are fixed for the native past, but the Transit is the interplay of the present with the native past.

Example: In a Natal Chart, Moon is positioned in Meena (Pisces). In the Gochara Chart, Venus is positioned in Leo. Hence counted from his Janma Rashi (Pisces), Venus is in the six house in Gochara. Similarly, Sun is in the 5th from his Janma Rashi. Rahu is in the 5th, Saturn in the 4th, Jupiter in the 6th, Moon in the 3rd, Ketu in the 11th, Mars in the 6th House. Similarly note the positions of these planets from Native's Lagna. Sun is in 3rd, Mercury is in 4th and so on. The effects produced by the 9 transiting planets, placed in various houses from the 9 natal planets, Janma Rashi and Lagna in natal chart makes total 10 x 9 x 12 = 1080 combinations as given in the Gochara Shashtra. The transit effects of the planets are very important. The Daily or Weekly Horoscope results are based on Gochara/Transit of Planets.

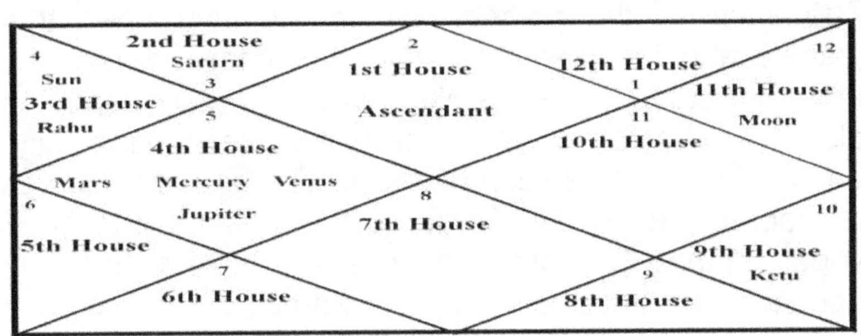

Fig 1: Natal Chart

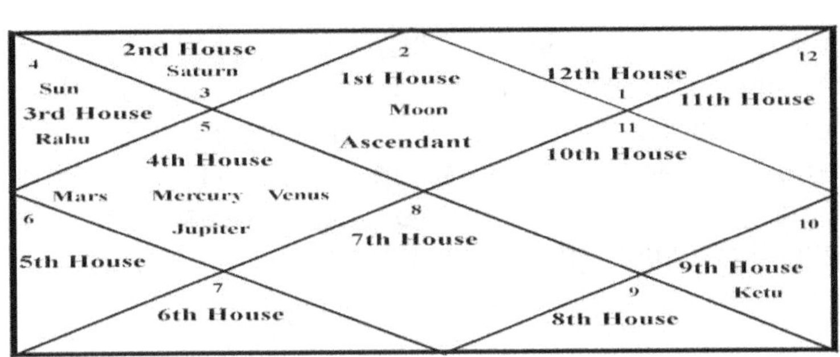

Fig 2: Transit (Gchara) Chart

Table 1: Planet Transit Period in one House

Planet	Sun	Moon	Mars	Mercury	Jupiter
Duration in one house	1 Month	2¼ Days	49 Days	1 Month	1 Year

Planet	Venus	Saturn	Rahu	Ketu
Duration in one house	1 Month	2 ½ Year	1 ½ Year	1 ½ Year

14.4 Predictions by Planet Transit (Gochara)

The effect of the planets will be felt only when the transit planets and the planets in the birth chart interact with each other. Actually, the Karaktva result of the aspect of natal planet will be felt when the transit planets, especially Saturn and Jupiter, receive such aspects. The age of the person must be kept in the mind while giving the timing of the events. The yoga (Planet Combination) present in the horoscope will become operative during the transit of the slow moving planets, Saturn, Jupiter, Rahu and Ketu. If aspect is from the same sign, then the result will be 100%. If there is aspect from the 2nd sign, the result is 75%, from the 3rd sign is 50%, from 5th sign 75%, from 7th sign 50%, from 9th sign 75%, from 11th sign 50% and from 12th sign 25%.

Jupiter: The Jupiter in the 2nd from natal Moon gives rewards, profit and gain of money. Jupiter will indicate the events according to the nature of the planet aspect from Natal Chart. Example: When natal Venus aspects transit Jupiter, the person will get married. In one case, Jupiter was transiting Meena Rashi, the native got married because Venus is in Cancer, being 5th sign from Meena. This result is applicable only to the unmarried people during the age of marriage timing but not during he is kid. If he is a married person, for the same aspect we have to predict that there will be a birth of daughter. When natal Mercury aspect transit Jupiter, the person will do well in studies.

Saturn: Saturn in the 2nd from the natal Moon gives sorrows and accidents. Saturn is Profession (Karma) Karaka; he represents the Professional fate of the person. Example: Suppose Saturn is transiting Mesha of Natal Chart. If Mars is posited in Mesha in the Natal Chart, it is 1st house, similarly Vrishabha is 2nd from Mesha, Mithuna is 3rd from Mesha, and so on, the person will be troubled by enemies but the percentage of the result will differ according to the aspect given as above. Example 1: In one case, when Saturn was transiting Dhanu Rashi, the native got appointment in Govt. Department, because Guru and Chandra both were in Simha Rashi, which was 9th sign from Dhanu. Therefore the transiting Sani located in Dhanu gets the aspect of Natal Jupiter and Moon located in Simha. When transit Saturn is aspect by natal Jupiter, the person will get a job or promotion. When transit Saturn is aspect by natal Moon, the person will have change of place. Therefore, the native got appointment in a different place.

Example 2: In one case, Saturn was transiting Kumbha Rashi, the native faced lot of difficulties. Since Natal Mars was aspecting transit Saturn from Simha Rashi which is 7th from Kumbha Rashi, Natal Ketu was also aspecting transit Saturn from Meena Rashi which was 2nd from Kumbha Rashi. But Natal Venus was aspecting over transit Saturn in Kumbha from 7th and gave her some money through other means.

14.5 Planet Transit over other Planet

Sun Transit over the following Natal planets:
While Sun Transit over the Natal planets, such as, Natal Sun: Success, vitality. Natal Moon: Travel to foreign, father gets blame. Natal Mars: Blood defect for the native, hindrance to the brothers. Natal Mercury: Father's gain in lands, success in commercial field, meetings with inmates. Natal Jupiter : Status, contact

with noble person, cooperation from cultured persons. Natal Venus: Marriage, fond of sex, wife suggestion, ill-health, financial difficulties. Natal Saturn: Lack of confidence, financial difficulties, unnecessary blame. Natal Rahu : Gain in speculation, laziness to father. Natal Ketu: worry through children, native's father will have divine thoughts.

Moon Transit over the following natal planets:

While Moon Transit over the Natal planets, such as, Natal Sun: Change of residence, good for operation, travel to father. Natal Moon: Opposite sex relations. Natal Mars: Mental unrest, Mother will become stubborn, travel for mail native's brother or travel for Female's husband. Natal Mercury: Lady uterine trouble, monitory loss in the field of commerce, unnecessary blame. Natal Jupiter: Travel, desire for learning, contract with women, happy celebrations. Natal Venus: Debts, ill health to wife. Natal Saturn: Change of place, love with old people, unnecessary blame, unnecessary expenditure of money, no mental peace. Natal Rahu: Women play vital role in life, change to mother mental hallucination for the native.

Natal Ketu: Difference of opinion with parents, ill health to mother.

Mars Transit over the following natal planets:

While Mars Transit over the Natal planets, such as, Natal Sun: Fever, mental tension, stubbornness, son gets wounded, problems due to anger, harassment by enemies in career, husband's behaviour is egoistic manner. Natal Moon: Mothers ill health, operation, travel, prospects for brother, mental anxiety, haste and fickle mindedness. Natal Mars: Accidents, increased energy. Natal Mercury: Trouble by enemies, unrest, and lack of mental peace, dispute with blood relatives, friends, and husband meets his intimate friends. Natal Jupiter: B.P, excess heat, danger from fire, stubbornness, hastiness, good time for husband. Natal Venus: Husbands financial gain, contacts with blood relations and intimate friends. Natal Saturn: Accidents, harassment to the native from enemies and financial loss. Natal Rahu: Over sex, death of brother. Natal Ketu: Desire for power.

Mercury Transit over the following natal planets:

When Mercury transit over the Natal planets, such as, Natal Sun: Friends help to father. Natal Moon: Blames, uneasiness through women or girlfriend, travel. Natal Mars: Dispute with brothers, friends, skin ailments, bone disorder, dispute with father in-law and maternal uncle. If the native is female husband meets friends. Natal Mercury: Strengthen all matters at birth. Natal

Venus: Gain of treasure, property, happiness and celebrations at home. Natal Saturn: Gain in business, gain of lands, and cooperation from friends. Natal Rahu: Good talent in communication, skin ailments. Natal Ketu: Unfavourable time.

Jupiter Transit over the following natal planets:

When Jupiter transits over the natal planets, such as, Natal Sun: Promotion, cooperation. Natal Moon: Change of residence, ill health due to cold. Natal Mars: Houses, status. Natal Mercury: Houses, gain of land, new knowledge. Natal Jupiter: Name and fame, birth of son. Natal Venus: Marriage, wealth, birth of daughter. Natal Saturn: Gets job, promotion, smooth going period in carrier. Natal Rahu: Operation, discontinuation of studies, death, abortion. Natal Ketu: Gives divine knowledge to the native, ill health to native.

Venue Transit over the following natal planets:

While Venus Transit over the Natal planets, such as, Natal Sun: Financial gain to father. Natal Moon: Loss of money, ill health to wife or daughter, mental unrest, financial dispute among the female inmates of the house. Natal Mars: Gain to brother, residence, wife may become pregnant, marriage, female native get cooperation from husband, financial gain to husband, hindrance to financial prospects. Natal Mercury: Treasure, gain of landed property. Natal Jupiter: Financial gain, misunderstanding in the family, acquiring luxurious goods, happiness and celebrations at home. Natal Venus: Fortune. Natal Saturn: Celebration at home, financial gain. Natal Rahu: Marriage, hard time to wife. Natal Ketu: Separation.

Saturn Transit over to the following natal planets:

While Saturn Transit over the Natal planets, such as, Natal Sun: Ill health to father, dispute between father and son, trouble by Govt. Natal Moon: Unnecessary blame, ill health to mother, unnecessary expenditure, mental unrest. Natal Mars: Trouble by enemies, unrest, gain of landed property; Natal Mercury: Gain of land, good time for education, Natal Jupiter: Change of profession, gets job or promotion, gastric trouble. Natal Venus: Marriage, acquiring property Natal Saturn: Troubles. Natal Rahu: Death in the house. Natal Ketu: Litigation, dispute, aimlessness, visiting holy places, financial loss.

Rahu Transit over the following natal planets:

While Rahu Transit over the Natal planets, such as, Natal Sun: Ill health problem to son. Natal Moon: Danger to mother, fear complex to native. Natal Mars: Accidents, blood defects, operation (surgery). Natal Mercury: Skin diseases, purchase of new vehicle, fear complex to native. Natal Jupiter: Accident, death in the family, black mark develops on the face. Natal Venus: Break in income, ill health to wife, hidden treasury. Natal Saturn: Performing lost rites, purchase of vehicle, laziness. Natal Rahu: Stomach disorder. Natal Ketu: Obstruction in good deeds.

Ketu Transit over the following natal planets:

While Ketu Transit over the Natal planets, such as, Natal Sun: Meeting holy persons. Natal Moon: Divine contemplation, ill health to mother, blood pressure. Natal Mars: Worries to brothers, blood pressure, and nerves debility to brother. Natal Mars: Dispute with girlfriend/boyfriend, litigation, poor memory. Natal Jupiter: Divine contemplation, nerves debility, ill health. Natal Venus: The native becomes victim to a lady. Natal Saturn: Litigation, dispute, aimlessness and quitting the job. Natal Rahu: Loose motions. Natal Ketu: Misunderstanding in family.

14.6 Planet Transit over Planet in Natal House

Sun Transit (Gochara) results

Sun takes one month to transit a sign. It is a very important planet in the Solar system and this transit has a strong impact on your life. The following are the effects of Sun's transit, as per Rashi (Moon-Sign), Lagna and other planets placed in birth chart. A full summary of a planets transit from all give a much better picture.

Sun Transit from Natal Moon (Rashi)

1st house: Loss of wealth & fame, illness, opposition with friends& relatives, tiresome journeys and fatigue, irritability, mental agony and anger.

2nd house: Bad company, headache, will be cheated, obstinacy, meanness, business will not yield results.

3rd house: Good health, happiness, acquisition of wealth & new position, destruction of enemies, gathering of friends and relatives, desired results.

4th house: Ill health, bad food, obstacles in journeys, troubles sexual enjoyment.
5th house: Laziness, sadness, mental worries, trouble from friends, heavy expenses & losses, embarrassment of all kinds.
6th house: Free from worries, good health, acquisition of wealth & clothes, destruction of enemies & desired results in work.
7th house: Illness to spouse & children, quarrel with spouse's relatives, humiliation, stomach troubles, tiresome & futile journeys, lack of enthusiasm.
8th house: Illness, fear, trouble from officials, arguments with enemies, tiresome journeys, bad news.
9th house: Depression, loss of money & goods, unnecessary quarrels, sorrow and mental worries.
10th house: Good health, gained in money, company of friends & relatives, success in undertakings, respect & honour from officials.
11th house: Advancement in job, honour, gains in money, happiness from spouse & children, good health.
12th house: Loss of position or change in place, heavy expenses, ill health, quarrels with friends & relatives, danger to life.

Sun Transit from Lagna

1st house: Loss of name friends and vehicles.
2nd house: Harmful for income.
3rd house: The destruction of enemies, desired results and gains.
4th house: Enjoyment with women.
5th house: Futile and tiresome journeys, quarrels.
6th house: Fame good health and no enemies.
7th house: Anger, fear of theft during journeys.
8th house: Trouble from enemies or sickness or scriptures.
9th house: Enmity and loss of position.
10th house: Gain of money land and cattle.
11th house: Gain of money land and cattle.
12th house: Futile work and unhappiness

: THANK YOU :

www.ingramcontent.com/pod-product-compliance
Lightning Source LLC
Chambersburg PA
CBHW080633180526

45168CB00008B/3149
* 9 7 8 1 5 1 2 1 1 2 9 6 2 *